# FLEXOELECTRICITY IN LIQUID CRYSTALS

## Theory, Experiments and Applications

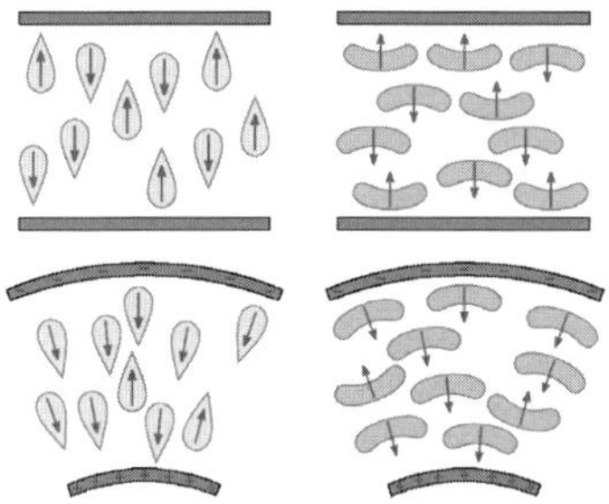

# FLEXOELECTRICITY IN LIQUID CRYSTALS

## Theory, Experiments and Applications

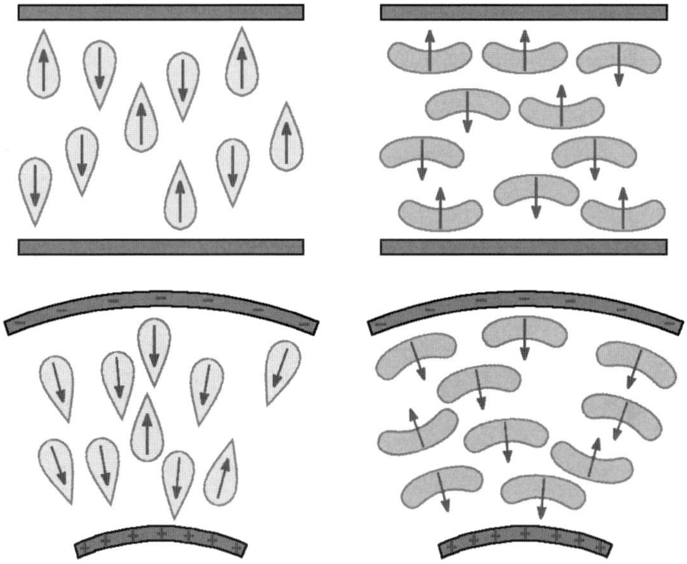


Edited by

**Ágnes Buka · Nándor Éber**

Institute for Solid State Physics and Optics,
Wigner Research Centre for Physics,
Hungarian Academy of Sciences, Hungary



ICP Imperial College Press

*Published by*

Imperial College Press
57 Shelton Street
Covent Garden
London WC2H 9HE

*Distributed by*

World Scientific Publishing Co. Pte. Ltd.
5 Toh Tuck Link, Singapore 596224
*USA office:* 27 Warren Street, Suite 401-402, Hackensack, NJ 07601
*UK office:* 57 Shelton Street, Covent Garden, London WC2H 9HE

**British Library Cataloguing-in-Publication Data**
A catalogue record for this book is available from the British Library.

**FLEXOELECTRICITY IN LIQUID CRYSTALS**
**Theory, Experiments and Applications**



ISBN 978-1-84816-799-5

Printed by FuIsland Offset Printing (S) Pte Ltd Singapore

# Preface

Flexoelectricity is a fascinating paradigm of phenomena allowed by symmetry in liquid crystals and there has been an ever increasing interest in it since its discovery in 1969 by R.B. Meyer.

Though it has been thoroughly studied theoretically from the beginning, only very few experimentalists chose it as a target of their investigations. For a long time it was regarded as a scientific curiosity only, as the understanding of the electro-optical phenomena, forming part of the mainstream of liquid crystal research in the 1970s and 1980s, did not require the presence of this type of cross-coupling between mechanical and electric properties.

Another reason for the moderate popularity of the field among experimentalists is that flexoelectricity-driven effects are rare; their detection mostly requires unusual surface (anchoring) conditions, moreover, the flexoelectric coefficients are small (in the range of pC/m), thus are not easily measurable and the results are often contradictory.

Despite all these problems, flexoelectricity has, from time to time, been found wanting when the results of an experiment could not fully be interpreted by the dielectric interaction alone; though its relevance or responsibility could not always be proved.

Nevertheless the results accumulated over decades have gradually made it clear that flexoelectricity may be of great importance in several phenomena (from biology through pattern formation to applications). The number of otherwise unexplained experimental facts that can be understood by adding flexoelectric effects to the standard description has grown steadily. The first displays utilizing flexoelectricity (the zenithal bistable device) have been commercialized; another display mode (the cholesteric flexoelectro-optic effect) is close to entering the market. Additionally, the introduction and spreading of bent-core molecules has provided another impetus for studying this phenomenon.

Now, about 40 years after its discovery, flexoelectricity is a mature field. Therefore it is timely to review the theoretical and experimental results obtained so far and to summarize what problems have been solved and what questions are waiting for answers. In order to accomplish this job we succeeded in persuading renowned experts in the field to contribute to this review.

Naturally, this volume could include only a limited number of topics, which have been selected according to the interests and fields of expertise of the contributors; thus we cannot claim that all aspects of flexoelectricity are fully addressed. Nevertheless, we think that the chapters provide a state-of-the-art comprehensive picture of the subject. Readers interested in more details will easily find in the references other works where related topics are discussed.

As supplementary material, at the end of the volume we have added an appendix that lists the measured values of the flexoelectric coefficients for a number of compounds that we could find in the literature. Hopefully this will be a helpful tool for those planning experiments or numerical simulations.

We are grateful to all contributors of the volume for accepting our invitation and for their cooperation during the editing process. We thank Katalin Fodor-Csorba, Věra Hamplova and Terézia Vojtylová for their help in finding the precise chemical names of the compounds listed in the Appendix. Finally we would like to express our thanks to all our co-workers at the Complex Fluids Department; they have provided an inspiring scientific atmosphere, which helped us through the difficulties of editing.

Ágnes Buka and Nándor Éber

Budapest, 2 August 2012

# Contents

# Introduction to Flexoelectricity: Its Discovery and Basic Concepts

Robert B. Meyer
*The Martin Fisher School of Physics, Brandeis University*
*Waltham, MA, 02454, USA*
*e-mail: meyer@brandeis.edu*

Flexoelectricity – what is it? How does it arise in liquid crystals? What are its consequences? What role does it play in liquid crystal phases, structures, and textures? How is it measured? What is its role, both realized and potential, in applications of liquid crystals? How was it discovered and what is its history in the context of the development of liquid crystal science and technology in the last 50 years? The name flexoelectricity clearly indicates the dual role of curvature distortions and electrical effects in liquid crystals, but just how are these two fundamental sets of concepts related by this phenomenon? This book attempts to lay out the answers to these questions, with a combination of broad reviews and focused insights into the role of flexoelectricity in the science and technology of liquid crystals. In this introduction there is first a little informal review of history along with some general comments on the fundamentals and the special challenges presented by this phenomenon, and then there is a brief sketch of the chapters of this book.

As a graduate student in the Division of Engineering and Applied Physics at Harvard University in the late 1960s, I was introduced to liquid crystals by David Turnbull, who suggested that I read Sir Frederick Charles Frank's 1958 paper on the theory of symmetry, elasticity, and defects in liquid crystals.[1] This paper introduced me to symmetry arguments as a powerful way of analyzing a system, and it made a deep impression on me. As I began working on liquid crystals, I soon came to realize that understanding molecular structure was crucial to understanding these systems, again an idea strongly suggested by Frank's paper. I was soon fortunate enough to set up a collaboration with a young organic chemist at Harvard, Dave

Dolphin, who synthesized new liquid crystals, and I became deeply interested in the details of molecular structure, especially the shape, electrical polarity, and chirality of these molecules. My research focused on field effects in nematics and cholesterics, especially the unwinding of the cholesteric helix by the field interacting with the anisotropy of the susceptibility in the uniaxial liquid crystal. With this set of interests and ideas in mind, I returned to Frank's paper,[1] and reviewed again his symmetry arguments, especially his argument that if the molecular organization of the nematic phase were intrinsically polar, that is, if rod-like molecules with one end different from the other were arranged to be all oriented one way, then that element of macroscopic symmetry would imply both spontaneous electrical polarization and a term in the elastic free energy density linear in splay distortion. The presence of that linear splay term would lead to spontaneous splay distortion of the polar nematic in its ground state, rather than the simple uniaxial structure observed in known nematics. The lack of that spontaneous splay argued strongly against the polar organization of the molecules in the nematic phase. While Frank focused on the nature of the ground state symmetry and properties, it was a small step for me to apply the same argument to excitations from the ground state, and see that electrical polarization parallel to the molecular long axis (defined by a unit vector **n** called the director) would induce splay, and vice versa. Looking again at Frank's arguments for eliminating terms in the elastic free energy density linear in bend, I saw the connection between polarization perpendicular to the nematic director and bend distortions, which break the rotational symmetry about the director. Reviewing the theory of piezoelectric crystals, I realized that what I was studying was a kind of second-order piezoelectricity involving certain second-order strains of the liquid crystal structure, that is, curvatures, rather than the first-order shear strains of piezoelectric crystals, which of course could not be sustained in the fluid phase. It took Pierre-Gilles de Gennes to see that this coupling of curvatures and electrical polarization deserved a new name, and he coined the term flexoelectricity.

Looking back on this discovery, I am struck by several things. First, it seems to me that Frank should have discovered this effect! His deep insights into symmetry and structure focused on the ground state symmetry, and not on excitations from that state. I was in the fortunate situation of thinking about a number of topics concerning liquid crystals, especially the coupling between their molecular structure and their response to electric fields. The stunning development of liquid crystal science and technology in the past 50 years has relied on the intimate interplay of knowledge and discovery in a wide range of sciences, including areas of physics, chemistry, biology, and engineering. The broader one's knowledge and interests, the better prepared one is to make significant discoveries, especially in fields like materials science, which are intrinsically interdisciplinary. Second, it is clear that returning to basic symmetry arguments is always worthwhile, and insights based on symmetry will have the most significant impact, since they are not limited to the odd quirks of a single material, but apply to whole classes of materials. And third, it is always worthwhile to review the work, and to try to understand the thinking, of great scientists like F.C. Frank. Their insights, and the thought processes that led to them, are a powerful lesson on how to think creatively in science.

A second kind of hindsight about this discovery is that I made a basic mistake in defining the flexoelectric coefficients for splay and bend. Because the flexoelectric effect is linear in curvatures and fields, in any scheme of definitions, there is no preference for the coefficients to be positive or negative on the average; either sign is physically possible. However, there is a reasonable geometric argument for the relative signs of the two coefficients for splay and bend distortions. For a simple cylindrical curvature of the director field, with the nematic director either radial (pure splay) or tangential (pure bend), the flexoelectric coefficients should be defined so that the curvature-induced polarization is either radial outward or radial inward for both splay and bend cases. Unfortunately, I chose definitions of the bend and splay vectors that were intrinsically opposite to one another, in this geometric sense, so that the curvature-induced polarizations were opposite in sign, relative to the radial direction, for cylindrical splay and bend curvatures of the director field. Using the currently accepted coefficients $e_1$ and $e_3$ based on my original paper, this means that the mean flexoelectric coefficient is $\bar{e} = (e_1 - e_3)/2$. This is not wrong mathematically, but it is confusing, and one often sees the expression $(e_1 - e_3)/2$ referred to as the *difference* between the flexoelectric coefficients, rather than their mean, which is surely confusing. To alleviate this problem, first I would define the splay and bend vectors so that their polar properties are consistently oriented the same way with respect to curvature. The splay vector should be $\mathbf{S} = \mathbf{n}(\nabla \cdot \mathbf{n})$ and the bend vector should be $\mathbf{B} = \mathbf{n} \times (\nabla \times \mathbf{n})$, which is opposite in sign to the bend vector defined in my first paper.[2] I would propose defining two new flexoelectric coefficients, $e_s$ and $e_b$ for the splay and bend cases, with $e_s = e_1$ and $e_b = -e_3$ of the previous definitions, related to the previously defined splay and bend vectors. The flexoelectric polarization would now be defined as $\mathbf{P_f} = e_s\mathbf{n}(\nabla \cdot \mathbf{n}) + e_b\mathbf{n} \times (\nabla \times \mathbf{n})$. Thus the mean flexoelectric coefficient would now be $\bar{e} = (e_s + e_b)/2$.

Returning again to the value of considering fundamental symmetry arguments, I must point out one of the startling fundamental discoveries about flexoelectricity, made eight years after the publication of the discovery of the effect. My initial picture of the molecular origin of the effect was based on the idea that the steric polarity of the molecule and its electrical polarity are intimately coupled in the structure of the molecule, and that this coupling is the link that resulted in flexoelectricity. Thus a cone-shaped molecule must possess an electrical dipole moment oriented either parallel or antiparallel to the axis of the cone, and a banana-shaped molecule must possess a transverse dipole moment oriented either toward or away from the centre of curvature of the banana. A splay distortion would tend to orient the cone molecules preferentially, and this would produce a net polarization. Likewise a bend curvature would preferentially orient the bananas, again producing electrical polarization. One could likewise understand the inverse effects of polarization-induced curvature. The only question was the magnitude of the couplings, which required some model of the interaction of an individual molecule with the curvature-distorted director field. Symmetry demanded both the internal couplings of steric polarity and electrical polarity in the molecular structure and the coupling of the molecular shape to the curvature field, but by chance either of these couplings could be very large or very small in a particular material. Based on this picture, I believed that molecules that were not intrinsically polar, both electrically and sterically, would not exhibit flexoelectricity. Most liquid crystal molecules are polar, but highly symmetric molecules that entirely lack any net polarity also form liquid crystals, and it seemed clear to me that they would not exhibit flexoelectricity. However, in 1977, Prost and Marcerou published a paper that overturned this idea.[3] They returned to the basic symmetry arguments about the structure of liquid crystals, relating to the most general description of the shape and charge distributions of molecules forming liquid crystals, and the structure of the distortion fields, splay and bend, which interact with the molecules. Highly symmetric molecules lacking a dipolar term in the description of their shape and charge distribution, but having a quadrupolar element to their structure, a natural requirement of any liquid crystal molecule, would interact with the curvature field to produce a spatial gradient of the quadrupole density, which is a dipole density! Thus even in these molecules curvature and electrical polarization are coupled to produce flexoelectricity. Again, the return to basic symmetry arguments proved to be a powerful approach to understanding material structures and phenomena.

From the start, it was clear that flexoelectricity involved some special challenges both for theory and for experiments. I have spoken of the "discovery" of flexoelectricity[2] in 1969 as an exercise in symmetry arguments and basic considerations of molecular structure and organization. But such "discoveries" in theory demand confirmation through observations of material properties and responses to applied fields, in this case, curvature fields and electric fields. Qualitatively, one must see that an applied electric field produces the expected curvatures, or conversely that curvatures produce the expected electrical polarizations, which must be detected in some way. Moreover, one wants to measure the strength of the effect, and relate it to the underlying structure of the molecules and their organization in the liquid crystal. This kind of quantitative knowledge is the basis for forming a solid understanding of any material and its properties, the kind of understanding that is necessary for progress, for instance in optimizing a material for a certain effect, or for designing an application of the effect to some practical purpose. Thus in the first paper, a rough estimate of the size of the effect was made, based on molecular properties, and two observations that might involve flexoelectricity were described. But these were only speculations to provide some initial guidance to possible experiments. They did not initially prove very useful, but they illustrated fundamental problems with the experimental aspects of flexoelectricity.

First, any electrical effect is immediately in competition with all other electrical effects, and must be carefully separated from them to be studied. For example, to study a curvature-induced polarization, one might consider producing a sample with a static curvature structure, say a layer of nematic confined between two cylindrical electrodes, with the director parallel or perpendicular to the electrodes. Flexoelectricity will produce space charges at the electrodes, but these would tend to be cancelled by free ionic charges in the material. Any fields produced by flexo-induced space charges would only propagate a distance of the order of the Debye screening length, so effects due to these fields would have to be detected on a microscopic scale in most materials.

Second, for observing curvature structures induced by an externally applied electric field, one must also consider the interaction of the field with the anisotropy of the polarizability of the medium, which tends to align the director either parallel or perpendicular to the applied field. This can suppress the flexoelectrically induced curvature in many geometries.

Third, because the flexoelectric terms in the free energy density are linear in the spatial derivatives, in a finite sample, they can often be integrated by parts as surface integrals rather than bulk integrals, so that the total flexoelectric energy in a sample is a function of the boundary conditions of the sample, and insensitive to continuous changes in the curvature structure within the sample. This means that the observation of a flexoelectrically induced structure often relies on weak anchoring of the director on the surfaces, or on the movement of topological defects into or through the sample, to change the topology of the interior of the sample contained by the strong anchoring boundary conditions. In the first case, it means that the observation of flexoelectric effects involves analysis combining knowledge of other effects, such as surface anchoring strength. In the second case, there is a threshold field to be crossed before the defects are generated or driven to new locations. Both of these restrictions make flexoelectricity difficult to study.

In general, flexoelectricity requires inhomogeneity in the sample being studied, and does not play a simply observed role in undistorted, single-domain samples in a homogeneous electric field, clearly the simplest sample geometry, and one commonly studied in liquid crystal experiments. It only plays a role when curvature distortions are involved or when there are inhomogeneous applied fields. This fundamental complication has been overcome in many clever experiments, so that there have now been many measurements of flexoelectricity in a range of materials.

For the development of engineering applications, flexoelectricity has some fundamental features to offer. Because it is linear in the applied electric field, it is fundamentally different from effects due to the anisotropy of the susceptibility in liquid crystals, which are quadratic in the field strength. Thus in a fixed geometry, the sign of the applied field matters, so positive and negative fields can produce quite different responses, either linear or non-linear in nature, depending on the geometry of the particular sample and electrodes. Moreover, the "difficulty" mentioned above, that certain flexoelectric responses require a change in sample topology, due to the motion of topological defects, can actually be more of a feature than a bug, leading to a sample response that locks in a change of topology after the external field is removed. This can be the basis of bistable or even multistable electrical switching phenomena, which are especially useful in display devices that retain information without continuing dissipation of electrical power.

The chapters of this book expand on the ideas mentioned here, as well as others that have arisen from the efforts of many talented scientists studying this rather unusual phenomenon over the past 40 years. To get the most out of this book, the reader should be familiar with the basics of liquid crystal phenomenology, including molecular structure, phases, textures, and typical sample geometries, as well as theoretical concepts, including curvature elasticity theory and some elements of the fluid dynamics of liquid crystals.

In the first chapter, Osipov presents a review of the molecular theory of the origins of the flexoelectric effect, with careful attention to how molecular shape and structural correlations determine the flexoelectric coefficients. In the second chapter, Madhusudana reviews the broad range of experimental methods that have been employed to measure the flexoelectric coefficients, discussing the many challenges this task has presented. Following these general reviews, in the third chapter, Jákli, Harden, and Éber discuss a particular measurement technique applied to an important new category of liquid crystal materials, the bent-core molecules, which can exhibit a giant flexoelectric effect. In the fourth chapter, Buka, Tóth-Katona, Éber, Krekhov, and Pesch discuss the role of flexoelectricity in pattern formation, especially addressing how this effect competes with other electro-dynamical effects leading to stripe patterns. Čepič, in Chapter 5, presents a clear argument for the role flexoelectricity plays in establishing the fundamental multi-layer ordering of the various polar smectic phases. In Chapter 6, Petrov traces the research on the role of flexoelectricity in the structure and molecular ordering of biological membranes, reviewing both theory and experimental results, including its role in nerve conduction and in the action of hair cells in hearing. In the final chapter, Rudquist and Lagerwall discuss in detail two important applications of flexoelectricity. The first is based on the flexoelectric response of a short-pitch cholesteric liquid crystal to a field applied perpendicular to the helix axis, which produces a fast linear electro-optical effect by rotating the average optical axis of the cholesteric. The second is based on the electric field-driven switching of a particular sample geometry between two topologically distinct states, by the field-induced motion of topological defects, an important feature of the non-linear flexoelectric response, as noted above. It is the basis of the first flexoelectric device to reach commercial production by the company ZBD, their acronym for "Zenithal Bistable Devices." Based on the foundations of our current knowledge about flexoelectricity reviewed in this book, it will be interesting to see how the science and technology of flexoelectricity develop over the next 40 years.

## References

1. F.C. Frank, On the theory of liquid crystals, *Discuss. Faraday Soc.* **25**(1), 19–28, (1958). doi: 10.1039/DF9582500019
2. R.B. Meyer, Piezoelectric effects in liquid crystals, *Phys. Rev. Lett.* **22**(18), 918–921, (1969). doi: 10.1103/PhysRevLett.22.918
3. J. Prost, and J.P. Marcerou, On the microscopic interpretation of flexoelectricity, *J. Phys. France* **38**(3), 315–324, (1977). doi: 10.1051/jphys:01977003803031500

# Chapter 1

# Molecular Theory of Flexoelectricity in Nematic Liquid Crystals

Mikhail A. Osipov

*Department of Mathematics and Statistics, University of Strathclyde*
*26 Richmond Street, Glasgow G1 1XH, UK*
*e-mail: m.osipov@strath.ac.uk*

The molecular-statistical theory of flexoelectricity in nematic liquid crystals is considered in detail and general expressions for the flexoelectric coefficients are obtained in terms of the direct correlation function. These general formulae are then used to express the flexoelectric coefficients in terms of molecular model parameters using the mean-field approximation. The relation between dipolar and quadrupolar flexoelectricity is discussed in detail and the effect of molecular longitudinal and transverse electric dipoles and the polarity of the molecular shape on the value of the flexoelectric coefficients is considered. The contribution from a strong dipole–dipole intermolecular correlation is also estimated, and the dependence of the value and sign of the difference of the flexocoefficients on various molecular parameters is analysed. The results of the theory are compared with existing experimental data. Finally we consider the original version of the mean-field theory of flexoelectricity, proposed by Ferrarini, which allows us to take into account the real molecular shape.

## Contents

## 1.1. Introduction

In this book the flexoelectric effect is mainly considered from the phenomenological point of view. At the same time it is very interesting and important to reveal the molecular origin of flexoelectricity and, in particular, to consider different types of intermolecular interactions that may be responsible for the dipolar ordering in a deformed liquid crystal, and to study the effects of intermolecular correlations and the molecular structure. This problem can only be solved using a molecular-statistical theory, which eventually allows us to express the flexoelectric coefficients in terms of molecular model parameters using various approximations.

The first microscopic interpretation of flexoelectricity in nematic liquid crystals was proposed by Meyer[1] and is illustrated in Fig. 1.1. We can readily see in Fig. 1.1 that in a nematic liquid crystal, composed of banana-shaped or cone-shaped molecules, the orientational deformation favours a dipolar ordering of the asymmetric molecules. Note that cone-like and banana-like molecules possess a polar molecular shape that can be specified by pronounced longitudinal and transverse steric dipoles, respectively. As illustrated in Fig. 1.1, in the nematic phase with splay deformation the opposite orientations of the longitudinal steric dipoles are no longer statistically equivalent, and as a result the macroscopic polarization appears in the direction parallel to the local director $\mathbf{n}$ if the molecules also possess longitudinal electric dipoles. In a similar way, transverse steric molecular dipoles of banana-shaped molecules should order in the nematic phase with bend deformation resulting in the corresponding ordering of transverse electric dipoles, which gives rise to the macroscopic polarization in the direction perpendicular to $\mathbf{n}$.

It should be noted, however, that the flexoelectric effect is not necessarily related to the ordering of molecular dipoles. Prost and Marcerou[2] proposed another microscopic mechanism of the flexoelectric effect, which requires neither the asymmetry of the molecular shape nor the permanent molecular dipole. The macroscopic polarization may simply appear in the direction of the gradient of average density of the molecular quadrupole moments. The quadrupole mechanism of flexoelectricity is more general because, in principle, it should manifest itself in any anisotropic material with a non-zero quadrupole density including solid crystals[3,4] and elastomers.

The first consistent statistical theory of the dipolar flexoeffect was developed by Straley[5] for an athermal nematic composed of polar hard rods. Helfrich[6] and Petrov and Derzhanski[7,8] also proposed explicit expressions

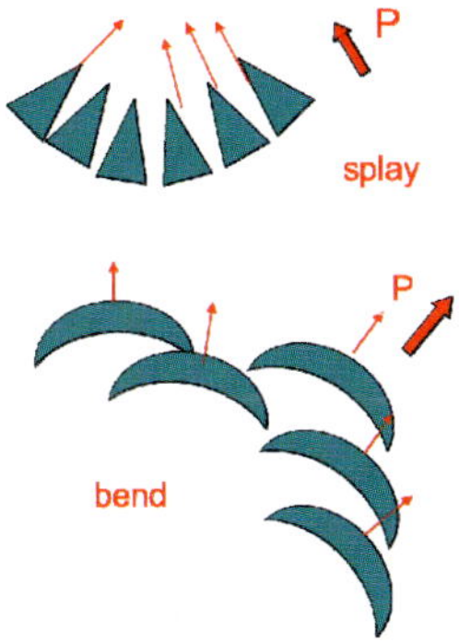


Fig. 1.1. Macroscopic polarization in the distorted nematic phase composed of bent-core molecules with transverse electric dipoles or cone-shaped molecules with longitudinal dipoles.

for the flexocoefficients in the nematic phase based on Meyer's model even before Straley's paper was published. These expressions, however, were obtained using a semi-phenomenological approach in which a microscopic molecular dipole is assumed to order in the effective potential created by macroscopic elastic forces.

During recent decades the molecular theory of flexoelectricity in nematic liquid crystals was developed further by various authors.[9–17] In particular, explicit expressions for the flexocoefficients were obtained[9] using the molecular-field approximation taking into account both steric repulsion and attraction between the molecules of polar shape. The influence of dipole–dipole correlations[10] and molecular flexibility[11] was later considered. Recently flexoelectric coefficients have been calculated numerically using the mean-field theory based on a simple surface intermolecular interaction model.[12,13] This approach allows us to take into consideration the real molecular shape and to evaluate the flexocoefficients for mesogenic molecules of different structures including dimers with flexible spacers.[13]

General expressions for the flexocoefficients of nematic liquid crystals have been obtained in terms of the direct correlation function using the powerful density functional approach.[14–16] These expressions have been used[17] to obtain some interesting numerical results using the Perkus–Yevic approximation for the pair correlation function. The results from the density functional theory have also been used in computer simulations of flexoelectricity using model bent-core molecules interacting via the Gay–Berne potential.[18] Alternative general expressions for the flexocoefficients have also been obtained using the linear response theory.[10] In this theory the flexoelectric coefficients are expressed as statistical averages of explicit microscopic functions evaluated in the homogeneous nematic phase, and the corresponding results have been used in computer simulations of flexocoefficients by a number of authors.[19–21]

Finally note that the flexoelectric effect is also important in the smectic phase although the corresponding molecular theory is at a rudimentary stage. Recently a molecular model for the conventional and the so-called discrete flexoelectric effect in tilted smectic phases has been proposed.[22]

This chapter is arranged as follows. In Section 1.2 we consider in more detail the dipolar and quadrupolar mechanisms of flexoelectricity, and in Section 1.3 we derive the general expressions for the flexocoefficients in terms of the direct pair correlation function. These results are used in Section 1.4 to obtain approximate expressions for the flexocoefficients in the molecular-field approximation taking into account both intermolecular repulsion and attraction. In that section we also consider the dependence of the flexocoefficients on the absolute value of the molecular dipole and on the orientation of the electric dipole with respect to the molecular long axes and the steric dipole. In Section 1.5 the effect of dipole–dipole correlations is analysed and in Section 1.6 we discuss the mean-field theory of flexoelectricity, which allows us to account for the real molecular shape.

## 1.2. Dipolar and Quadrupolar Flexoelectricity

From the microscopic point of view the macroscopic polarization in a uniaxial nematic liquid crystal medium is given by the following expansion:

$$\mathbf{P}(\mathbf{r}) = \rho_0 \langle \mathbf{d} \rangle + \nabla \cdot \langle \mathbf{q}(\mathbf{r}) \rangle + ..., \tag{1.1}$$

where the first term is a contribution from the average molecular dipole $\langle \mathbf{d} \rangle$ and the second term is a gradient of the average quadrupole density tensor $\langle \mathbf{q}(\mathbf{r}) \rangle$, which can be expressed as:

$$\langle \mathbf{d} \rangle = \int \mathbf{d} f(\mathbf{a} \cdot \mathbf{n}) d\mathbf{b} d^2\mathbf{a}, \tag{1.2}$$

and

$$\langle q_{\alpha\beta} \rangle = \rho_0 \int q_{\alpha\beta} f(\mathbf{a} \cdot \mathbf{n}) d\mathbf{b} d^2\mathbf{a}, \tag{1.3}$$

where $\rho_0$ is the number density of molecules, the unit vectors $\mathbf{a}$ and $\mathbf{b}$ are in the direction of the long and short molecular axes, respectively, $\mathbf{d}$ is the molecular permanent dipole, $f(\mathbf{a} \cdot \mathbf{n})$ is the orientational distribution function of the uniaxial nematic phase and $q_{\alpha\beta}$ is the molecular quadrupole moment. Here $d^2\mathbf{a}$ denotes integration over all orientations of the long molecular axis $\mathbf{a}$ and $d\mathbf{b}$ denotes integration over the orientation of the short molecular axis $\mathbf{b}$ about the fixed axis $\mathbf{a}$. In terms of angles $d^2\mathbf{a} = d\cos\beta d\phi$ and $d\mathbf{b} = d\psi$ where $\beta$ is the polar angle between the long axis $\mathbf{a}$ and the director $\mathbf{n}$, $\phi$ is the corresponding azimuthal angle and the angle $\psi$ specifies the orientation of the short axis $\mathbf{b}$ about the long axis $\mathbf{a}$.

The anisotropic quadrupole tensor of the biaxial molecule can be expressed as:

$$q_{\alpha\beta} = \Delta q(a_\alpha a_\beta - \frac{1}{3}\delta_{\alpha\beta}) + q_b(b_\alpha b_\beta - c_\alpha c_\beta), \tag{1.4}$$

where the unit vector $\mathbf{c}$ is in the direction of the second short molecular axis, $\Delta q$ is the uniaxial and $q_b$ is the biaxial quadrupole moment.

The average quadrupole moment can now be written in the form:

$$\langle q_{\alpha\beta}\rangle = \rho_0 \left(\Delta q S + q_b D\right) \left(n_\alpha n_\beta - \frac{1}{3}\delta_{\alpha\beta}\right), \tag{1.5}$$

where $S$ is the nematic order parameter, $D$ is the orientational order parameter, which describes the ordering of the short molecular axis in the uniaxial nematic phase, and $\mathbf{n}$ is the director.

Note that the average molecular quadrupole is non-zero in the homogeneous nematic phase. At the same time the quadrupolar contribution to the macroscopic polarization vanishes in the homogeneous phase because it is determined by the gradient of the quadrupole density. In contrast, the average molecular dipole $\langle\mathbf{d}\rangle$ vanishes in the homogeneous phase because the orientational distribution function $f(\mathbf{a} \cdot \mathbf{n})$ is non-polar, i.e. it is invariant under the sign reversal of the molecular long axis $\mathbf{a}$ and the short axis $\mathbf{b}$. Mathematically this means that the orientational distribution function depends only on the squares of the scalar products $(\mathbf{a} \cdot \mathbf{n})$ and $(\mathbf{b} \cdot \mathbf{n})$, i.e. $f(\mathbf{a}, \mathbf{b}, \mathbf{n}) = f((\mathbf{a} \cdot \mathbf{n})^2, (\mathbf{b} \cdot \mathbf{n})^2)$.

In the distorted liquid crystal the orientational distribution function is modified, and for a small orientational deformation it contains a small correction $h(\mathbf{a}, \mathbf{b}, \mathbf{n})$, which is generally polar; different orientations of the molecular dipole are not statistically equivalent in the distorted phase:

$$f(\mathbf{a}, \mathbf{b}, \mathbf{n}) = f_0(\mathbf{a}, \mathbf{b}, \mathbf{n}) \left(1 + h(\mathbf{a}, \mathbf{b}, \mathbf{n})\right), \tag{1.6}$$

where $f_0(\mathbf{a}, \mathbf{b}, \mathbf{n})$ is the orientational distribution function of the undistorted nematic phase and the correction $h(\mathbf{a}, \mathbf{b}, \mathbf{n}) \propto \nabla_\alpha n_\beta$ is supposed to be small and proportional to the gradients of the director.

Then the average molecular dipole in the distorted nematic phase can be expressed as:

$$\langle \mathbf{d} \rangle = \int \mathbf{d} f_0(\mathbf{a}, \mathbf{b}, \mathbf{n}) h(\mathbf{a}, \mathbf{b}, \mathbf{n}) d\mathbf{b} d^2\mathbf{a}, \tag{1.7}$$

where the molecular dipole is

$$\mathbf{d} = d_{\parallel}\mathbf{a} + d_{\perp}\mathbf{b}. \tag{1.8}$$

Here the polar correction to the orientational distribution function $h(\mathbf{a}, \mathbf{b}, \mathbf{n})$ should be determined using a corresponding molecular-statistical theory.

According to Eq. (1.1) the quadrupole contribution to the flexoelectric polarization is determined by the gradient of the average quadrupole density, which can be written in the following form using Eq. (1.5):

$$P_\alpha(r) = \nabla_\beta \langle q_{\alpha,\beta}(\mathbf{r}) \rangle = \rho_0 \left( \Delta q S + q_b D \right) \nabla_\beta (n_\alpha n_\beta). \tag{1.9}$$

Equation (1.9) can be rewritten in the conventional form:

$$\mathbf{P} = e_1 \mathbf{n}(\nabla \cdot \mathbf{n}) + e_3 (\mathbf{n} \cdot \nabla)\mathbf{n}, \tag{1.10}$$

where

$$e_1 = e_3 = \rho_0 \left( \Delta q S + q_b D \right). \tag{1.11}$$

Thus the quadrupole mechanism yields very simple expressions for the flexoelectric coefficients, which are proportional to the nematic order parameter $S$ in the first approximation (usually the parameter $D$ is much smaller than $S$). In addition, the difference of the flexoelectric coefficients appears to be equal to zero, i.e. $\Delta e = e_1 - e_3 = 0$ if the quadrupole contribution alone is taken into account.

At the same time experimental facts indicate that the difference between the flexocoefficients is non-zero and even rather large for a number of nematic materials, and it strongly depends on the absolute value and the orientation of the permanent dipole within the molecular structure.[23–25] Moreover, the difference between the flexocoefficients determines the flexoelectro-optic effect, which has been extensively studied experimentally.[26,27] There exist also some other experimental data which, in principle, allowed us to distinguish between dipolar and quadrupolar flexoelectricity.[28] This can be considered as an argument in favour of the dipolar interpretation of the flexoelectric effect. On the other hand, the actual ratio of the dipolar and quadrupolar contributions to the flexocoefficients of particular nematic materials remains unknown. It is only possible to speculate that the quadrupolar contribution is not expected to be sensitive to the details of the molecular structure because it is determined by the quadrupole moment of the whole molecule, while the dipolar contribution should be very sensitive to those changes of the molecular structure, which affect the electric and steric dipoles.

## 1.3. Density Functional Theory of Flexoelectricity

A molecular-statistical theory of the flexoelectric effect in the nematic phase can be derived in a general way using the density-functional approach to the theory of liquid crystals.[29–31] In this approach, the free energy of a liquid crystal, $F$, is a functional of the density $\rho_0(\omega) = \rho_0 f(\omega)$ where $f(\omega)$ is the orientational distribution function. The general structure of the functional $F(\rho)$ is not known, but the functional derivatives are known and are related to the direct correlation functions of the nematic phase.

The free energy of the distorted nematic liquid crystal can generally be written as a sum of two terms:

$$F = W + H, \tag{1.12}$$

where $W$ is the free energy of the system without intermolecular interactions. Without the external field the ideal gas free energy $W$ can be expressed as

$$W = \rho_0 k_B T \int f(\omega) \ln [f(\omega)\Lambda] \, d\omega, \tag{1.13}$$

where $\omega = (\mathbf{a}, \mathbf{b}, \mathbf{r})$ and $\Lambda = \text{const}$.

The reduced free energy $H(\rho)$ is the generating functional for the direct correlation functions of the system:

$$\delta\beta H/\delta\rho(\omega) = -C_1(\omega) = -\ln\rho(\omega) + \beta\mu + \text{const}, \tag{1.14}$$

$$\delta^2\beta H/\delta\rho(\omega_1)\delta\rho(\omega_2) = -C_2(\omega_1, \omega_2), \tag{1.15}$$

where $\beta = 1/(k_B T)$, $\mu$ is the chemical potential and $C_2(\omega_1, \omega_2)$ is the direct pair correlation function of the liquid crystal. The direct correlation function is related to the full pair correlation function $g_2(\omega_1, \omega_2)$ by the Ornstein–Zernike equation:

$$g_2(\omega_1, \omega_2) = C_2(\omega_1, \omega_2) + \int g_2(\omega_1, \omega_3) C_2(\omega_2, \omega_3) f(\omega_3) d\omega_3.$$

In this equation the first term describes the so-called direct correlations between the molecules '1' and '2' while the second term describes the correlations via a third particle. The direct correlation function $C_2(\omega_1, \omega_2)$ generally describes short-range intermolecular correlations and decays rapidly with increasing intermolecular separation while the full pair correlation function $g_2(\omega_1, \omega_2)$ normally has a long-range tail. Note that from a mathematical point of view $C_1(\omega)$ and $C_2(\omega_1, \omega_2)$ are functions of $\omega$ and, simultaneously, can be considered as functionals of the density $\rho(\omega)$. For example, the functional derivative $\delta C_1(\omega_1)/\delta\rho(\omega_2) = C_2(\omega_1, \omega_2)$. Thus it is more precise to use the forms $C_1(\omega, [\rho(\omega)])$ and $C_2(\omega_1, \omega_2, [\rho(\omega_1)], [\rho(\omega_2)])$, which will be used below.

The one-particle distribution function of the distorted liquid crystal can be determined by minimization of the free energy with respect to $f(\omega)$ taking into account the normalization condition. We obtain

$$\delta\left(F + \lambda \int \rho(\omega) d\omega\right) \delta\rho(\omega) = -K_B T C_1(\omega, [\rho(\omega)]) + k_B T \ln f(\omega) + \Lambda + \lambda = 0, \tag{1.16}$$

where the Lagrange multiplier $\lambda$ is determined from the normalization condition.

Let us now assume that the orientational deformation is small, which corresponds to the linear flexoeffect considered in this chapter. Then the one-particle distribution function of the distorted nematic liquid crystal is given by Eq. (1.6):

$$f(\omega) = f_0(\omega)\,(1 + h(\omega))\,, \tag{1.17}$$

where the function $h$ is small and proportional to the orientational deformation $\nabla_\alpha n_\beta$, and where $f_0(\omega) = f_0((\mathbf{a}\cdot\mathbf{n})^2, (\mathbf{b}\cdot\mathbf{n})^2)$ is the orientational distribution function of the undistorted nematic phase, which, however, depends on the inhomogeneous director $\mathbf{n}(\mathbf{r})$.

Let us now use the functional Taylor expansion of the function $C_1(\omega, [\rho(\omega)])$ on the right-hand side of Eq. (1.16) about the local distribution $f_0(\omega, \mathbf{n}_1)$ at the point $\mathbf{r}_1$:

$$C_1(\omega, [\rho(\omega_1)]) = C_1(\omega, [\rho_0(\omega_1)])$$

$$+ \rho_0 \int C_2(\omega_1, \omega_2, [\rho_0(\omega_1)], [\rho_0(\omega_2)])\,(f(\omega_2, \mathbf{n}_2) - f_0(\omega_2, \mathbf{n}_1))\,d\omega_2 + ...., \tag{1.18}$$

where $\mathbf{n}_1 = \mathbf{n}(\mathbf{r}_1)$ and $\mathbf{n}_2 = \mathbf{n}(\mathbf{r}_2)$. Note that the higher-order terms in the expansion (which are determined by higher-order direct correlation functions) do not contribute to the flexoelectric effect because they are proportional to higher powers of $(f(\omega_2, \mathbf{n}_2) - f_0(\omega_2, \mathbf{n}_1))$ and hence to higher powers of the orientational deformation.

The distribution function $f(\omega_2, \mathbf{n}_2)$ in Eq. (1.18) is given by the general Eq. (1.17) where the unperturbed distribution $f_0(\omega_2) = f_0(\mathbf{a}_2, \mathbf{b}_2, \mathbf{n}_2)$ depends on the local director $\mathbf{n}(\mathbf{r}_2)$ at the point $\mathbf{r}_2$. The function $f_0(\mathbf{a}_2, \mathbf{b}_2, \mathbf{n}(\mathbf{r}_2))$ can now be expanded in gradients of the director and to the first order (introducing $\mathbf{r}_{12} = \mathbf{r}_2 - \mathbf{r}_1$):

$$f_0(\mathbf{a}_2, \mathbf{b}_2, \mathbf{n}(\mathbf{r}_2)) \approx f_0(\mathbf{a}_2, \mathbf{b}_2, \mathbf{n}(\mathbf{r}_1)) + (\mathbf{r}_{12} \cdot \nabla) f_0(\mathbf{a}_2, \mathbf{b}_2, \mathbf{n}(\mathbf{r}_1)).$$

Substituting this gradient expansion into Eq. (1.17) we obtain

$$(f(\omega) - f_0(\omega)) = (\mathbf{r}_{12} \cdot \nabla) f_0((\mathbf{a}\cdot\mathbf{n})^2, (\mathbf{b}\cdot\mathbf{n})^2) + f_0((\mathbf{a}\cdot\mathbf{n})^2, (\mathbf{b}\cdot\mathbf{n})^2) h(\omega)\,. \tag{1.19}$$

Substitution of Eq. (1.19) into Eq. (1.18) and then into the general Eq. (1.16) and with the expansion of $\ln f$ in Eq. (1.16) to the first order in the small function $h$, yields the following integral equation for the correction $h(\omega)$:

$$\begin{aligned} h(\mathbf{x}_1) = &-\rho_0 \int C_2(\mathbf{x}_1, \mathbf{x}_2, \mathbf{r}_{12})(\mathbf{r}_{12} \cdot \nabla) f_0((\mathbf{a}\cdot\mathbf{n})^2, (\mathbf{b}\cdot\mathbf{n})^2) d\mathbf{x}_1 d\mathbf{x}_2 d\mathbf{r}_{12} \\ &- \rho_0 \int C_2(\mathbf{x}_1, \mathbf{x}_2, \mathbf{r}_{12}) f_0((\mathbf{a}\cdot\mathbf{n})^2, (\mathbf{b}\cdot\mathbf{n})^2) h(\mathbf{x}_2) d\mathbf{x}_1 d\mathbf{x}_2 d\mathbf{r}_{12}, \end{aligned} \tag{1.20}$$

where $\mathbf{x} = (\mathbf{a}, \mathbf{b})$.

In the general part the correction $h(\mathbf{a}, \mathbf{b})$ is a sum of the 'symmetric' and the 'antisymmetric' parts:

$$h(\mathbf{a}, \mathbf{b}) = h_s(\mathbf{a}, \mathbf{b}) + h_a(\mathbf{a}, \mathbf{b}), \tag{1.21}$$

where the symmetric part $h_s(\mathbf{a}, \mathbf{b})$ is non-polar, i.e. it is invariant under the transformation $\mathbf{a} \to -\mathbf{a}$ and $\mathbf{b} \to -\mathbf{b}$, while the antisymmetric part $h_a(\mathbf{a}, \mathbf{b})$ changes sign under such a transformation. The flexoelectric effect is determined only by the antisymmetric function $h_a(\mathbf{a}, \mathbf{b})$, which is sensitive to the direction of the molecular dipole.

Equation (1.20) can now be split into two independent equations for the functions $h_s(\mathbf{a}, \mathbf{b})$ and $h_a(\mathbf{a}, \mathbf{b})$ because the last term in Eq. (1.20) vanishes when $h(\mathbf{a}, \mathbf{b}) = h_s(\mathbf{a}, \mathbf{b})$. We are interested only in the equation for the antisymmetric function, which can be written in the following simple form:

$$h(\mathbf{x}_1) = -\rho_0 \int C_2(\mathbf{x}_1, \mathbf{x}_2, \mathbf{r}_{12})(\mathbf{r}_{12} \cdot \nabla) f_0((\mathbf{a} \cdot \mathbf{n})^2, (\mathbf{b} \cdot \mathbf{n})^2) d\mathbf{x}_1 d\mathbf{x}_2 d\mathbf{r}_{12}. \tag{1.22}$$

Substituting Eq. (1.22) into the general Eqs (1.1) and (1.7) we obtain the following expression for the flexoelectric polarization:

$$P_\alpha = e_{\alpha\beta\gamma\delta} n_\gamma \nabla_\beta n_\delta, \tag{1.23}$$

where the tensor of flexoelectric coefficients reads

$$e_{\alpha\beta\gamma\delta} = -2\rho_0^2 kT \int d\mathbf{x}_1 d\mathbf{x}_2 \left( C_2(\mathbf{x}_1, \mathbf{x}_2, \mathbf{r}) r_\beta d\mathbf{r} \right) f_0(\mathbf{x}_1, \mathbf{n}) \times$$

$$\left( \frac{\partial f_0}{\partial (\mathbf{a}_2 \cdot \mathbf{n})^2} a_{2\gamma} a_{2\delta} + \frac{\partial f_0}{\partial (\mathbf{b}_2 \cdot \mathbf{n})^2} b_{2\gamma} b_{2\delta} \right). \tag{1.24}$$

The flexoelectric polarization can now be written in the conventional form:

$$\mathbf{P} = e_1 \mathbf{n}(\nabla \cdot \mathbf{n}) + e_3 (\mathbf{n} \cdot \nabla)\mathbf{n}, \tag{1.25}$$

where

$$e_1 = 2e_{xzzx} = 2e_{yzzy}; \quad e_3 = 2e_{xzxz} = 2e_{yzyz}. \tag{1.26}$$

Note that the expressions for the flexoelectric coefficients, obtained using the density functional approach, are very general as they take into account all contributions linear in the gradients of the director. Thus all approximate expressions, used in the literature to establish a relation between the flexocoefficients and the molecular model parameters, can be obtained from these general formulae by using an appropriate approximation for the direct correlation function. In the following two sections we will consider the results obtained using the molecular-field approximation and taking into account dipole–dipole correlations, respectively.

## 1.4. Influence of Polar Molecular Shape on the Flexocoefficients

Simple expressions for the flexoelectric coefficients in the nematic phase can be obtained[9] using the generalized van der Waals approximation. In this approximation the intermolecular attraction is taken into account in the mean-field approximation while the steric repulsion is accounted for by taking into consideration short-range steric correlations via the excluded volume effect. Then the free energy of the nematic phase can be written in the form:

$$F = \frac{1}{2}\rho_0^2 \int f_0(\mathbf{x}_1) f_0(\mathbf{x}_2) V(\mathbf{x}_1, \mathbf{x}_2, \mathbf{r}_{12}) \left(\Theta(\xi_{12} - r_{12}) + 1\right) d\mathbf{x}_1 d\mathbf{x}_2 d\mathbf{r}_{12}$$

$$-\frac{1}{2}kT\rho_0^2 \int f_0(\mathbf{x}_1) f_0(\mathbf{x}_2) \Theta(\xi_{12} - r_{12}) d\mathbf{x}_1 d\mathbf{x}_2 d\mathbf{r}_{12} + \rho kT \int f_0(\mathbf{x}_1) \ln f_0(\mathbf{x}_1) d\mathbf{x}_1, \tag{1.27}$$

The first term in Eq. (1.27) is the internal energy of the system where $V(\mathbf{x}_1, \mathbf{x}_2, \mathbf{r}_{12})$ is the pair intermolecular potential and $\Theta(\xi_{12} - r_{12})$ is the step function. Here $\xi_{12}$ is the closest distance of approach for the two molecules '1' and '2' with fixed relative orientation. $\Theta(\xi_{12} - r_{12}) = -1$ if $r_{12} < \xi_{12}$, i.e. if the molecules penetrate each other, and $\Theta(\xi_{12} - r_{12}) = 0$ if $r_{12} \geq \xi_{12}$. The second term in Eq. (1.27) is the so-called packing entropy and the third term is the orientational entropy. Thus in the first term the integration over $r_{12}$ is restricted to $r_{12} \geq \xi_{12}$, which takes into account that the molecules cannot penetrate each other.

In this approximation the direct correlation function in the nematic phase can be obtained using Eq. (1.15):

$$C_2(\mathbf{x}_1, \mathbf{x}_2, \mathbf{r}_{12}) = \Theta(\xi_{12} - r_{12}) \left(\beta V(\mathbf{x}_1, \mathbf{x}_2, \mathbf{r}_{12}) - 1\right). \tag{1.28}$$

The direct correlation function can now be substituted into the general Eqs (1.24) and (1.26) to obtain the particular expressions for the flexocoefficients.

The results of Straley[5] can be obtained by neglecting the pair attraction interaction potential $V(\mathbf{x}_1, \mathbf{x}_2, \mathbf{r}_{12})$ in the equation for the direct correlation function. Indeed, the Straley theory of flexoelectricity was developed for the system of hard polar rods, while for thermotropic liquid crystals both the molecular shape and the intermolecular attraction are important.

As mentioned above, the dipolar flexoelectric coefficients are determined by the polar molecular shape, which can be characterized by the molecular steric dipole. For a molecule having the shape of a truncated cone, as shown in Fig. 1.2, the steric dipole is in the direction of the long molecular axis **a** and is proportional to the cone angle $\gamma_{\parallel}$, while for a bent-rod molecule the steric dipole is parallel to the short axis **b** and is proportional to the bend angle $\gamma_{\perp}$. The relation between the flexocoefficients and the molecular shape is determined by the distance of closest approach $\xi_{12} = \xi_{12}(\mathbf{x}_1, \mathbf{x}_2, \mathbf{r}_{12})$, which reflects the polarity of the shape.

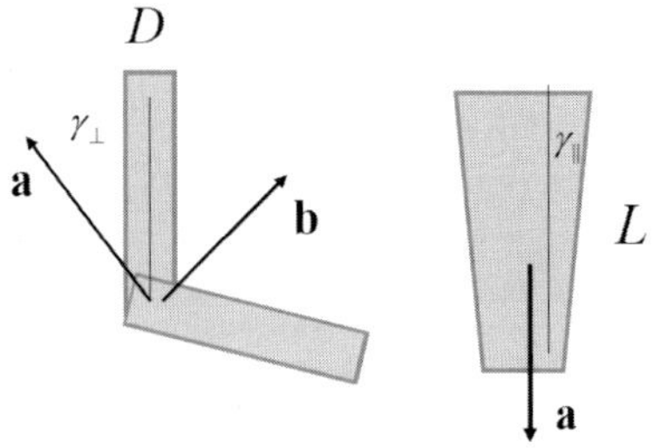


Fig. 1.2. Bent-rod and truncated-cone models for a polar molecular shape specified by the bend angle $\gamma_\perp$ and the cone angle $\gamma_\|$, respectively.

At small angles $\gamma \ll 1$ the function $\xi_{12}(\mathbf{x}_1, \mathbf{x}_2, \mathbf{r}_{12})$ can be expanded in powers of $\gamma$:

$$\xi_{12} \approx \xi_{12}^0 + \gamma_\perp \xi_\perp + \gamma_\| \xi_\| , \tag{1.29}$$

where $\xi_{12}^0$ is the closest distance of approach for cylindrically symmetric molecules and the functions $\xi_\perp(1,2)$ and $\xi_\|(1,2)$ are the coefficients of the expansion.

The pair direct correlation function can then be approximately expressed as:

$$C_2(\mathbf{x}_1, \mathbf{x}_2, \mathbf{r}_{12}) = \Theta(\xi_{12}^0 - r_{12})\,(\beta V(\mathbf{x}_1, \mathbf{x}_2, \mathbf{r}_{12}) - 1)$$

$$+ \delta(\xi_{12}^0 - r_{12})\left(\beta V(\mathbf{x}_1, \mathbf{x}_2, \xi_{12}^0) - 1\right)\left(\gamma_\perp \xi_\perp + \gamma_\| \xi_\|\right). \tag{1.30}$$

In the present section we consider only the flexoelectric effect relating to the polar molecular shape, and we therefore assume that the pair attraction interaction potential $V(\mathbf{x}_1, \mathbf{x}_2, \mathbf{r}_{12})$ is even in $\mathbf{a}_1, \mathbf{a}_2, \mathbf{b}_1, \mathbf{b}_2$. In this case the first term in Eq. (1.30) does not contribute to the flexoelectric coefficients, which are determined mainly by steric dipoles.

In the general case the attraction interaction potential may also contain a polar contribution, which is determined mainly by the polar part of the dispersion and the multipole interaction energies. It can be shown,[9] however, that the corresponding contribution makes only a small correction to the value of the flexocoefficients.

Flexoelectric coefficients are mainly determined by the function $\xi_{12}$, which specifies the molecular shape, and the pair attraction interaction potential $V(1,2)$. In the general case neither of these functions can be written in a simple analytical form, which makes it very difficult to evaluate the flexoelectric coefficients using the general formulae. At the same time it is possible to obtain some reasonable estimates of the coefficients in terms of the parameters of the molecular shape and the nematic order parameter. As shown by Osipov,[9] the flexoelectric coefficients can be estimated as:

$$e_1 = \frac{\rho_0^2}{kT} DL^2 S \left(S(\gamma_\parallel d_\parallel \lambda_1 + \gamma_\perp d_\perp \kappa_1) + \gamma_\parallel d_\parallel \lambda_2 + \gamma_\perp d_\perp \kappa_2\right),$$

$$e_3 = \frac{\rho_0^2}{kT} DL^2 S \left(S(\gamma_\parallel d_\parallel \lambda_1 - \gamma_\perp d_\perp \kappa_3) + \gamma_\parallel d_\parallel \lambda_2 + \gamma_\perp d_\perp \kappa_2\right), \qquad (1.31)$$

where $S$ is the nematic order parameter and where the coefficients are estimated as

$$\lambda_1 \sim \frac{1}{2} 10^{-2} (A + B - kT),$$

$$\lambda_2 \sim 10^{-2} (A + 2B - kT),$$

$$\kappa_1 \sim 10^{-1} (2A + B - 2kT),$$

$$\kappa_2 \sim 10^{-1} \left(A + \frac{1}{2} B - kT\right),$$

$$\kappa_3 \sim 10^{-1} (A + B - kT). \qquad (1.32)$$

The parameter $A$ is the isotropic part of the total attraction interaction potential while the parameter $B$ is the anisotropic part of the same potential. These parameters are defined in the following way.

The pair interaction potential integrated over the intermolecular vector $\mathbf{r}_{12}$ and the short molecular axes $\mathbf{b}_1$ and $\mathbf{b}_2$ only depends on the coupling between the long axes $(\mathbf{a}_1 \cdot \mathbf{a}_2)$, and thus can be expanded using Legendre polynomials $P_n(\mathbf{a}_1 \cdot \mathbf{a}_2)$. Taking into account the first non-polar term we obtain:

$$\langle V(1,2) \rangle = A + BP_2(\mathbf{a}_1 \cdot \mathbf{a}_2) + \ldots. \qquad (1.33)$$

The parameters $A$ and $B$ can readily be calculated if the explicit form of the potential $V(1,2)$ is known. At the same time, it is particularly interesting to estimate the coefficients $A$ and $B$ for typical mesogenic molecules.

It follows from Eqs (1.31) and (1.32) that the predominant contribution to the flexoelectric coefficients is determined by the isotropic intermolecular attraction modulated by the polar molecular shape. Indeed, in the general case the maximum attraction interaction energy $V(R) \gg kT$ where $R$ is the equilibrium distance between the two molecules. It follows then that $A \gg$

$kT$. This conclusion is supported by the theory of Gelbart and Barboy,[32] who showed that the most important anisotropic interaction in nematic liquid crystals is the isotropic intermolecular attraction modulated by the anisotropic molecular shape.

As mentioned above, the first expressions for the flexoelectric coefficients were obtained by Helfrich[6] and Petrov and Derzhanski[7,8] while a systematic molecular-statistical theory was developed later by Straley.[5] The results of these two approaches were compared by Marcerou and Prost[2] who concluded that the theories of Helfrich and Petrov and Derzhanski and of Straley describe different mechanisms for the dipolar flexoelectric effect because Straley's theory yields values for the flexocoefficients that are two orders of magnitude smaller than the experimental ones, and which therefore can be neglected.

The general mean-field results, presented in this section, enable us to clarify this problem. It should be noted that Straley's theory was developed for a system of rigid rods and thus it takes into consideration only a short-range steric repulsion between molecules. On the other hand, in the theory of Helfrich and Petrov and Derzhanski the flexocoefficients are expressed in terms of Frank elastic constants, which, in turn, are determined by both the intermolecular attraction and repulsion. The relation between the two contributions can be clarified using Eqs (1.31) and (1.32), which can be used to obtain the following estimate of the flexoelectric coefficients:

$$e \sim \frac{\rho^2}{10kT} DL^2 Sd\gamma\,(A + B - kT)\,. \tag{1.34}$$

Using the values $\rho_0 = 10^{27}$ m$^{-3}$, $D = 0.5$ nm, $L = 3$ nm, $S = 1/2$ and $d = 1$ D $= 3.3356 \times 10^{-30}$ Cm, we obtain $e \sim 16 \times 10^{-2}(A + B - kT)/kT$ pC/m. Here $B$ is the characteristic anisotropic intermolecular interaction energy and hence $B \sim kT_{NI}$ where $T_{NI}$ is the nematic-isotropic transition temperature. The parameter $A$ is the isotropic part of the attraction interaction energy and thus we expect that $A \gg kT$. The interaction energies of typical mesogenic molecules have been calculated numerically.[33,34] Using these results of Baran and Les[33] we obtain $A \sim 5kT$ or from Tokita *et al.*[34] $A \sim 10kT$. Taking the value $A = 10kT$ we estimate $e \sim 2$ pC/m, which is close to the experimental values between 1 pC/m and 10 pC/m. In fact, our estimate corresponds to the one obtained[2] using the Petrov and Derzhanski and Helfrich theory. On the other hand, steric repulsion, considered by Straley, actually gives a much smaller contribution, $e < 10^{-1}$ pC/m, which can be obtained by setting $A = B = 0$.

Let us now discuss the approximate expressions for the flexoelectric coefficients, Eq. (1.31), in more detail. Firstly, note that the expressions for both coefficients $e_1$ and $e_3$ contain terms proportional to both $S$ and $S^2$. It has been assumed in the literature[6,8] that the dipolar contribution to the flexoelectric coefficients is always proportional to $S^2$ while the quadrupole contribution[2] is proportional to $S$, and even the method of separation between the dipolar and quadrupolar flexoelectric effect has been proposed[2,28] based on these preliminary results. The results of the consistent molecular theory presented in this section allow us to conclude that the relation $e \sim S^2$ for the dipolar contribution is due to the shortcomings of the semi-phenomenological approach.[6,8] The results of this section also cast some doubt on the quantitative ratio of the dipolar and quadrupole contributions[28] based on a comparison of the two terms in the expression $e = e_0 S + e_2 S^2$. At the same time, the absence of the linear term in $S$ in the dependence $e(S)$ for a number of nematic materials[28] still points to the predominant role of the quadrupole flexoeffect for those materials.

Secondly, it follows from Eqs (1.31) and (1.32) that the longitudinal molecular dipole $d_{\|}$ provides a much smaller contribution to the flexocoefficients than the transverse dipole $d_{\perp}$, since $\lambda/\kappa \sim 10^{-1}$. Thus we conclude that the dipole flexoeffect is expected to be important only for molecules with large transverse dipoles. Note that the significant dipole flexoeffect has indeed been determined[28] for nematics composed of molecules with large transverse dipoles. For cyanobiphenyl liquid crystals Marcerou and Prost did not find any dipolar flexoelectric effect,[28] which may be determined not only by the tendency to form dimers with antiparallel dipoles but also by a relatively small contribution from transverse molecular dipoles to the flexoelectric coefficients.

The difference of flexoelectric coefficients $e^* = e_1 - e_3$ plays an important role in many cases including, in particular, the so-called flexoelectro-optic effect. As shown in Section 1.2, $e^* \approx 0$ for the purely quadrupole flexoeffect. On the other hand, experimental data indicate that the difference $e^*$ is generally not small. Indeed, according to some estimates[1,23] the difference $e^* \approx 5.6$ pC/m in the liquid crystal **BMAOB** (4-n-butyl-4′-methoxyazoxybenzene) and $e^* \approx 3$ pC/m in **MBBA** (N-(4-metoxybenzylidene)-4′-butylaniline).[24] Direct measurements[25] of the difference of the flexocoefficients $e^*$ also yield values of the order of 3 pC/m for 4-octyl-4′-cyanobiphenyl and 4-n-octyloxy-4′-cyanobiphenyl (**8CB** and **8OCB**, respectively). Therefore the difference of the flexocoefficients in the nematic phase should be due mainly to the dipolar flexoeffect. This conclusion is also supported by the strong effect of the absolute value and the orientation of the molecular dipole on the difference $e^*$ as directly observed.[24]

One obtains the following expression for the difference of flexocoefficients from Eqs (1.31):

$$e^* = e_1 - e_3 = \frac{\rho^2}{kT} DL^2 S d_\perp \gamma_\perp (\kappa_1 + \kappa_3) . \tag{1.35}$$

Thus the difference of the flexocoefficients is determined only by the transverse electric and steric dipoles. For molecules with $d_\perp = 0$ the difference $e^*$ should be small, while in the general case the difference $e^*$ is of the same order of magnitude as the sum $e_1 + e_3$. The latter conclusion can be verified experimentally. Unfortunately, the available experimental data are too scanty to be directly compared with Eq. (1.35).

The expressions for the flexoelectric coefficients presented in this section are derived using the molecular-field approximation. Therefore, care should be taken in the description of nematic liquid crystals composed of strongly polar molecules. In such liquid crystal materials (for example, cyanobiphenyls) the flexoelectric coefficients may be strongly affected by the short-range dipole–dipole correlations, which are considered in the following section.

## 1.5. Influence of Dipole–Dipole Correlations

The general expressions for the flexoelectric coefficients obtained using the density functional approach can be used to estimate the contributions from different types of intermolecular correlations, including short-range dipole–dipole correlations. We use the following approximation for the direct pair correlation function:

$$C_2(\mathbf{x}_1, \mathbf{x}_2, \mathbf{r}_{12}) = \Theta(r_{12} - \xi_{12}) \exp\left(-\beta V(\mathbf{x}_1, \mathbf{x}_2, \mathbf{r}_{12})\right) , \tag{1.36}$$

where the step function $\Theta(r_{12} - \xi_{12})$ describes the excluded volume effect and where $V(\mathbf{x}_1, \mathbf{x}_2, \mathbf{r}_{12})$ is the total intermolecular attraction interaction potential:

$$V(1,2) = V_{disp}(1,2) + V_{dd}(1,2). \tag{1.37}$$

Here $V_{dd}(1,2)$ is the energy of the electrostatic dipole–dipole interaction

$$V_{dd}(1,2) = r_{12}^{-3} \left(\mathbf{d}_1 \cdot \mathbf{d}_2 - 3(\mathbf{d}_1 \cdot \mathbf{u}_{12})(\mathbf{d}_2 \cdot \mathbf{u}_{12})\right) , \tag{1.38}$$

where $\mathbf{u}_{12} = \mathbf{r}_{12}/r_{12}$, and where $V_{disp}(1,2)$ is the remaining part of the attraction interaction energy, which is mainly determined by the dispersion intermolecular interaction.

Now the direct correlation function may be expanded using powers of the potential $V(1,2)$ and to the second order:

$$C_2(\mathbf{x}_1, \mathbf{x}_2, \mathbf{r}_{12}) = \Theta(r_{12} - \xi_{12})(1 - \beta V_{disp}(1,2) - \beta V_{dd}(1,2)$$

$$+ \frac{1}{2}\beta^2 \left(V_{disp}^2(1,2) + V_{dd}^2(1,2) + 2V_{disp}(1,2)V_{dd}(1,2)\right) + ...). \qquad (1.39)$$

Note that $\beta V_{dd}(1,2) < 1$. Indeed, the dimensionless parameter $\beta \rho_0 d^2 \approx 1/3$ for a very large dipole $d = 4$ D and $\rho_0 = 10^{27}$ m$^{-3}$. Thus for a description of the dipole–dipole correlations we need only to keep the first terms in the expansion.

The linear terms in the expansion Eq. (1.39) do not contribute to the flexoelectric coefficients because the dipole–dipole interaction potential is odd both in $\mathbf{d}_1$ and $\mathbf{d}_2$ and hence the corresponding contributions vanish after averaging over the orientation of the molecular axes. Thus it is necessary to take into account the quadratic terms in the expansion of the direct correlation function. Then the contribution from the dipole–dipole correlations to the flexocoefficients can be written in the form:

$$\Delta e_{\alpha\beta\gamma\delta} = -\beta\rho^2 \int \left(V_{dd}^2(\mathbf{x}_1, \mathbf{x}_2, \mathbf{r})\Theta(r_{12} - \xi_{12}) r_\beta d\mathbf{r}\right) \times$$

$$f_0(\mathbf{x}_1, \mathbf{n}) \left(\frac{\partial f_0}{\partial(\mathbf{a}_2 \cdot \mathbf{n})^2} a_{2\gamma} a_{2\delta} + \frac{\partial f_0}{\partial(\mathbf{b}_2 \cdot \mathbf{n})^2} b_{2\gamma} b_{2\delta}\right) d\mathbf{x}_1 d\mathbf{x}_2, \qquad (1.40)$$

where $f_0(\mathbf{x}_1, \mathbf{n}) = f_0((\mathbf{a}_1 \cdot \mathbf{n})^2, (\mathbf{b} \cdot \mathbf{n})^2)$. For simplicity the integration in Eq. (1.40) can be performed for perfect orientational order.

Let us now consider a system of bent-core molecules with the shape of slightly bent ellipsoids of revolution of length $L$ and diameter $D$. The steric dipole of such a molecule is determined by the average bend angle $\gamma$. For small $\gamma$ the closest distance of approach for the molecules '1' and '2' can be expressed as:

$$\xi_{12}^2(\mathbf{x}_1, \mathbf{x}_2, \mathbf{u}_{12}) = \xi_{12}^0(\mathbf{a}_1, \mathbf{a}_2, \mathbf{u}_{12}) + \Delta\xi(\mathbf{a}_1, \mathbf{b}_1, \mathbf{a}_2, \mathbf{b}_2, \mathbf{u}_{12}), \qquad (1.41)$$

where the function $\xi_{12}^0$ is independent of $\gamma$ while $\Delta\xi \to 0$ as $\gamma \to 0$.

For perfect orientational order of long molecular axes, $\mathbf{a}_1 = \mathbf{a}_2 = \mathbf{n}$, where $\mathbf{n}$ is the director, and hence we obtain:[10]

$$\Delta\xi(\mathbf{a}_1, \mathbf{b}_1, \mathbf{a}_2, \mathbf{b}_2, \mathbf{u}_{12}) = L\gamma\left((\mathbf{b}_1 \cdot \mathbf{u}_{12}) + (\mathbf{b}_2 \cdot \mathbf{u}_{12})\right)|(\mathbf{n} \times \mathbf{u}_{12})|/2. \quad (1.42)$$

The closest distance of approach for two uniaxial parallel ellipsoids can be expressed in the analytical form:

$$\xi^0(\mathbf{n}\cdot\mathbf{u}) = D\left(1-\left(1-\frac{D^2}{L^2}\right)(\mathbf{n}\cdot\mathbf{u})^2\right)^{-1/2}. \qquad (1.43)$$

For small angles $\gamma \ll 1$ the step function can be expanded as

$$\Theta(r_{12}-\xi_{12}) = \Theta(r_{12}-\xi^0(\mathbf{n}\cdot\mathbf{u}))-\delta(r_{12}-\xi^0(\mathbf{n}\cdot\mathbf{u}))\Delta\xi(\mathbf{b}_1,\mathbf{b}_2,\mathbf{n},\mathbf{u}). \quad (1.44)$$

Substituting Eqs (1.42) and (1.43) into Eq. (1.44) and then into the general Eq. (1.40) we obtain the following expressions for the contributions from the dipole–dipole correlations to the flexoelectric coefficients of the nematic phase:[10]

$$\Delta e_1 \approx -\frac{1}{2(kT)^2}\pi\rho_0^2 d_\parallel^2 d_\perp LD^{-2}\gamma(\alpha_1 d_\parallel^2 - \alpha_2 d_\perp^2),$$

$$\Delta e_3 \approx -\frac{1}{2(kT)^2}\pi\rho_0^2 d_\perp LD^{-2}\gamma(\kappa_2 d_\perp^2 - \kappa_1 d_\parallel^2)(d_\parallel^2 - d_\perp^2), \qquad (1.45)$$

where $\alpha_1 \approx 5.6, \alpha_2 \approx 4, \kappa_1 \approx 5$ and $\kappa_2 \approx 1.8$.

It follows from Eq. (1.45) that the contribution from the dipole–dipole correlations strongly depends on the value of the molecular dipoles, i.e. $\Delta \sim d_\perp d_\parallel^4$ if $d_\parallel > d_\perp$. Therefore the correlation contribution can be substantial in liquid crystal materials composed of strongly polar molecules. The effect of dipole–dipole correlations on the flexoelectric coefficients has been discussed in experimental papers.[24,25]

Let us now estimate the corrections $\Delta e_1$ and $\Delta e_3$ for the nematic liquid crystal **8OCB**, which has been studied experimentally.[24] Note that in **8OCB** $d_\parallel \approx 4$ D and $d_\perp \approx 1$ D. Using the values $\rho_0 = 10^{27}$ m$^{-3}$, $D = 0.5$ nm, $L = 2.2$ nm and $T = 300$ K we obtain $\Delta e_1 \sim -3\gamma$ pC/m and $\Delta e_3 \sim -2.6\gamma$ pC/m. Assuming that the bend angle $\gamma \sim 1$ gives the estimate $\Delta e \sim 3$ pC/m, i.e. the correlation contribution is of the same order as the flexocoefficients themselves. Thus we conclude that the dipole–dipole correlations make a substantial contribution to the flexoelectric coefficients of strongly polar nematic liquid crystals and therefore they cannot generally be neglected.

Note that the corrections $\Delta e_1$ and $\Delta e_3$ have opposite signs when $d_\parallel > d_\perp$ ($\Delta e_1 < 0$ and $\Delta e_3 > 0$). Thus the contribution to the difference of the flexocoefficients appears to be negative and anomalously large: $\Delta e^* = \Delta e_1 - \Delta e_3 \sim -6$ pC/m. Indeed, it has been found[24] that the difference

$e^* \approx 3$ pC/m in the nematic liquid crystal **8CB** and $e^* \approx -15$ pC/m in the nematic material **8OCB**. Note that the molecules of the **8OCB** material have an additional transverse dipole $d_\perp \approx 1$ D, which is absent in the **8CB** molecule. Thus the experimentally observed large negative contribution to the difference in the flexocoefficients $e^*$ in the nematic liquid crystal **8OCB** may be determined by dipole–dipole correlations.

It should be noted that dipole–dipole correlations may contribute to the flexoelectric coefficients only when mesogenic molecules have both a longitudinal and a sufficiently large transverse dipole. This may explain why the correlation contribution seems to be very important for oxycyanobiphenyls (and not for cyanobiphenyls, which do not possess any transverse dipoles). For weakly polar molecules $d_\parallel \sim d_\perp \sim 1$ D, and in this case the contribution from the dipole–dipole correlations is two orders of magnitude smaller than for **8OCB** and can be neglected.

## 1.6. Influence of Real Molecular Shape

The molecular-statistical theory of flexoelectricity, presented in the previous sections, does not allow us to establish a direct relation between the flexocoefficients and the details of a particular molecular structure (except for permanent electric and steric dipoles) because the theory is based on simple model interaction potentials. A different version of the mean-field theory, which takes into consideration the real molecular shape, has recently been proposed by Ferrarini *et al.*[12,13] This approach is based on the following simple surface mean-field potential:

$$U_{MF}(i) = k_B T \nu \int_\sigma P_2(\mathbf{s} \cdot \mathbf{n}) d\sigma, \tag{1.46}$$

where the parameter $\nu$ determines the strength of the surface interaction, the unit vector $\mathbf{s}$ is the normal to the molecular surface and the integration is performed over the whole molecular surface.

The one-particle distribution function of the nematic phase is expressed in terms of the mean-field potential Eq. (1.46) in the following way

$$f(\omega) = \frac{1}{Z} \exp\left(-\frac{1}{k_B T} U_{MF}(\omega)\right), \tag{1.47}$$

where $Z$ is the normalization factor.

The mean-field potential describes a coupling between the surface element $d\sigma$ and the local director $\mathbf{n}$, and it can be described as a coupling between the director **n** and the molecular surface tensor

$$S_{\alpha\beta} = \int_{\sigma} s_{\alpha} s_{\beta} d\sigma.$$

The molecular surface tensor is a symmetric second-rank tensor, which can be written in the diagonal form:

$$S_{\alpha\beta} = s_1 a_\alpha a_\beta + s_2 b_\alpha b_\beta + s_3 c_\alpha c_\beta, \tag{1.48}$$

where the unit vectors $\mathbf{a}, \mathbf{b}$ and $\mathbf{c}$ are the principal axes of the tensor $S_{\alpha\beta}$. These axes define the molecular frame. Here the principal molecular axis **a** corresponds to the largest eigenvalue $s_1$. The surface mean-field potential can now be expressed as:

$$U_{MF} = k_B T\nu s_0 + k_B T\nu \Delta s \left( a_\alpha a_\beta - \frac{1}{3}\delta_{\alpha\beta} \right) + k_B T\nu \Delta s_\perp \left( b_\alpha b_\beta - c_\alpha c_\beta \right), \tag{1.49}$$

where

$$s_0 = \frac{1}{3}(s_1 + s_2 + s_3),$$

$$\Delta s = s_1 - \frac{1}{2}(s_2 + s_3),$$

$$\Delta s_\perp = s_2 - s_3. \tag{1.50}$$

In the distorted nematic the director distribution is inhomogeneous, and hence $\mathbf{n} = \mathbf{n}(\mathbf{r})$ in Eq. (1.46). Taking into account that the length scale of the director variation is much larger than the molecular size we can apply the gradient expansion of the director $\mathbf{n}(\mathbf{r})$ with respect to some point $\mathbf{R}_0$ inside the molecule:

$$\mathbf{n}(\mathbf{r}) = \mathbf{n}(\mathbf{R}_0) + ((\mathbf{r} - \mathbf{R}_0) \cdot \nabla)\, \mathbf{n}(\mathbf{R}_0) + ... \tag{1.51}$$

taking into account only the linear term.

Then the mean-field potential can be approximately expressed as:

$$U_{MF}(\mathbf{R}_0, \mathbf{a}, \mathbf{b}) = kT\nu \int_{\sigma} P_2(\mathbf{s} \cdot \mathbf{n}) d\sigma + 3 \sum_{\alpha,\beta,\gamma} n_\gamma \nabla_\beta n_\alpha kT\nu \int_{\sigma} r_\beta s_\gamma s_\alpha d\sigma, \tag{1.52}$$

where **r** is the position of the elementary surface element in the molecular frame.

Substituting Eq. (1.52) for the mean-field potential into Eq. (1.47) for the orientational distribution function gives the following expression for the correction $h(\omega)$ to the one-particle distribution function in a distorted nematic liquid crystal:

$$h(\mathbf{r}, \mathbf{a}, \mathbf{b}) = -3\nu \sum_{\alpha,\beta,\gamma} n_\gamma \nabla_\beta n_\alpha \int_\sigma r_\beta s_\gamma s_\alpha d\sigma. \tag{1.53}$$

The flexoelectric polarization is given by the same general Eq. (1.1) where the quadrupole contribution can again be written in the form of Eqs (1.9)–(1.11) with

$$\Delta q = (\mathbf{a} \cdot \mathbf{q} \cdot \mathbf{a}) - \frac{1}{2} ((\mathbf{b} \cdot \mathbf{q} \cdot \mathbf{b}) + (\mathbf{c} \cdot \mathbf{q} \cdot \mathbf{c})),$$

$$\Delta q_b = (\mathbf{b} \cdot \mathbf{q} \cdot \mathbf{b}) - (\mathbf{c} \cdot \mathbf{q} \cdot \mathbf{c}), \tag{1.54}$$

where we have taken into account that for molecules of sufficiently low symmetry the principal axes of the molecular quadrupole tensor do not necessarily coincide with the principal axes of the surface tensor $S_{\alpha\beta}$. In this case the quadrupolar flexocoefficients are determined by the diagonal elements of the quadrupole tensor in the molecular frame $(\mathbf{a}, \mathbf{b}, \mathbf{c})$.

The dipolar contribution to the flexoelectric polarization can now be written in the form:

$$\mathbf{P}_{dip} = -3\rho_0 \nu \sum_{\alpha,\beta,\gamma} n_\gamma \nabla_\beta n_\alpha \int_\sigma \langle \mathbf{d} r_\beta s_\gamma s_\alpha \rangle_0 d\sigma, \tag{1.55}$$

where $\mathbf{d}$ is the total molecular permanent dipole and the brackets $\langle (...) \rangle_0$ denote the average for the orientational distribution function $f_0(i)$ for the undistorted nematic liquid crystal.

Equation (1.55) for the dipolar flexoelectric polarization can again be rewritten in the conventional form where the flexoelectric coefficients are expressed as:

$$e_1 = -3\rho_0 \nu \int_\sigma \langle d_z r_x s_x s_z \rangle_0 d\sigma,$$

$$e_3 = -3\rho_0 \nu \int_\sigma \langle d_x r_z s_x s_z \rangle_0 d\sigma, \tag{1.56}$$

where the $z$ axis is parallel to the local director $\mathbf{n}$.

This approach has been used to calculate the flexoelectric coefficients numerically for several typical mesogens taking into account their real molecular surface and selected molecular conformations.[12] Numerical results have also been obtained for a number of nematics composed of molecular dimers that exhibit large values of the flexoelectric coefficients.[13] The results coincide with the experimental data at least within an order of magnitude, which makes this approach very promising. On the other hand, the approach is based on the oversimplified surface interaction potential, which does not account for any specific interactions between particular molecular fragments including, for example, localized dipoles, effective charges or strongly polarizable bonds. Strong orientational correlations including the dipole–dipole ones are also not taken into consideration.

We conclude that the existing molecular theory of flexoelectricity is far from being complete. Any future more advanced theory should combine a more sophisticated molecular-statistical approach, which will allow us to take into account at least short-range intermolecular correlations, and more realistic intermolecular interaction potentials, which reflect the real molecular structure.

## References


1. R.B. Meyer, Piezoelectric effects in liquid crystals, *Phys. Rev. Lett.* **22**(18), 918–921, (1969). doi: 10.1103/PhysRevLett.22.918
2. J. Prost and J.P. Marcerou, On the microscopic interpretation of flexoelectricity, *J. Phys. France* **38**(3), 315–324, (1977). doi: 10.1051/jphys:01977003803031500
3. A.K. Tagantsev, Pyroelectric, piezoelectric, flexoelectric, and thermal polarization effects in ionic crystals, *Sov. Phys. Uspekhi*, **30**(7), 588–603, (1987). doi: 10.1070/PU1987v030n07ABEH002926
4. V.L. Indenbom, E.B. Loginov and M.A. Osipov, On the microscopic origin of the flexoelectric effect in the smectic C liquid crystals, *Sov. Phys. Crystallogr.* **26**, 1157–1160, (1981).
5. J.P. Straley, Theory of piezoelectricity in nematic liquid crystals, and of the cholesteric ordering, *Phys. Rev. A* **14**(5), 1835–1841, (1976). doi: 10.1103/PhysRevA.14.1835
6. W. Helfrich, The strength of piezoelectricity in liquid crystals, *Z. Naturforsch.* **26A**(5), 833–835, (1971).
7. A.I. Derzhanski and A.G. Petrov, Piezoelectricity in nematic liquid crystals, *Acta Phys. Polon. A* **55**(5), 747–754, (1979).
8. A.G. Petrov and A.I. Derzhanski, A possible relationship between the dielectric permeability and the piezoelectric properties of nematic liquid crystals, *Phys. Lett.* **34A**(7), 427–428, (1971). doi: 10.1016/0375-9601(71)90951-0
9. M.A. Osipov, Molecular theory of flexoelectric effect in nematic liquid crystals, *Sov. Phys. JETP* **58**(6), 1167–1171, (1983).
10. M.A. Osipov and V.B. Nemtsov, On the statistical theory of the flexoelectric effect in liquid crystals, *Sov. Phys. Crystallogr.* **31**, 125–130, (1986).

11. M.A. Osipov, The order parameter dependence of the flexoelectric coefficients in nematic liquid crystals, *J. Physique Lett.* **45**(16), 823–826, (1984). doi: 10.1051/jphyslet:019840045016082300
12. A. Ferrarini, Shape model for the molecular interpretation of the flexoelectric effect, *Phys. Rev. E* **64**(2), 021710/1–11, (2001). doi: 10.1103/PhysRevE.64.021710
13. A. Ferrarini, C. Greco and G.R. Luckhurst, On the flexoelectric coefficients of liquid crystal monomers and dimers: A computational methodology bridging length-scales, *J. Mater. Chem.* **17**(11), 1039–1042, (2007). doi: 10.1039/B618928H
14. Y. Singh and U.P. Singh, Density-functional theory of the flexoelectric effect in nematic liquids, *Phys. Rev. A* **39**(8), 4254–4262, (1989). doi: 10.1103/PhysRevA.39.4254
15. S.A. Pikin, M.A. Osipov and E.M. Terentjev, The polarization properties of liquid crystals, *Soviet Scientific Reviews, Harwood Acad. Publ.* **11 A**, 193–283, (1989).
16. A.M. Somoza and P. Tarazona, Density functional theory of the elastic constants of a nematic liquid crystal, *Mol. Phys.* **72**(4), 911–926, (1991). doi: 10.1080/00268979100100641
17. A.V. Zahkarov and R.Y. Dong, The flexoelectric effect in nematic liquid crystals: A statistical-mechanical approach, *Eur. Phys. J. E* **6**(1), 3–6, (2001). doi: 10.1007/s101890170021
18. J. Stelzer, R. Berardi and C. Zannoni, Flexoelectric effects in liquid crystals formed by pear-shaped molecules. A computer simulation study, *Chem. Phys. Lett.* **299**(1), 9–16, (1999). doi: 10.1016/S0009-2614(98)01262-7
19. D.L. Cheung, S.J. Clark and M.R. Wilson, Calculation of flexoelectric coefficients for a nematic liquid crystal by atomistic simulations, *J. Chem. Phys.* **121**(18), 9131–9139, (2004). doi: 10.1063/1.1802231
20. J. Stelzer, R. Berardi and C. Zannoni, Flexoelectric coefficients for model pear shaped molecules from Monte-Carlo simulations, *Mol. Cryst. Liq. Cryst. A* **352**(1), 187–194, (2000). doi: 10.1080/10587250008023176
21. J.L. Billeter and R.A. Pelcovits, Molecular shape and flexoelectricity, *Liq. Cryst.* **27**(9), 1151–1160, (2000). doi: 10.1080/02678290050121999
22. A.V. Emelyanenko and M.A. Osipov, Theoretical model for the discrete flexoelectric effect and a description for the sequence of intermediate smectic phases with increasing periodicity, *Phys. Rev. E* **68**(5), 051703/1–16, (2003). doi: 10.1103/PhysRevE.68.051703
23. M.I. Barnik, L.M. Blinov, A.N. Trufanov and B.A. Umanski, Flexo-electric domains in liquid crystals, *J. Phys. France* **39**(4), 417–422, (1978). doi: 10.1051/jphys:01978003904041700
24. I. Dozov, Ph. Martinot Lagarde and G. Durand, Flexoelectrically controlled twist of texture in a nematic liquid crystal, *J. Physique Lett.* **43**(10), L-365–L-369, (1982). doi: 10.1051/jphyslet:019820043010036500
25. I. Dozov, Ph. Martinot Lagarde and G. Durand, Conformational flexoelectricity in nematic liquid crystals, *J. Physique Lett.* **44**(19), L-817–L-822, (1983). doi: 10.1051/jphyslet:019830044019081700

26. J.S. Patel and R.B. Meyer, Flexoelectric electro-optics of a cholesteric liquid crystal, *Phys. Rev. Lett.* **58**(15), 1538–1540, (1987). doi: 10.1103/PhysRevLett.58.1538
27. P. Rudquist, M. Buivydas, L. Komitov and S.T. Lagerwall, Linear electro-optic effect based on flexoelectricity in a cholesteric with sign change of dielectric anisotropy, *J. Appl. Phys.* **76**(12), 7778–7783, (1994). doi: 10.1063/1.358523
28. J.P. Marcerou and J. Prost, The different aspects of flexoelectricity in nematics, *Mol. Cryst. Liq. Cryst.* **58**(3-4), 259–284, (1980). doi: 10.1080/00268948008082127
29. Y. Singh, Molecular theory of liquid crystals: Application to the nematic phase, *Phys. Rev. A* **30**(1), 583–593, (1984). doi: 10.1103/PhysRevA.30.583
30. T.J. Sluckin and P. Shukla, Molecular field theory of nematics: Density functional approach. I. Bulk effects, *J. Phys. A: Math. Gen.* **16**(7), 1539–1554, (1983). doi: 10.1088/0305-4470/16/7/030
31. M.A. Osipov, Molecular theories of liquid crystals. In eds. D. Demus, J. Goodby, G.W. Gray, H.-W. Spies and V. Vill, *Handbook of Liquid Crystals. Vol.1, 2nd edition*, Wiley-VCH, Weinheim, 1998. pp. 39–71.
32. W.M. Gelbart and B. Barboy, A van der Waals picture of the isotropic-nematic liquid crystal phase transition, *Accts. Chem. Res.* **13**(8), 290–296, (1980). doi: 10.1021/ar50152a007
33. J.W. Baran and A. Les, On the interaction energy between some nematogenic molecules, *Mol. Cryst. Liq. Cryst.* **54**(3-4), 273–288, (1979). doi: 10.1080/00268947908084860
34. K. Tokita, K. Fujimura, S. Kondo and M. Takeda, The pair intermolecular interaction energy of the two nematogenic liquid crystals, *Mol. Cryst. Liq. Cryst.* **64**(5-6), 171–176, (1981). doi: 10.1080/01406568108072524

# Chapter 2

# Flexoelectro-optics and Measurements of Flexocoefficients

N.V. Madhusudana

*Raman Research Institute*
*C.V. Raman Avenue, Bangalore 560080, India*
*e-mail: nvmadhu@rri.res.in*

This chapter is concerned with experimental measurements of flexoelectric coefficients. After a brief introduction to flexoelectricity in nematic liquid crystals, some applications exploiting the flexoelectric effect and the influence of this effect on electrohydrodynamic instabilities are pointed out. Flexoelectricity arises in samples with a splay-bend distortion in the director field and as such its measurement is not as direct as for dielectric constants. The theoretical background needed to analyse electro-optic experiments and extract the flexocoefficients is outlined in Section 2.2. Various experimental techniques that have been developed are described in Section 2.3. These involve cells in which the alignment of the nematic director is homeotropic, or planar or hybrid. In the first case, the interdigitated electrode technique is particularly noteworthy, as it has been used to establish several features of flexoelectricity: (1) the effect can arise purely from the quadrupolar nature of the medium, and (2) the dipolar contribution relaxes at a relatively low frequency. It has also been used to study the second sound resonance in smectic A liquid crystals and measure the compression modulus. For measuring the flexocoefficients $(e_1 - e_3)$ and $(e_1 + e_3)$, hybrid-aligned nematic cells have been used extensively. AC techniques avoid problems associated with ionic impurities, but require elaborate numerical fitting of the data. Some observations on the published measurements of flexocoefficients are made in Section 2.4, which ends with a few concluding remarks.

## Contents

## 2.1. Introduction

Electric polarization in condensed matter is a subject that has attracted the attention of both fundamental and applied researchers over the past couple of centuries. Indeed the very first theory of nematic liquid crystals was based on the assumption that they are ferroelectric liquids.[1] The *apolar* nature of the nematic director **n** is of course now well established. Even when the mesogenic molecules have strong longitudinal dipoles, near neighbours will have an antiparallel orientation to lower the electrostatic interaction energy,[2] ensuring the apolar character of **n**. The molecules can have *shape polarity* as well, either along longitudinal (pear-shaped, with an intrinsic splay) or transverse (banana-shaped, with an intrinsic bend) directions. In a well-aligned nematic with a spatially uniform orientation of **n**, the orientations of these shape dipoles also get averaged out. If the medium is subjected to splay or bend curvature distortions of the director field, the relevant shape dipoles can no longer be expected to be averaged out, for better packing. Meyer[3] realized that if the electric and shape dipoles of the molecules are aligned, these distortions of the **n**–field lead to macroscopic electric polarizations of the medium. There is some similarity of these distortion-induced polarizations with the phenomenon of piezoelectricity found in certain classes of crystals. Using Meyer's notation, the flexoelectric polarization is written as[4]

$$\mathbf{P_f} = e_1\mathbf{n}(\nabla \cdot \mathbf{n}) + e_3(\nabla \times \mathbf{n}) \times \mathbf{n}. \tag{2.1}$$

It is clear that the two polar vectors respect the apolar nature of **n**. There is also an obvious analogy of the above mechanism with the orientation polarization of a liquid dielectric, which was used by Helfrich[5] to relate the two flexocoefficients with molecular properties. The intrinsic splay or bend can be related to an appropriate angle and molecular dimensions. The relevant component of the electric dipole moment and the curvature elastic constant, *viz.*, $\mu_{\|}$ and the splay constant $K_1$ or $\mu_{\perp}$ and the bend constant $K_3$ figure in the estimation of the flexocoefficients. Nematic liquid crystals made of banana-shaped molecules have been studied only recently, and a comparison of the experimental measurements with the Helfrich formula leads to interesting inferences, as will be mentioned later in this chapter, and covered more thoroughly in the companion Chapter 3 by Jákli *et al.*[6]

The orientationally ordered nematic has a non-zero quadrupole density, irrespective of the shape or polar nature of the constituent molecules. Prost and Marcerou[7] realized that, as a spatial gradient in the quadrupole density is equivalent to polarization, all nematics have non-zero flexocoefficients. In other words, flexoelectricity is a *universal* property of all nematic liquid crystals. As the nematic is a prime example of *soft* condensed matter, the magnitude of the flexocoefficient can be simply estimated as

$$e \sim \text{elementary charge/molecular length} \sim 10^{-11}\text{C/m}\,.$$

We shall see later that this is of the correct order in many nematics with simple rod-like molecules.

The flexoelectric effect can be expected to have a discernible influence wherever there is splay-bend distortion in the director field. The distortion may occur due to boundary conditions, as for example in nematic drops suspended in an immiscible liquid with either tangential or normal orientation of $\mathbf{n}$ at the interface.[3] On the other hand, the distortion can be induced by an external field, like a magnetic field above the Freedericksz threshold. An applied electric field can have a particularly important effect, as it directly couples to the flexoelectric polarization. We should, however, remember that a nematic is a fluid made of organic compounds and unless special precautions are taken the medium will have a substantial amount of ionic impurities. These screen out the polarization, and on length scales beyond that corresponding to the (Debye) screening ($\lambda_\text{D}$), the flexoelectric effect has no significant physical influence. Often $\lambda_\text{D} \leq 0.1$ $\mu$m, while the sample thickness is a few micrometers, and flexoelectric effects are usually ignored in the analysis of experiments. A well-known counterexample is the flexoelectro-optic effect in cholesteric liquid crystals,[8] in which the relevant length is the pitch of the helical structure. There has been much effort to optimize the physical parameters, including enhancement of the flexoelectric coefficients with the hope of exploiting the effect in applications. An in-plane field applied to a short-pitch cholesteric with the helix axis oriented perpendicular to the cell walls also leads to useful electro-optic effects.[9] The *Zenithal Bistable Device* (ZBD), which is already in commercial production, exploits the polarity-dependent flexoelectric coupling with an external electric field to switch between two optically distinct states.[10] The technical applications of the flexoelectric effect are covered in the companion Chapter 7 by Rudquist and Lagerwall.[11] *Electrohydrodynamic* (EHD) instabilities in nematic liquid crystals are driven by the action of an external electric field on charge densities in the medium, which *requires* the latter to have sufficiently large concentrations of ionic impurities.[12] The Carr–Helfrich mechanism works for a nematic with negative dielectric anisotropy and positive conductivity anisotropy, indicated by the symbol (-,+). As we realized some time ago,[13] flexoelectric polarization and its divergence make important contributions to the phenomenon, like generating oblique rolls in which the wave vector of the instability makes a non-zero angle with the undistorted director, and enable even (+,+) nematics, in which the Carr–Helfrich mechanism does not work, to exhibit EHD instabilities.[14] This topic is covered in detail in the companion Chapter 4 by Buka *et al.*[15] Liquid Crystal Displays (LCDs) are now commercially highly successful and the nematic mixtures used in the devices are thoroughly purified of ionic contamination to enhance the working life of the displays. Further, the device thickness is reduced to $\leq 4\ \mu$m, so that the switching times are fast enough. It is clear that flexoelectric effects should become important in the electro-optic response of the device, which inevitably involves splay-bend distortions of the director field.

Thus, measurements of flexoelectric coefficients, and relating them with molecular properties, are important both intrinsically and for the interpretation of various experiments of fundamental and practical interest. One possible idea is to start from the definition given by Eq. (2.1) and create suitable splay and bend distortions in the director field by appropriate boundary conditions and to measure the resulting flexoelectric polarization. The polarization produces a surface charge density, which may possibly be measured using an electrometer. A static distortion that produces a DC electrical signal is not suitable, due to the screening problem mentioned above. It is only recently that a technique for producing an oscillatory distortion of the required type has been developed and the resulting AC signal used to measure $e_3$, the bend flexocoefficient in a nematic made of banana-shaped molecules, apart from those of some other materials with rod-like molecules.[16] This will be discussed in detail in Chapter 3 by Jákli *et al.*[6]

Practically all other methods developed for the measurement of flexocoefficients are 'indirect'. These exploit the fact that the polarization resulting from the splay-bend distortion couples linearly with an applied electric field **E**. This contributes to the total free energy of the sample, and hence alters the distortion of the director field compared to that in the absence of flexoelectric polarization. An external electric field of course acts on the dielectric anisotropy ($\Delta\varepsilon$) of the nematic, which, like the orientational order itself, is a second-rank tensor. Indeed $E^2$ is conjugate to the order and the highly successful LCD industry exploits a few different electro-optic effects that result from this coupling. The indirect methods of measuring the flexocoefficients have to take into account the dielectric coupling, which dominates at higher field strengths. In this chapter I will summarize the results from a few different techniques that have been developed to measure the flexocoefficients. Petrov[17] reviewed the measured data up to 2000. More recently, Elston and Parry-Jones[18] briefly reviewed the subject, in connection with a meeting held at Oxford in 2006. Some aspects of the theoretical background are covered in the next section.

## 2.2. Theoretical Background

Since the flexoelectric effect is associated with curvature distortions of the director field it seems natural to expect that the splay and bend elastic constants themselves may have contributions from flexoelectricity. The shape polarity of the molecules invoked by Meyer will have a direct mechanical influence independently of flexoelectricity and can be expected to lower the relevant elastic constants.[19,20] The flexoelectric polarization will generate an electrostatic self-energy and hence make an independent contribution to the elastic constants. In the absence of any external field, the electric displacement $\mathbf{D} = 0$ and the flexoelectric polarization generates an internal field $\mathbf{E} = -\mathbf{P}/\varepsilon_0$, where $\varepsilon_0$ is the vacuum dielectric constant. Considering only a director deformation confined to a plane, and described by a polar angle $\theta(z)$, and in the absence of ionic screening, the energy density due to a splay-bend deformation reads as[21]

$$F = \frac{1}{2}\left\{K_1 \sin^2\theta + K_3 \cos^2\theta + \frac{(e_1+e_3)^2 \cos^2\theta \sin^2\theta}{\varepsilon_0(\varepsilon_{\parallel}\cos^2\theta + \varepsilon_{\perp}\sin^2\theta)}\right\}\left(\frac{d\theta}{dz}\right)^2, \quad (2.2)$$

where the subscripts $\parallel$ and $\perp$ refer to directions in relation to the director.

It is clear that the positive self-energy due to the polarization effectively *increases* the elastic constant of the nematic depending on the magnitude of the *sum* of the flexocoefficients $(e_1 + e_3)$. Ionic screening reduces this enhancement if the director distortion occurs over a length that is longer than the screening length. A well-known experimental technique for measuring the elastic constants of nematics exploits the Freedericksz transition in the alignment of the director brought about by an external magnetic or electric field beyond a threshold value.[4] In some geometries, the director field has splay-bend distortion above the threshold value and electro-optic effects associated with this distortion are exploited in all commercial LCDs. Deuling showed that while the threshold field itself is unaffected by the flexoelectric properties of the medium, the distortion above the threshold experiences the additional stiffening mentioned above.[22,23] Brown and Mottram[24] analysed the AC electric field-induced distortion in an initially planar-aligned cell and showed that the distortion profile can be fitted quite well by appropriately lowering the bend elastic constant to compensate for the flexoelectric stiffening. The fitted $K_3$ shows an inverse parabolic relationship with $|e_1 + e_3|$, and thus such an experiment cannot be used to uniquely determine either of them. They also showed that a sufficiently large number density of ions can screen out the flexoelectric polarization, and the maximum distortion angle at the midplane can be restored to the value which would be obtained in the absence of the flexoelectric effect.[25]

It is worthwhile recalling that the traceless and symmetric tensor order parameter of the nematic is given by[4]

$$Q_{ij} = S\left(n_i n_j - \frac{1}{3}\delta_{ij}\right), \tag{2.3}$$

where $S$ is the magnitude of the order parameter and $\delta_{ij}$ is the Kronecker delta. A Landau–de Gennes free energy expression including spatial gradients in $Q_{ij}$ can be used to define both the elastic and flexoelectric properties, and, as can be guessed from Eq. (2.3), $S$ can have spatial gradients for a fixed direction of the director $\mathbf{n}$. In the latter case, we get 'order electric' polarization in the medium,[26] with two independent components whose coefficients are related to $e_1$ and $e_3$. The spatial gradients in $S$ can be very important near the nematic-isotropic interface and can account for the tilt angle of $\mathbf{n}$ often found at this interface. We will not discuss order electricity any further.

In this chapter, we will be describing different electro-optic techniques that have been developed specifically for measurement of the flexocoefficients. In these techniques, an electric field is applied to a nematic liquid crystal in a cell with well-defined boundary treatments. The net torque on the medium, which involves dielectric, elastic and flexoelectric components, is set to zero both in the bulk and at the two surfaces in order to find the equilibrium director configuration if a DC field is applied. In AC techniques, the above torques are balanced with the viscous torque. The optical properties of the medium are in turn calculated for the distorted profiles and compared with experimental measurements. The flexoelectric, and in some methods other, parameters can be adjusted to fit the experimental and calculated optical properties. Thus these measurements involve optimizing a fair number of unknown parameters and the errors in the measurements of flexocoefficients can be $\sim 20\%$ or higher.

The torque acting on the director is given by $\mathbf{n} \times \mathbf{h}$, where $\mathbf{h}$ is called the molecular field,[4] which can be derived from the Euler–Lagrange equation. The energy density corresponding to the flexoelectric polarization is given by $-\mathbf{P}_{\mathbf{fl}} \cdot \mathbf{E}$ and the molecular field can be expressed in the form[17]

$$\mathbf{h}_{\mathbf{fl}} = (e_1 - e_3)\,[\mathbf{E}(\nabla \cdot \mathbf{n}) - (\nabla \mathbf{n}) \cdot \mathbf{E}] - (e_1 + e_3)\mathbf{n} \cdot \nabla \mathbf{E}. \tag{2.4}$$

In the bulk, only the term involving $(e_1 - e_3)$ contributes to the flexoelectric torque under a uniform external field $\mathbf{E}$. Further, if the field is applied along the $z$ axis between two conducting plates, and the director is confined to one plane (say the $xz$ plane) and its orientation $\theta(z)$ depends only on $z$, the term inside the square brackets of Eq. (2.4) becomes zero. The only contribution to the flexotorque arises from the $(e_1 + e_3)$ term, which is non-zero if the *field gradient* $\nabla \mathbf{E}$ has a non-zero value. The dielectric anisotropy $(\Delta\varepsilon)$ can give rise to a field gradient if $\theta(z)$ is non-uniform in the bulk. If $\Delta\varepsilon = 0$, there is no bulk flexotorque. However, the director at the bounding surface is subject to a flexotorque given by

$$\Gamma_s = (e_1 + e_3) \sin \theta_s \cos \theta_s E, \tag{2.5}$$

where the subscript $s$ indicates surface values. It again depends only on the sum of the flexocoefficients and vanishes for $\theta_s = 0$ or $\pi/2$. If the director is very strongly anchored at the surface, the flexotorque is not effective, and the flexoelectric effect can be ignored in the analysis.

If the director distortion is not constrained as in the above, and can depend on an additional coordinate ($x$ or $y$), or if it is not confined to a plane, the bulk flexotorque is not zero, and these geometries can be used to measure the flexocoefficients, as we shall describe later.

Several molecular theories have been developed to relate the flexocoefficients with the molecular properties and the orientational order parameter, as described in the companion Chapter 1 by Osipov.[27] We just note that the predicted dependencies on the order parameter $S$ are as follows:

$$e_1 = aS + bS^2,$$

$$e_3 = aS + cS^2. \tag{2.6}$$

The linear dependence on $S$ arises from both the dipolar and quadrupolar contributions, whereas the quadratic dependence is entirely due to the dipolar effect. Thus the difference $(e_1 - e_3)$ is expected to arise only from the dipolar effect.

## 2.3. Experimental Techniques

The first electro-optic experiment that was designed to measure flexocoefficients was due to Schmidt *et al.*[28] They used a homeotropically aligned N-(4-metoxybenzylidene)-4′-butylaniline (**MBBA**) sample with weak surface anchoring and applied a transverse DC field. As the sample has negative dielectric anisotropy, the director distortion is entirely due to the flexo-torque at the surfaces. In this geometry, when the tilt angle is small, the surface torque balance yields $d\theta/dz = e_3 E/K_3$, depending only on $e_3$. The resulting optical path difference $\Delta l \propto d^3$, where $d$ is the sample thickness. This dependence was verified experimentally and $e_3$ was estimated to be $\approx +10^{-12}$ C/m at 22 °C.

Helfrich[29] also pointed out that a homeotropic sample with strong anchoring at one surface and weak anchoring at the other can exhibit a polarity-dependent flexoelectro-optic effect for a field applied *along* the undistorted director. Indeed there are different types of possible DC flexoelectro-optic effects depending on whether the initial orientation of the director is planar or homeotropic, and depending on the relative anchoring energies at the two surfaces, the sign of dielectric anisotropy and the direction of the applied field.[30] Even if both surfaces have equally weak homeotropic anchoring, a polar instability can occur at one of the surfaces depending on the sign of the applied field[31] beyond a threshold value, provided that the assumed positive $\Delta\varepsilon < (e_1 + e_3)^2/(\varepsilon_0 K)$. However, we have to note that any surface is intrinsically polar. The two end groups of most nematogenic molecules are different, and the different affinities of the two molecular ends on any surface producing homeotropic alignment can lead to a surface polarization $(p_s)$. Indeed in 4-pentyl-4′-cyanobiphenyl (**5CB**), $p_s$ is an order of magnitude larger than $(e_1 + e_3)$.[32,33] Further, a preferential adsorption of ions with one sign can give rise to a strong enough field gradient at the surface to contribute to a director instability,[34] and effectively alter the anchoring energy. Indeed high-field electrolytic processes can give rise to such field gradients and produce electro-optic responses in the kilohertz frequency range[35] in **5CB**. If the surfactant used for the homeotropic anchoring can desorb from the surface, its concentration will have a gradient in the cell. The flexocoefficient can itself have a spatial gradient, leading to a new type of bulk torque,[36] which has to be taken into account in interpreting the experimental results for such samples. The influence of ions can also be quite complicated,[37] and there are no reliable measurements of flexocoefficients using samples with an initially homeotropic alignment using direct electro-optic techniques.

(a)
(b)
(c)
(d)
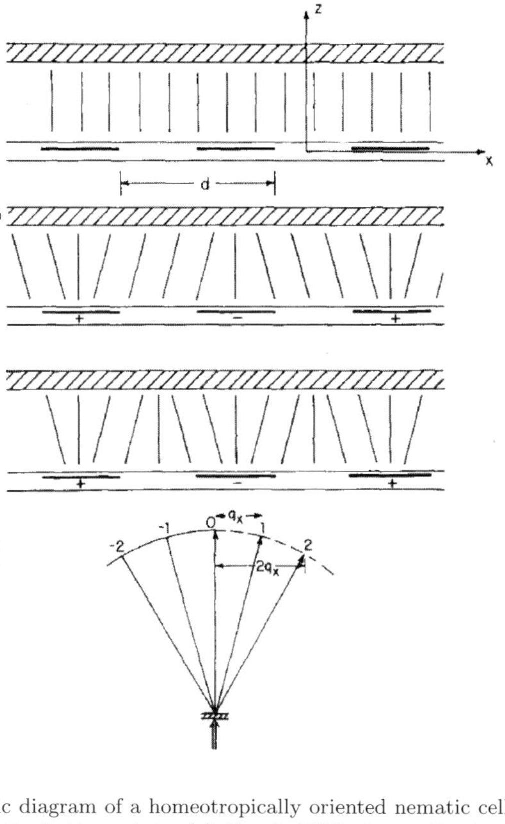


Fig. 2.1. Schematic diagram of a homeotropically oriented nematic cell with interdigitated electrodes on the bottom plate. (a) $E = 0$. With the application of potentials of opposite signs on adjacent electrodes, the director distortion has a periodicity (b) equal to $2d$ where $d$ is the electrode spacing, due to the flexoeffect depending linearly on the field, and (c) equal to $d$ due to the dielectric coupling depending quadratically on the field. (d) The corresponding scattering beams will be, respectively, along $\pm 1$ and $\pm 2$ with the incident beam along 0. (Reprinted with permission from Prost and Pershan,[38] copyright 1976, American Institute of Physics.)

Prost and Pershan[38] developed an interdigitated electrode technique (Fig. 2.1), which overcomes all the problems mentioned above. The field-induced director distortion generates a phase grating whose spacing due to the flexoelectric effect is twice that of the electrodes ($d$), while that due to the dielectric coupling equals $d$. The bulk flexotorque arises due to the field gradient $dE_x/dz$ which is obviously non-zero in the experimental geometry. The authors further show that the effective anchoring energy is small, so that $(e_1 + e_3)$ is measured. If there is a surface polarization $p_s$, the surface torque depends on $e_1 + e_3 + p_s$. The experiment is conducted with an applied sinusoidal electric field and the light scattered by the phase grating generated by the flexo-distortion is synchronously detected. The AC technique overcomes the problem due to ionic impurities and the synchronous detection is sensitive to director deviations $\sim 1°$, allowing for some simplifications in the analysis. Nevertheless, the analysis is quite involved, requiring the electric field profile and taking into account the elastic, dielectric and viscous contributions in relating the intensity of the relevant light scattering to the flexocoefficient. Subsequently, using a refinement of the theoretical analysis and from measurements of a few compounds,[39] a number of important conclusions were drawn:

(1) In a compound with perfectly symmetric molecules, *viz.*, 4,4′-dihexyloxytolan (**HOT**) with zero net molecular dipole moment, there is no dipolar contribution, and $(e_1 + e_3)$ was found to vary as $S$, as expected from Eq. (2.6).

(2) In **MBBA**, the flexo-polarization was found to be screened out by ionic impurities at low frequencies.

(3) In 4-n-octyloxy-4′-cyanobiphenyl (**8OCB**), which has a strongly polar cyano end group, $(e_1 + e_3)$ was again found to vary as $S$ in the higher temperature regions of the nematic range, showing that the dipolar contribution was essentially zero. The antiparallel associations between the neighbouring dipolar molecules[2] can give rise to molecular pairs with negligible dipole moments, but substantial quadrupole moments, accounting for this trend.[39] Further, as the temperature was lowered towards the nematic-to-smectic A transition point, the divergence of the bend elastic constant $K_3$ resulted in a corresponding reduction in the intensity of scattered light.

(4) For p-pentylphenyl 4-(p-pentylbenzoyloxy)-3-chlorobenzoate, the relatively weak molecular dipole moment does not result in a strong enough antiparallel interaction between neighbouring molecules, and the dipolar contribution to $(e_1 + e_3)$ was estimated to be $\sim 30\%$, which relaxed at the corresponding dielectric relaxation frequency. The technique is sensitive enough to measure the flexocoefficient arising from the electric field-induced orientational order above the nematic-to-isotropic transition point $T_{NI}$, which diverges as $T_{NI}$ is approached.[40] The method can be used

for measuring the flexocoefficient of a smectic A (SmA) liquid crystal as well, which has an additional contribution from the gradients in the layer displacement vector **u**. More interestingly, as an important *academic application* of the flexoeffect, the technique was used to locate the second sound resonance leading to an accurate measurement of the layer compression modulus $B$.[41] Further, $B$ was found to tend to vanish as the frequency approached zero, as was predicted theoretically based on the anharmonic elasticity of a smectic liquid crystal.

Recently, a very different type of experiment has been conducted using an interdigitated pattern etched on a transparent indium-tin-oxide (ITO) electrode, and aligning the commercial mixture **E7** (with a large positive $\Delta\varepsilon$) between a structured electrode and one without interdigitation such that the director is parallel to the plates but orthogonal to the fingers of the interdigitated electrodes.[42] A Mach–Zehnder interferometer was used to give a fringe pattern corresponding to the optical path of a He-Ne laser beam polarized parallel to **n**. An applied 0.2 Hz square-wave signal shifted the position of the fringes in the gaps between the electrodes, as the director orientation is distorted. A high-speed digital recording shows that the distortion increases after every reversal of the field, attains a maximum value and dies down to zero after a couple of seconds. Based on an independent measurement of the transient capacitance with a similar field profile, it is argued that ion migration screens out the field in each half cycle, and that the field, which is larger than the applied one immediately after the reversal, decays exponentially. The optical experiment with the interdigitated electrodes shows that the peak in the fringe shift differs considerably between the two signs of the applied field, showing the flexoelectric contribution to the director profile. The Euler–Lagrange relations corresponding to both the director tilt angle $\theta$ and the electric potential $V$ were solved numerically to fit the spatiotemporal profiles of the fringe pattern. Using the known values of the rotational viscosity and the dielectric and elastic constants, $(e_1 + e_3)$ was estimated at room temperature.

It is clear that the above geometry just corresponds to that for Freedericksz transition if there is no interdigitation on either of the electrodes. If the surface anchoring is weak, the flexotorque [Eq. (2.5)] enhances the tilt angle at one surface, while lowering it at the other, depending on the sign of the applied DC electric field and that of $(e_1+e_3)$. If the anchoring strengths $(A)$ at the two surfaces are equal, it is clear that an applied AC voltage will not generate an electro-optic signal at the applied frequency. However, if the anchoring strengths are different, such a signal is produced,[43] and can be used to measure $(e_1 + e_3)$. Interestingly, as the dielectric anisotropy is reduced to extremely low values, the flexoelectric interaction with an external DC field dominates, and the energy is actually lowered by periodic structures whose wave vectors are orthogonal to the field.

The first such possibility was pointed out by Meyer in the original paper on the subject,[3] in which he showed, assuming that the anchoring strength $A = 0$, $\Delta\varepsilon = 0$ and $e_1 = -e_3$, a DC field can generate a periodic structure along the direction of the field-free orientation of the director **n**. Under the field, **n** develops alternate bend-rich and splay-rich distortions.

Though this type of periodic structure with multiple arches of the director is difficult to generate in a nematic, it is already present in a cholesteric liquid crystal when viewed in a plane whose normal makes an oblique angle with the helical axis. The flexoelectric effect changes the periodicity of this structure under a DC field applied normal to the helical axis,[8] effectively rotating the latter. This can be used in turn to measure $(e_1 - e_3)$.[9]

For a planar-oriented nematic sample with a *low* value of $\Delta\varepsilon$, the flexoelectric effect readily gives rise to another type of periodic structure under a DC field above a threshold value,[44] even when the director is strongly anchored. The wave vector lies along the *normal* to the $\mathbf{n} - \mathbf{E}$ plane, as had been experimentally found earlier.[45] The director distortion involves both the zenithal angle $\theta$ and the azimuthal angle $\phi$, which depend on both $z$ and $y$ coordinates. Thus, referring back to Eq. (2.4), near the threshold, only the $(e_1 - e_3) = e^*$ term, which is acted upon by $E$, is significant. Assuming solutions of the form

$$\theta = \theta_0 \cos(qy) \cos(\pi z/d),$$

$$\phi = \phi_0 \sin(qy) \cos(\pi z/d), \tag{2.7}$$

the threshold values of voltage and the wave vector are given by

$$U_{th} = \frac{2\pi K}{|e^*|(1+\mu)} \text{ and } q_{th} = \frac{\pi}{d}\sqrt{\frac{1-\mu}{1+\mu}}, \tag{2.8}$$

where $\mu = (\varepsilon_0 \Delta\varepsilon K / e^{*2})$ in the one-elastic-constant approximation with $K = K_1 = K_2$. If $\Delta\varepsilon$ is negative, the $\theta$ distortion is not favoured and as the magnitude of $\Delta\varepsilon$ is increased, the threshold voltage as well as the wave vector tend to $\infty$ as $\mu$ approaches $-1$. In any case, if such materials have conducting impurities, the DC field gives rise to an EHD instability[12] beyond a threshold voltage, as we mentioned earlier. In materials with positive $\Delta\varepsilon$, the flexo threshold voltage is lower than that of the Freedericksz transition, as the energy density of the former is linear in the distortion angles near the threshold. However, as $\mu$ approaches 1, the wave vector of the flexo-instability tends to 0, and beyond that, only the Freedericksz transition can be seen. These trends have been seen in experiments[46] and were used to make the first measurements of $e^*$. The theoretical analysis becomes quite complicated when the elastic anisotropy is taken into account.[47] As the periodic instability involves twist deformation, larger values of $K_1/K_2$ lead to a divergence of $q_{th}$ for smaller negative values of $\mu$ in materials with negative $\Delta\varepsilon$, while greatly increasing the value of $\mu$ at which $q_{th}$ goes to zero for materials with positive $\Delta\varepsilon$.[15,47]

We have hitherto considered techniques that employed initially uniform orientations of the director. In the Hybrid-Aligned Nematic (HAN) cell, the surface treatments ensure that the director is aligned parallel to the surface (along the $x$ axis) at one of the plates, and normal to the surface (along the $z$ axis) at the other one. The director profile will have a splay-bend distortion involving $\theta(z)$ in the cell, leading to a flexo-polarization $\mathbf{P}(z)$. An external electric field can now couple to $\mathbf{P}$, changing the distortion profile, leading to measurements of the flexocoefficients.

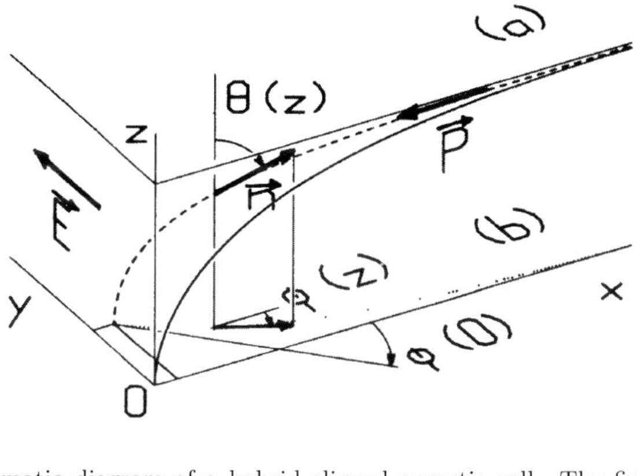


Fig. 2.2. Schematic diagram of a hybrid-aligned nematic cell. The field-free director (shown by the continuous curved line) has a splay-bend curvature distortion in the $xz$ plane. A DC field applied along the $y$ axis rotates the polarization and the director (shown by the curved dashed line) acquires a $\phi(z)$ profile. (Reproduced from Dozov *et al.*,[48] with the permission of EDP Sciences, http://publications.edpsciences.org.)

Dozov *et al.*[48] proposed a very simple technique of measuring $e^*$ by noting that a DC electric field applied along the $y$ axis rotates $\mathbf{P}$, imparting an azimuthal angle $\phi(z)$ on the director field [see Eq. (2.4) and Fig. 2.2].

For small values of $E$, the dielectric coupling ($\propto E^2$) can be ignored, and the Euler–Lagrange equations corresponding to $\theta(z)$ and $\phi(z)$ can be used to get a very simple relation for the maximum twist angle $\phi(0)$ occurring close to the homeotropically aligned plate:

$$\phi(0) = -\frac{e^* E d}{\pi K} \tag{2.9}$$

where $d$ is the sample thickness and as usual $K$ is the single curvature elastic constant. If a light beam polarized along $x$ enters the cell at the planar-aligned plate, the polarization is rotated by $\phi(0)$ as it traverses the cell whose thickness d is large enough for the Mauguin criterion[4] to be satisfied. An analyser can now be used to measure $\phi(0)$ and hence $e^*$. This simple technique has been used to measure $e^*$ for a large number of nematic liquid crystals.[49–51] If the orientation of the analyser is not altered from that of the crossed setting, the intensity of transmitted light increases from zero as the field is increased. Detailed numerical calculation of the transmitted intensity as a function of the field using the known values of all the elastic constants, the dielectric anisotropy, the refractive indices and the finite anchoring energies at the two surfaces for **MBBA** shows that the $e^*$ value is larger than that found by using the simplifying approximations.[51]

If the external electric field is applied between the bounding plates of a HAN cell, a bulk flexotorque is generated only if there is a field gradient [see Eq. (2.4)]. In a sample free of ions, the $z$ component of the electric displacement $D_z$ is a constant in this geometry and a field gradient results if the nematic has a non-zero value of $\Delta\varepsilon$. In addition, if the anchoring strength at one or both surfaces is not very high, the tilt angle can deviate from 0 or $\pi/2$ rad at the plates treated for planar and homeotropic anchorings, respectively. The flexoelectric effect then contributes to the surface torques as well [see Eq. (2.5)]. Both the bulk and surface torques depend on $(e_1 + e_3)$, which is often measured using this geometry.

In a HAN cell in which the anchoring energy is very weak at the surface treated for homeotropic alignment and quite strong at the planar one, **MBBA** was shown to exhibit a very large *linear* flexoelectro-optic effect[52] when viewed between crossed polarizers set at 45° to the plane containing the director. At low voltages, the dielectric contribution can be ignored, and the observation can be assumed to be entirely due to the surface effect, allowing for a determination of the sign, which is negative, and an estimation of the magnitude of $(e_1 + e_3)$. The DC field is essentially screened up to $\sim \pm 1$ V, which corresponds to the redox potential of the material, beyond which the molecule exchanges an electron with the electrode. Similar results are obtained if the planar alignment has a weak anchoring strength, whereas the homeotropic one is strong.[53] An improved analysis was achieved by assuming that the medium has a high density of ions, and consequently the local electric field $\mathbf{E}(z)$ to be determined by a constancy of the current density across the sample. The authors analysed the data collected for **MBBA** samples for a few different cell thicknesses, and confirmed the negative sign of $(e_1 + e_3)$, but obtained a somewhat reduced magnitude compared to the earlier estimation. However, the influence of ions is very complicated when a DC field is applied, and a rigorous analysis including the ion distribution in the HAN cell, the dielectric anisotropy, the flexoeffect and the elasticity leads to a pair of coupled non-linear equations, which can be solved only numerically.[54] The parameters can be adjusted to quantitatively fit one set of the experimental data collected by Madhusudana and Durand[52] with a moderate value of the anchoring strength for the homeotropic alignment. The authors found that with the influence of the ion distribution taken into account, the best fit with the data leads to a fairly large $(e_1 + e_3)$ of positive sign for **MBBA**! Another experiment[51] used the sign of the voltage shift (in the electro-optic response about 0 V) arising from the flexo-polarization in the HAN cell, to obtain the negative sign, which was confirmed by a subsequent detailed analysis.[55]

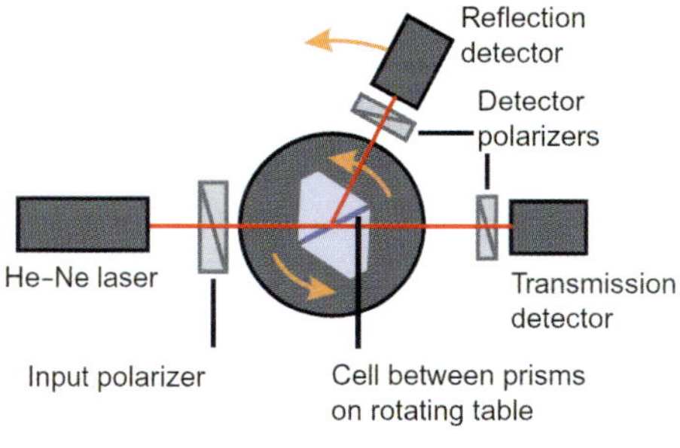


Fig. 2.3. Experimental layout for the fully leaky guided mode technique. (Reproduced from Taphouse *et al.*,[57] with the permission of the Institute of Physics.)

The fully leaky guided mode technique makes use of a thin ($\sim 3\ \mu$m) HAN sample between glass plates with a refractive index lower than both the principal indices of the nematic.[56,57] A $p$ or $s$ polarized He-Ne laser beam, with the electric vector in the plane of incidence or orthogonal to it, respectively, incident in an appropriate range of incident angles $\beta$ (see

Fig. 2.3) exhibits a few minima and maxima of intensity in both polarized (*pp*, etc.) and depolarized (*ps*, etc.) components of both reflected (R) and transmitted (T) beams as functions of $\beta$. A detailed numerical analysis of the experimental data can be used to find the director profile $\theta(z)$ in the cell. The profile is altered by applying a DC voltage to the cell, which is reversed after 250 ms. The differences in the equilibrium profiles obtained for the two polarities of the electric field after the transients have died down (in practice, about 200 ms after the field of the given sign is applied) can be used to measure $(e_1 + e_3)$ by using the torque balance relations. For **E7**, a commercial mixture with components having molecules with the strongly polar cyano end groups, a DC offset was needed to fit the data,[56] and was attributed to a surface polarization at the homeotropic surface. For **ZLI-4788-000**, another commercial mixture with negative dielectric anisotropy, the offset was attributed to a surface density ($\sigma$) of adsorbed charges.[57] Indeed in this case, the authors found degeneracy in the values of $\sigma$ and a positive value of $(e_1 + e_3)$ to give a good fitting, and the sign of $(e_1 + e_3)$ could even be reversed by increasing the assumed value of $\sigma$. In the final investigation, the authors also studied the dynamics of the director reorientation after the field was either applied or reversed, and no combination of $\sigma$ and $(e_1+e_3)$ could reproduce the data. It was necessary to include the motion of (a relatively small concentration of) ions apart from the usual Leslie viscosities of the nematic to understand the dynamics.[58] A relatively large value of $\sigma$ and small negative value of $(e_1 + e_3)$ were found to give the best reproduction of the experimental results. It may be noted that the authors assumed strong anchoring at both surfaces in the analysis, though the sample thickness was only $\sim 3$ $\mu$m.

As all the above measurements using DC fields have the problems discussed earlier, several different techniques using AC fields have been developed. Blinov *et al.*[59] used a surface treatment to give $\sim 72°$ pretilt angle to enhance the surface flexotorque [see Eq. (2.5)]. An applied AC voltage of frequency $f$ generates surface oscillations of the director, which decay exponentially with distance from the surface, the decay length varying inversely as the square root of $f$. The average surface modulation is measured using the evanescent wave, which penetrates the medium when a laser beam incident at a suitable angle on a prism-shaped substrate with a refractive index that is higher than both the ordinary and extraordinary indices of the nematic is totally internally reflected. The first harmonic optical signal (of frequency $f$) is generated by the flexoelectric coupling and is measured using a lock-in amplifier. The frequency dependence of the amplitude and the phase angle of the signal can be related to the viscoelastic properties and the anchoring energy to extract $(e_1 + e_3)$. Using a mixture of **MBBA** and 4-heptyl-4′-cyanophenyl benzoate to adjust $\Delta\varepsilon$ to a value $\approx 0$, $(e_1+e_3)$ is found to be about half of the value obtained by other techniques, and attributed to a reduced value of the nematic order parameter resulting from the rough corrugations in the SiO coating needed to give $\theta_0 = 72°$. The sign was determined by using DC electric pulses, and the positive sign quoted in the paper again is uncertain for the reasons discussed earlier.

A simpler AC technique makes use of a HAN cell.[60,61] A sample with positive $\Delta\varepsilon$ is used in a cell with the anchoring energy at the surface treated for homeotropic alignment being weaker than that at the one treated for planar alignment. The tilt angle is smaller than $\pi/2$ rad at the former surface and an applied AC voltage can generate a significant flexotorque. The sample is placed between crossed polarizers set at 45° to the plane containing the director. An AC signal with a frequency $f \sim 100$ Hz, which is high enough to avoid the problems due to ion motion, is applied between the plates. The experiment consists of measuring the DC level, the amplitudes and phases of the first and second harmonics of the optical signal (at $f$ and at $2f$, respectively) using an appropriate multimeter and lock-in amplifier. As the voltage is increased, the dielectric coupling ensures that the average tilt angle in the cell tends towards $\pi/2$ rad, thus reducing the DC component of the optical path difference. The signal at $f$ initially increases with voltage, both due to the field gradient in the bulk arising from the non-zero value of $\Delta\varepsilon$, and the surface contribution from the surface treated for homeotropic alignment. As the voltage is increased beyond some value, both these contributions decrease as the tilt angle at that surface and in most of the bulk approaches $\pi/2$ rad. However, the director develops a very large gradient near the planar surface, which in turn increases the tilt angle and hence the flexotorque at that surface. Thus the first harmonic signal starts to increase again. The detailed torque balance relations for both the in-phase and out-of-phase $\theta(z)$ profiles in the bulk, as well as at the two surfaces can be worked out, including contributions from viscous, elastic, dielectric and flexoelectric effects. The DC level and the amplitude and phase angle of the $f$ component of the optical signal can now be calculated. The elastic and dielectric constants and the refractive indices are usually measured by independent techniques. By treating the rotational viscosity $\gamma$, the anchoring energies at the two surfaces and $(e_1 + e_3)$ as adjustable parameters, a non-linear least-squares fitting procedure has been developed for optimizing the values of the unknown parameters to obtain an overall agreement between the voltage dependences of the calculated and measured values of the DC level and the amplitude and phase angle of the $f$ signal. The experimental data for 4-heptyl-4′-cyanocyclohexylcyclohexane (**CCH7**), which is essentially non-aromatic and thus relatively free of ionic impurities, has been used to obtain the temperature dependences of the four parameters mentioned above. We may note that both the sign and magnitude of $(e_1 + e_3)$ were obtained using this technique.[61]

There is yet another direction in which an external electric field can be applied to a HAN cell, *viz.*, along the $x$ axis in the plane containing the distorted director.[62] If the $x$ component of the flexo-polarization $\mathbf{P}$ is oriented in a direction opposite to the DC field, $P_x$ starts to rotate towards $\mathbf{E}$ beyond a threshold value of $E$, which varies as $1/e^*$. This can be used to measure $e^*$.

The intrinsic asymmetry of the HAN cell with the two surfaces subjected to different treatments leads to some problems mentioned earlier: there is surface polarization at the wall treated for homeotropic alignment if the molecules have longitudinal dipoles, and there is the possibility of different surface densities of the adsorbed charges at the two plates, apart from the net flexo-polarization. These generate intrinsic voltages in the cell, which have to be taken into account especially when an external DC field is applied. Even when an AC field is used, the intrinsic field can influence the overall response of the director to some extent. One way of overcoming these problems is to use similarly treated surfaces at both walls. The Pi-cell has such a structure, with a pretilt angle of $\sim 40°$ at both surfaces, which are arranged to give a splay distortion of the director, which exhibits only an $x$ component of flexo-polarization. A large enough electric field applied between the plates reorients the nematic with positive $\Delta\varepsilon$ to get an essentially vertical (V) alignment. A DC field will result in a larger tilt angle at one plate compared to that at the other, if the anchoring energy is not strong. The opposite field reverses the magnitudes of the tilt angles at the two surfaces. This does not change the optical path difference for a beam travelling normal to the cell, but if the beam enters at an angle to the vertical, the two director configurations generate different path differences. Kischka *et al.*[63] used 100 Hz AC voltages, monitoring the DC, $f$ and $2f$ optical signals and analysing the voltage-dependent data as was done by Warrier and Madhusudana,[61] and obtained a fairly large positive value of $(e_1 + e_3)$ for **E7**.

A twisted nematic (TN) does not by itself exhibit flexo-polarization, but a DC electric field applied normal to the helical axis generates a tilt to gain energy due to flexoelectric coupling.[8,9] A TN cell can be obtained by appropriate surface treatments in a non-chiral nematic. By applying the DC field in the plane of the cell, the director develops a tilt angle, whose sense reverses if the field is reversed. By sending two laser beams at ±45° to the normal to the plane of the cell (i.e., the $z$ axis) as well as the electric field (i.e., the $x$ axis), the variations of the transmitted intensities between suitably oriented crossed polarizers as functions of both positive and negative fields can be measured[63] (see Fig. 2.4). They can also be numerically calculated using the known values of the elastic and dielectric constants, the refractive indices and a trial value of $e^*$. The sign and magnitude of the flexocoefficient is adjusted to give the best agreement with the measurements. For **E7**, $e^*$ was also found to have a positive sign and was about a third of the magnitude compared to that of $(e_1 + e_3)$. In fact, in the above geometry, even if there is no twist in the director field, when a DC electric field is applied normal to the undistorted director the usual Freedericksz transition for twist distortion is pre-empted by a distortion involving both tilt and azimuthal angles of the director to gain energy due to the resulting flexo-polarization.[64]

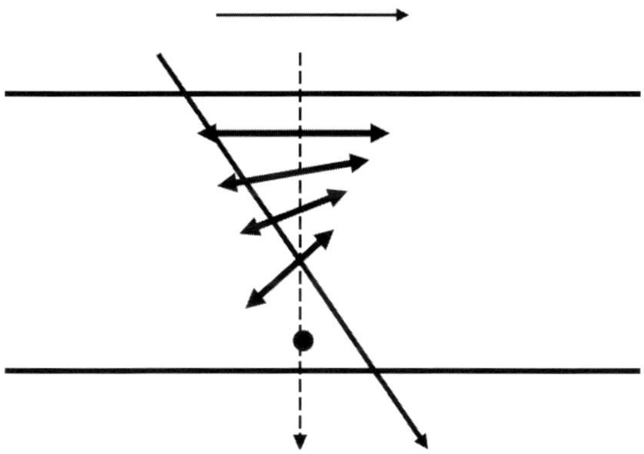

Fig. 2.4. A transverse electric field, indicated by the arrow pointed towards the right at the top of the figure, tilts the apolar director as shown by double headed arrows in a specific direction due to the flexoelectric effect on a 90° twisted nematic cell. The tilting direction reverses if the field direction is reversed. The transmitted intensity measured with a polarized light beam traversing the cell vertically as indicated by the dashed line will be identical in the two cases. On the other hand, with an oblique beam, the transmitted intensities for the two tilted director structures will be different, and can be used to measure the flexocoefficient (adapted from Kischka *et al.*[63]).

If a chiralized nematic with a long pitch ($= 2\pi/q$) and positive $\Delta\varepsilon$ is filled in a thin ($\sim 3\ \mu$m) cell whose walls are treated to produce a strong planar alignment of the director, the helix becomes unwound. An AC electric field of frequency $f$ applied between the plates beyond the Freedericksz threshold ($E_{thr}$) value generates a $\theta(z,t)$ profile, which in turn gives rise to a DC as well as to an oscillatory optical signal (at $2f$), which have maximum values for a setting of the initial $\mathbf{n}$ at 45° to the crossed polarizers. In addition, the cell also gives rise to an $f$ signal again for $E > E_{thr}$, which has a maximum value for the initial $\mathbf{n}$ set at 22.5° to the polarizer, which is a signature of $\phi(z)$ oscillations of the director.[65] Further, this signal is found to be $\propto q$, changing sign with that of $q$, when the sample temperature is varied across the helix inversion temperature in an appropriate mixture. Indeed the experiment was set up to look for this signal, which was taken to be a signature of the symmetry-permitted electromechanical coupling in chiral liquid crystals.[4,65] Recently, this experiment has been reinvestigated by Dequidt and Oswald,[66] who showed that the electric field dependence of the $f$-signal is better described by the flexoelectric effect, the electric field developing a spatial gradient due to the $\theta(z,t)$ profile above $E_{thr}$. Thus this experiment, which makes use of a simple symmetric cell structure, can be used to measure $(e_1 + e_3)$, though again it requires a detailed numerical analysis and the field-dependent signal depends on the three elastic constants, the dielectric constants, the rotational viscosity and the refractive indices, as well as on $(e_1 + e_3)$.

## 2.4. Some Remarks on the Experimental Results

From the above survey of various experimental techniques, we can make some general observations.

(1) As the measurements of $e^*$ require only a uniform electric field, the torque on the flexo-polarization by the external DC field and hence the sign of $e^*$ is unambiguous. If precautions are taken, such as applying an AC field for a short time before the measuring DC field, even the magnitude may be reasonable. However, the distortions generally involve both the zenithal and azimuthal angles of the director, and most samples have non-zero values of the dielectric anisotropy. The light beam passes through a fairly complicated structure in the cell before the transmitted intensity is measured. A detailed modelling of the experiment and using accurate values of all the other material properties of the nematic are necessary for the extraction of a good value of $e^*$. In most experiments that are reported, many simplifying approximations are made, and thus the quoted values are

subject to substantial errors, usually $\sim \pm 30$ to 50%.

(2) The measurements of $(e_1 + e_3)$ are subject to even more serious uncertainties. As the relevant flexotorque involves field gradients in the bulk or surface contributions, some additional factors contribute to the measurements using DC fields: surface polarization due to dipolar molecules of the nematic, adsorbed ions on the surfaces, ions moving under the action of the space-varying field, a part of which may be caused by the flexo-polarization itself, contribute to the response of the cell. Indeed as we have mentioned earlier, the data can be fitted by different combinations of these parameters, thus making even the quoted sign of $(e_1 + e_3)$ sometimes ambiguous. AC techniques using better geometries for the cell and detailed modelling of the response of the nematic to the field have started improving the reliability of the results.

All the data show that the flexocoefficients for the many different single components as well as multicomponent mixtures of nematogens with monomeric rod-like molecules are $\sim 10^{-11}$ C/m, the measurements being made well below the nematic-isotropic transition temperature. This agrees well with the estimate made on the basis of simple physical considerations, *viz.*, as the ratio of an elementary charge over a molecular dimension.

There have been studies on the influence of a small concentration of solute molecules on the flexocoefficients of a host nematic. A swallow-tailed molecule has two chains attached at one end, thus giving it the shape of a wedge. By dissolving such a compound in a host nematogen, the $e^*$ value was found to increase linearly with concentration up to $\sim 4.5$ mole % of the solute.[67] A number of compounds have been specifically synthesized to enhance the $e^*$ values of commercial nematic mixtures.[68] Even a small concentration of the solutes with bent-shaped molecules or dimers can increase $e^*$ for **E7** by a factor of three or four, showing the importance of molecular shape as originally envisaged by Meyer. However, there was no clear correlation with the lateral dipole moments of the molecules. Swallow-tailed molecules with the highly polar cyano groups attached at the narrower ends of the molecules also enhance $e^*$ for **E7**,[69] though by a somewhat smaller factor compared to the bent-shaped molecules. The $e^*$ value of another commercial mixture **MLC6437-000** shows a modest increase (decrease) by doping it with rod-like molecules with positive (negative) $\Delta\varepsilon$.[70] A large number of mesogens with bent-core (BC) molecules have been synthesized during the last decade. Doping only $\approx 5\%$ of such molecules in **8OCB** enhances the $e^*$ value by a large factor, indicating that the molecular flexocoefficient of the BC molecules is an order of magnitude larger than that of the rod-like molecules,[71] as can be estimated from molecular models.[5] Increasing the concentration further enhances $e^*$ non-linearly, indicating the formation of clusters of BC molecules. Indeed, the flexocoefficient of a typical compound with pure BC molecules is three orders of magnitude larger than that of rod-like molecules.[16]

There are only a few measurements on the temperature dependences of the flexocoefficients of pure compounds. The interdigitated electrode technique shows that $(e_1 + e_3)S/K$ is independent of temperature for a non-polar symmetric tolane compound,[39] consistent with the quadrupolar origin of the flexocoefficient in this case. A similar trend is also found in the highly polar compound **8OCB**, the ratio decreasing only as the nematic-smectic $A_d$ transition point is approached, reflecting the divergence of $K_3$. Surprisingly, $e^*$ for **8OCB**, which has a negative sign,[49] is also found to be $\propto S$, even though the essentially dipolar origin of $e^*$ is expected to lead to an $S^2$ dependence [see Eq. (2.6)]. The main contribution to $e^*$ for **8OCB** arises from the transverse dipole moment of the alkoxy group. The *antiparallel pair* can take bent conformations, increasing the magnitude of $e^*$, which actually becomes enhanced at higher temperatures as the flexibility of the chains and hence the ease of forming bent conformations also increases. This enhanced contribution in turn decreases the reduction of $e^*$ with temperature because of the order parameter $S$. In 4-octyl-4′-cyanobiphenyl (**8CB**) without the oxygen atom, $e^*$ has a positive sign, and its magnitude is about five times smaller than that of **8OCB**.[49] When the phenyl ring attached to the alkyl chain is replaced by a cyclohexane ring (to get 4-cyanophenyl-trans-4′-n-alkylcyclohexane compounds identified by the symbol **PCHn**), $e^*$ which is $\propto S^2$ retains a positive sign but decreases in magnitude[50] compared to that of the corresponding **nCB**. However, if the other phenyl ring is also replaced by a cyclohexane ring (to get **CCHn**), $e^*/K$ becomes negative in sign, and increases with temperature as with **8OCB**. In **8CB**, **PCH7** and **CCH7**, the lateral components of the dipoles also arise from the cyano groups. The orientation of these components with respect to the bent structures of the antiparallel pairs, however, changes sign between **PCH7** and **CCH7** compounds due to the non-planar structure of the cyclohexane ring.[50] The greater flexibility of the non-aromatic **CCH7** molecule gives rise to the observed temperature dependence. The sign of $(e_1 + e_3)$ for **CCH7** remains positive[61] as in the case of all compounds with cyano end groups, the antiparallel pairing producing similar quadrupole moments in such compounds. $(e_1 + e_3)/S$ remains practically constant with temperature in the lower nematic range for **CCH7**, but decreases in the higher temperature range, the increased flexibility now *lowering* the quadrupole moment of the antiparallel pair of non-aromatic molecules.

In conclusion, as flexoelectricity is associated with splay-bend distortion of the director field, measurements of the relevant coefficients are not as straightforward as in the case of dielectric constants or refractive indices. The medium is an organic liquid and ionic impurities usually screen out the polarization. AC electro-optic techniques have been developed to overcome this problem, but the extraction of the flexocoefficients from the electro-optic data involves a fairly complicated numerical analysis. Independently measured values of the elastic constants, dielectric constants, refractive indices and rotational viscosity are input parameters in the fitting procedure, apart from the usually unknown surface properties such as the anchoring energies, ion adsorption and surface polarization, which have to be treated as fitting parameters. The absolute values of flexocoefficients quoted in the literature have substantial errors though the magnitude agrees with theoretical estimates. With a greater appreciation of the important role played by flexoelectric effects in both applications and various phenomena under the action of electric fields, recently there have been a few attempts to address some of the issues mentioned above. It would be preferable to have at least two very independent measurement techniques in which these issues can be addressed, so that the measured values can be cross-checked.

## References

1. M. Born, Über anisotrope Flüssigkeiten. [1.]: Versuch einer Theorie der flüssigen Kristalle und des elektrischen Kerr-Effekts in Flüssigkeiten, *Sitz. Preuss. Akad. Wiss.* **30**, 614–650, (1916).
2. N.V. Madhusudana and S. Chandrasekhar, The role of permanent dipoles in nematic order, *Pramana Suppl.* **1**, 57–68, (1975).
3. R.B. Meyer, Piezoelectric effects in liquid crystals, *Phys. Rev. Lett.* **22**(18), 918–921, (1969). doi: 10.1103/PhysRevLett.22.918
4. P.G. de Gennes and J. Prost, *The Physics of Liquid Crystals, 2nd edition*, Clarendon Press, Oxford, 1993.
5. W. Helfrich, The strength of piezoelectricity in liquid crystals, *Z. Naturforsch.* **26a**, 833–835, (1971).
6. A. Jákli, J. Harden and N. Éber, Chapter 3. Flexoelectricity of bent-core molecules, In eds. Á. Buka and N. Éber, *Flexoelectricity in Liquid Crystals. Theory, Experiments and Applications*, Imperial College Press, London, 2012. pp. 61–99.

7. J. Prost and J.P. Marcerou, On the microscopic interpretation of flexoelectricity, *J. Phys. France* **38**(3), 315–324, (1977). doi: 10.1051/jphys:01977003803031500
8. J.S. Patel and R.B. Meyer, Flexoelectric electro-optics of a cholesteric liquid crystal, *Phys. Rev. Lett.* **58**(15), 1538–1540, (1987). doi: 10.1103/PhysRevLett.58.1538
9. F. Castles, S.M. Morris and H.J. Coles, Flexoelectro-optic properties of chiral nematic liquid crystals in the uniform standing helix configuration, *Phys. Rev. E* **80**(3), 031709/1–9, (2009). doi: 10.1103/PhysRevE.80.031709
10. J.C. Jones, G.P. Bryan-Brown, E.L. Wood, A. Graham, P. Brett and J.R. Hughes, Novel bistable liquid crystal displays based on grating alignment, *Proc. SPIE* **3955**, 84–93, (2000). doi: 10.1117/12.379969
11. P. Rudquist and S.T. Lagerwall, Chapter 7. Applications of flexoelectricity. In eds. Á. Buka and N. Éber, *Flexoelectricity in Liquid Crystals. Theory, Experiments and Applications*, Imperial College Press, London, 2012. pp. 211–247.
12. W. Helfrich, Conduction-induced alignment of nematic liquid crystals: basic model and stability considerations, *J. Chem. Phys.* **51**(9), 4092–4105, (1969). doi: 10.1063/1.1672632
13. N.V. Madhusudana, V.A. Raghunathan and K.R. Sumathy, Flexoelectric origin of oblique-roll electrohydrodynamic instability in nematics, *Pramana-J. Phys.* **28**(3), L311–L316, (1987).
14. N.V. Madhusudana and V.A. Raghunathan, Influence of flexoelectricity on electrohydrodynamic instabilities in nematics, *Liq. Cryst.* **5**(6), 1789–1812, (1989). doi: 10.1080/02678298908045689
15. Á. Buka, T. Tóth-Katona, N. Éber, A. Krekhov and W. Pesch, Chapter 4. The role of flexoelectricity in pattern formation. In eds. Á. Buka and N. Éber, *Flexoelectricity in Liquid Crystals. Theory, Experiments and Applications*, Imperial College Press, London, 2012. pp. 101–135.
16. J. Harden, B. Mbanga, N. Éber, K. Fodor-Csorba, S. Sprunt, J.T. Gleeson and A. Jákli, Giant flexoelectricity of bent-core nematic liquid crystals, *Phys. Rev. Lett.* **97**(15), 157802/1-4, (2006). doi: 10.1103/PhysRevLett.97.157802
17. A.G. Petrov, Measurements and interpretation of flexoelectricity. In eds. D.A. Dunmur, A. Fukuda and G.R. Luckhurst, *Physical Properties of Liquid Crystals, Vol 1: Nematics*, Inspec, London, 2001. pp. 251–264.
18. S.J. Elston and L.A. Parry-Jones, in the Abstracts book, *Conference on Flexoelectricity in Liquid Crystals*, Oxford, September 2006, pp. 21–29.
19. M.R. Dodge, C. Rosenblatt, R.G. Petchek, M.E. Neubert and M.E. Walsh, Bend elasticity of mixtures of V-shaped molecules in ordinary nematogens, *Phys. Rev. E* **62**(4), 5056–5063, (2000). doi: 10.1103/PhysRevE.62.5056
20. B. Kundu, R. Pratibha and N.V. Madhusudana, Anomalous temperature dependence of elastic constants in the nematic phase of binary mixtures made of rodlike and bent-core molecules, *Phys. Rev. Lett.* **99**(24), 247802/1–4, (2007). doi: 10.1103/PhysRevLett.99.247802

21. W. Helfrich, Inherent bounds to the elasticity and flexoelectricity of liquid crystals, *Mol. Cryst. Liq. Cryst.* **26**(1-2), 1–5, (1974). doi: 10.1080/15421407408084818
22. H.J. Deuling, The piezo-electric effect in nematic layers, *Mol. Cryst. Liq. Cryst.* **26**(3-4), 281–284, (1974). doi: 10.1080/15421407408083106
23. H.J. Deuling, On a method to measure the flexoelectric coefficients of nematic liquid crystals, *Solid State Commun.* **14**(11), 1073–1074, (1974). doi: 10.1016/0038-1098(74)90275-0
24. C.V. Brown and N.J. Mottram, Influence of flexoelectricity above the nematic Freedericksz transition, *Phys. Rev. E* **68**(3), 031702/1–5, (2003). doi: 10.1103/PhysRevE.68.031702
25. A.A.T. Smith, C.V. Brown and N.J. Mottram, Theoretical analysis of the magnetic Freedericksz transition in the presence of flexoelectricity and ionic contamination, *Phys. Rev. E* **75**(4), 041704/1–8, (2007). doi: 10.1103/PhysRevE.75.041704
26. G. Barbero, I. Dozov, J.F. Palierne and G. Durand, Order electricity and surface orientation in nematic liquid crystals, *Phys. Rev. Lett.* **56**(19), 2056–2059, (1986). doi: 10.1103/PhysRevLett.56.2056
27. M.A. Osipov, Chapter 1. Molecular Theory of flexoelectricity in nematic liquid crystals. In eds. Á. Buka and N. Éber, *Flexoelectricity in Liquid Crystals. Theory, Experiments and Applications*, Imperial College Press, London, 2012. pp. 9–32.
28. D. Schmidt, M. Schadt and W. Helfrich, Liquid-crystalline curvature electricity: the bending mode of MBBA, *Z. Naturforsch.* **27a**, 277–280, (1972).
29. W. Helfrich, Polarity-dependent electro-optical effect of nematic liquid crystals, *Appl. Phys. Lett.* **24**(10), 451–452, (1974). doi: 10.1063/1.1655006
30. A. Derzhanski, A.G. Petrov and M.D. Mitov, One-dimensional dielectric-flexoelectric deformations in nematic layers, *J. Phys. France* **39**(3), 273–285, (1978). doi: 10.1051/jphys:01978003903027300
31. M. Monkade, Ph. Martinot-Lagarde and G. Durand, Electric polar surface instability in nematic liquid crystals, *Europhys. Lett.* **2**(4), 299–305, (1986). doi: 10.1209/0295-5075/2/4/007
32. O.D. Lavrentovich, V.G. Nazarenko, V.V. Sergan and G. Durand, Dielectric quenching of the electric polar surface instability in a nematic liquid crystal, *Phys. Rev. A* **45**(10), R6969–R6972, (1992). doi: 10.1103/PhysRevA.45.R6969
33. V.G. Nazarenko, R. Klouda and O.D. Lavrentovich, Unipolar electro-optic effect in a nematic cell, *Phys. Rev. E* **57**(1), R36–R38, (1998). doi: 10.1103/PhysRevE.57.R36
34. G. Barbero, L.R. Evangelista and N.V. Madhusudana, Effect of surface electric field on the anchoring of nematic liquid crystals, *Eur. Phys. J. B* **1**(3), 327–331, (1998). doi: 10.1007/s100510050190
35. G. Basappa and N.V. Madhusudana, New type of high field electrooptic response in nematics, *Proc. SPIE* **4147**, 116–125, (2000). doi: 10.1117/12.385668

36. S. Ponti, G. Barbero, A. Strigazzi, Y. Marinov and A. Petrov, Surface energy dissipation in homeotropic nematic layers: the role of flexoelectricity and surfactant desorption, *Mol. Cryst. Liq. Cryst.* **420**(1), 55–72, (2004). doi: 10.1080/15421400490478164
37. M. Buczkowska and G. Derfel, Analysis of deformations of flexoelectric homeotropic liquid crystal layers with various anchoring strengths, *Opto-Electronics Review* **19**(1), 56–60, (2011). doi: 10.2478/s11772-010-0065-0
38. J. Prost and P.S. Pershan, Flexoelectricity in nematic and smectic-A liquid crystals, *J. Appl. Phys.* **47**(6), 2298–2312, (1976). doi: 10.1063/1.323021
39. J.P. Marcerou and J. Prost, The different aspects of flexoelectricity in nematics, *Mol. Cryst. Liq. Cryst.* **58**(3-4), 259–284, (1980). doi: 10.1080/00268948008082127
40. J.P. Marcerou and J. Prost, Flexoelectricity in isotropic phases, *Phys. Lett.* **66A**(3), 218–220 (1978). doi: 10.1016/0375-9601(78)90662-X
41. J.C. Rouillon, M. Benzekri, T. Claverie, J.P. Marcerou, H.T. Nguyen and J. Prost, Experimental evidence for the breakdown of conventional elasticity in smectic-A, *Liq. Cryst.*, **16**(6), 1065–1072, (1994). doi: 10.1080/02678299408027875
42. C.L. Trabi, A.A.T. Smith and C.V. Brown, Interferometric method for the measurement of the flexoelectric polarisation using nematic liquid crystal phase gratings, *Mol. Cryst. Liq. Cryst.* **509**(1), 378–392, (2009). doi: 10.1080/15421400903066125
43. S.D. Lee and J.S. Patel, Symmetry-breaking effect of interfacial interactions on electro-optical properties of liquid crystals, *Phys. Rev. Lett.* **65**(1), 56–59, (1990). doi: 10.1103/PhysRevLett.65.56
44. Y.P. Bobylev and S.A. Pikin, Threshold piezoelectric instability in a liquid crystal, *Sov. Phys. JETP* **45**(1), 195–201, (1977).
45. L.K. Vistin, A new electrostructural effect in liquid crystals of nematic type, *Sov. Phys. Crystallogr.* **15**, 514–518, (1970).
46. M.I. Barnik, L.M. Blinov, A.N. Trufanov and B.A. Umanski, Flexo-electric domains in liquid crystals, *J. Phys. France* **39**(4), 417–422, (1978). doi: 10.1051/jphys:01978003904041700
47. A. Krekhov, W. Pesch and Á. Buka, Flexoelectricity and pattern formation in nematic liquid crystals, *Phys. Rev. E* **83**(5), 051706/1–13, (2011). doi: 10.1103/PhysRevE.83.051706
48. I. Dozov, Ph. Martinot Lagarde and G. Durand, Flexoelectrically controlled twist of texture in a nematic liquid crystal, *J. Physique Lett.* **43**(10), L-365–L-369, (1982). doi: 10.1051/jphyslet:019820043010036500
49. I. Dozov, Ph. Martinot Lagarde and G. Durand, Conformational flexoelectricity in nematic liquid crystals, *J. Physique Lett.* **44**(19), L-817–L-822, (1983). doi: 10.1051/jphyslet:019830044019081700
50. P.R. Maheswara Murthy, V.A. Raghunathan and N.V. Madhusudana, Experimental determination of the flexoelectric coefficients of some nematic liquid crystals, *Liq. Cryst.* **14**(2), 483–496, (1993). doi: 10.1080/02678299308027664

51. T. Takanishi, S. Hashidate, S. Nishijou, M. Usui, M. Kimura and T. Akahane, Novel measurement method for flexoelectric coefficients of nematic liquid crystals, *Jpn. J. Appl. Phys. Part 1.* **37**(4A), 1865–1869, (1998). doi: 10.1143/JJAP.37.1865
52. N.V. Madhusudana and G. Durand, Linear flexo-electro-optic effect in a hybrid aligned nematic liquid crystal cell, *J. Physique Lett.* **46**(5), L-195–L-200, (1985). doi: 10.1051/jphyslet:01985004605019500
53. B. Valenti, C. Bertoni, G. Barbero, P. Taverna-Valabrega and R. Bartolino, Flexoelectricity in the hybrid aligned nematic cell, *Mol. Cryst. Liq. Cryst.* **146**(1), 307–320, (1987). doi: 10.1080/00268948708071820
54. S. Ponti, P. Ziherl, C. Ferrero and S. Zumer, Flexoelectro-optic effect in a hybrid nematic liquid crystal cell, *Liq. Cryst.* **26**(8), 1171–1177, (1999). doi: 10.1080/026782999204192
55. N.T. Kirkman, T. Stirner and W.E. Hagston, Continuum modelling of hybrid-aligned nematic liquid crystal cells: optical response and flexoelectricity-induced voltage shift, *Liq. Cryst.* **30**(9), 1115–1122, (2003). doi: 10.1080/0267829031000159456
56. S.A. Jewell and J.R. Sambles, Fully leaky guided mode study of the flexoelectric effect and surface polarization in hybrid aligned nematic cells, *J. Appl. Phys.* **92**(1), 19–24, (2002). doi: 10.1063/1.1483392
57. T.S. Taphouse, S.L. Cornford, J.E. Birkett and J.R. Sambles, Time-resolved sign-dependent switching in a hybrid aligned nematic liquid crystal cell, *New J. Phys.* **10**(8), 083045/1–15, (2008). doi: 10.1088/1367-2630/10/8/083045
58. S.L. Cornford, T.S. Taphouse and J.R. Sambles, Analysis of the sign-dependent switching observed in a hybrid aligned nematic cell, *New J. Phys.* **11**(1), 013045/1–14, (2009), doi: 10.1088/1367-2630/11/1/013045
59. L.M. Blinov, G. Durand and S.V. Yablonsky, Curvature oscillations and linear electro-optical effect in a surface layer of a nematic liquid crystal, *J. Phys. II France* **2**(5), 1287–1300, (1992). doi: 10.1051/jp2:1992200
60. P.R. Maheswara Murthy, V.A. Raghunathan and N.V. Madhusudana, An AC electro-optic technique of comparing the two anchoring energies of a hybrid aligned nematic cell, *Liq. Cryst.* **14**(4), 1107–1124, (1993). doi: 10.1080/02678299308027819
61. S.R. Warrier and N.V. Madhusudana, An AC electrooptic technique for measuring the flexoelectric coefficient $(e_1+e_3)$ and anchoring energies of nematics, *J. Phys. II France* **7**(12), 1789–1803, (1997). doi: 10.1051/jp2:1997215
62. N.V. Madhusudana, J.F. Palierne, Ph. Martinot-Lagarde and G. Durand, Twist instability of a flexoelectric nematic domain in an external field, *Phys. Rev. A* **30**, 2153R–2154R, (1984). doi: 10.1103/PhysRevA.30.2153
63. C. Kischka, L.A. Parry-Jones, S.J. Elston and E.P. Raynes, Measurement of the flexoelectric coefficients $e_1$ and $e_3$ in nematic liquid crystals, *Mol. Cryst. Liq. Cryst.* **480**(1), 103–110, (2008). doi: 10.1080/15421400701825706
64. V.A. Raghunathan and N.V. Madhusudana, A new threshold flexoelectric instability in nematic liquid crystals, *Mol. Cryst. Liq. Cryst. Lett.* **6**, 103–111, (1989).

65. H.P. Padmini and N.V. Madhusudana, Electromechanical effect in cholesteric mixtures with a compensation temperature, *Liq. Cryst.* **14**(2), 497–511, (1993). doi: 10.1080/02678299308027665
66. A. Dequidt and P. Oswald, Does the electric Lehmann effect exist in cholesteric liquid crystals? *Eur. Phys. J. E* **24**(2), 157–166, (2007). doi: 10.1140/epje/i2007-10226-0
67. Y. Marinova, J. Kosmopoulos, W. Weissflog, A.G. Petrov and D.J. Photinos, Flexoelectricity of wedge-like molecules in nematic mixtures, *Mol. Cryst. Liq. Cryst.* **357**(1), 221–228, (2001). doi: 10.1080/10587250108028255
68. J.H. Wild, K. Bartle, N.T. Kirkman, S.M. Kelly, M. O'Neill, T. Stirner and R.P. Tuffin, Synthesis and investigation of nematic liquid crystals with flexoelectric properties, *Chem. Mater.* **17**(25), 6354–6360, (2005). doi: 10.1021/cm051682y
69. J.H. Wild, K. Bartle, M. O'Neill, S.M. Kelly and R.P. Tuffin, Synthesis and mesomorphic behaviour of wedge-shaped nematic liquid crystals with flexoelectric properties, *Liq. Cryst.* **33**(6), 635–644, (2006). doi: 10.1080/02678290600665131
70. N. Aziz, S.M. Kelly, W. Duffy and M. Goulding, Rod-shaped dopants for flexoelectric nematic mixtures, *Liq. Cryst.* **36**(5), 503–520 (2009). doi: 10.1080/02678290903031643
71. B. Kundu, A. Roy, R. Pratibha and N.V. Madhusudana, Flexoelectric studies on mixtures of compounds made of rodlike and bent-core molecules, *Appl. Phys. Lett.* **95**(8), 081902/1-3, (2009). doi: 10.1063/1.3213544

# Chapter 3

# Flexoelectricity of Bent-core Molecules

Antal Jákli[♯], John Harden[♯] and Nándor Éber[§]

[♯] *Liquid Crystal Institute, Kent State University*
*Kent, Ohio 44242, USA*
*e-mail: ajakli@kent.edu*

[§] *Institute for Solid State Physics and Optics*
*Wigner Research Centre for Physics*
*Hungarian Academy of Sciences*
*P.O. Box 49, H-1525 Budapest, Hungary*
*e-mail: eber@szfki.hu*

Bent-core (BC) liquid crystals (LC) have a tendency to form polar smectic ordering even in their nematic phase, which has a great influence on their flexoelectric properties. A measuring technique based on flexing sandwich cells is introduced, which allows us to determine the flexocoefficients by the direct flexoelectric effect. We report on giant flexoelectricity found in several bent-core nematics by this method, as well as by detecting the electric field-induced flexing of a cell (a manifestation of giant converse flexoelectricity). Conventional (indirect) measurements of flexocoefficients on BC nematics, which might not detect giant flexoelectricity, are also reviewed. We address the physical origin of giant flexoelectricity and attempt to explain the apparent discrepancy found between results obtained by different measurement techniques. Finally, we show that giant flexoelectricity is observable in BC elastomers too, which may open the way for practical applications.

## Contents

## 3.1. Introduction

### 3.1.1. *Bent-core (banana-shaped) liquid crystals*

Although the first synthesis of bent-shaped liquid crystals was reported 80 years ago by Vorländer and Apel,[1] their potential for producing liquid crystalline phases was not widely accepted until the early 1990s. It was the synthesis and characterization by Matsunaga *et al.*[2,3] that showed that not just rod-shape (calamitic) or discotic molecules can form mesophases, but bent-core (BC) molecules may do so as well. This discovery prompted a surge of research, which attracted a large number of liquid crystal scientists to explore this area. Soon it became clear that a relatively tiny change in the molecular shape leads to huge differences in physical properties. First of all, the translational symmetry is broken and the simplest symmetric and bare BC molecules prefer smectic layering. In addition, the tight molecular packing requirement (dictated by the entropy) leads to macroscopic polarization provided that the electric dipoles along the kink direction are not zero. Such a situation is shown in Fig. 3.1, where the axis along the end-to-end distance of the molecules (the primary director $\mathbf{n}$) is parallel to the smectic layer normal $\mathbf{k}$. Since $\mathbf{n}\|\mathbf{k}$ and the layers are polar, this structure can be designated SmAP. Although the first observations found ferroelectric polarization switching,[4,5] soon several groups reported that the switching is instead antiferroelectric.[6–8] This indicates that one has to consider two-layer units, where the layer polarization $\mathbf{P}$ can be either parallel or antiparallel in the subsequent layers, corresponding to ferroelectric ($\mathrm{SmAP_F}$) or antiferroelectric ($\mathrm{SmAP_A}$) structures.

The three orthogonal unit vectors characterizing these structures ($\mathbf{n}$, $\mathbf{m}$ and $\mathbf{P}$) are the following: $\mathbf{n}$ is the unit vector along the long axis; $\mathbf{m}$ is normal to the molecular plane and the layer polarization $\mathbf{P}$ is along the kink direction.

Experimental results, which followed the publications of the Tokyo Institute of Technology[4,5] also showed that the molecular planes denoted by

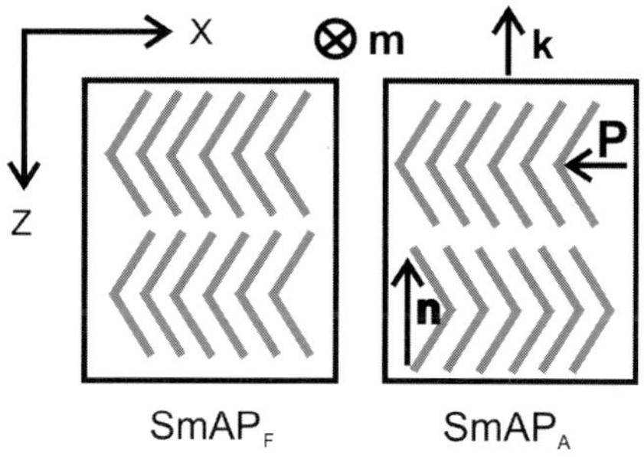


Fig. 3.1. Schematic structure of the ferroelectric $SmAP_F$ and antiferroelectric $SmAP_A$ phases of bent-core molecules.

vector **m** in Fig. 3.1 are tilted and should rather be called the polar smectic C (SmCP) phase. The Boulder group was the first to realize that these polar tilted layers are actually chiral, even if the constituent molecules do not possess any chiral carbon or any other type of stereo centre.[6] This situation is illustrated in Fig. 3.2.

Just as the electric polarization in the subsequent layers can be parallel or antiparallel, the chirality in the subsequent layers is not necessarily the same but can alternate. When the neighbouring layers have the same handedness, the structure is called 'homochiral', and when the layer chirality alternates, the structure is usually designated as 'racemic'; although actually could be named as 'antichiral' to distinguish it from the well-known racemic mixtures with equal numbers of left- and right-handed molecules. It was found later that racemic and homochiral states have very similar free energies, and not just their polarization, but also the layer chirality can be changed; for example by the application of electric fields of different waveforms. Specifically, it was found[9] that a triangular waveform tends to induce antichiral structures whereas rectangular fields lead to homochiral structures. Later examples were found for direct flipping of both the chirality and the polarity at the same time.[10–14]

The situations where the molecular planes are tilted with respect to the layer normal (i.e., when **m** is not perpendicular to **k**) are shown in the upper row of Fig. 3.3. In the plane determined by the polarization **P** and the layer normal **k** (the polar plane) this tilt is illustrated by a nail at the end of the directors that is closer to the observer. Depending on whether the tilt directions are parallel or antiparallel we refer to 'synclinic' (s) and 'anticlinic' (a) structures. Combining these different situations with the 'Ferroelectric' (F) and 'Antiferroelectric' (A) layer polarizations we can

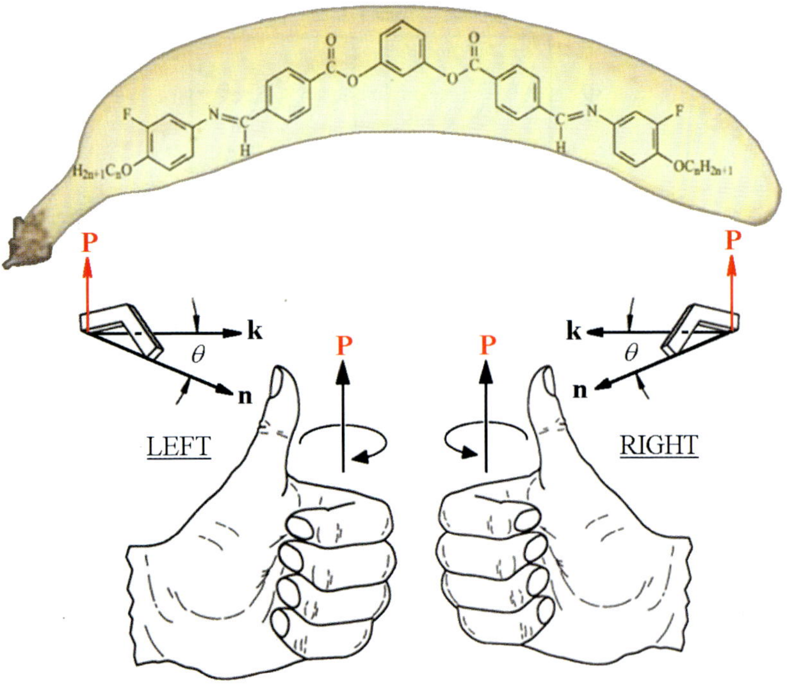


Fig. 3.2. Illustration of layer chirality of bent-core liquid crystals (an example of a typical molecular structure is shown in the upper picture) using the left and right hands.

have four different subphases: $\mathrm{SmC_sP_F}$, $\mathrm{SmC_aP_A}$, $\mathrm{SmC_sP_A}$ and $\mathrm{SmC_aP_F}$. A similar notation was introduced by Link *et al.*[6] and is widely used in the literature. Note that the SmCP layers have only a two-fold symmetry axis around **P** ($C_2$ symmetry), i.e. they have the same symmetry as that of chiral SmC* materials.

In principle, we can also envision that the director **n** only is tilted with respect to the layer normal **k**, without any change in **m**. Such tilting is usually called 'leaning' to distinguish it from the tilt of the molecular plane. In this situation the layers remain achiral, but the successive layers can be either 'synleaning' or 'antileaning'. We call the leaning positive or negative, if the top or the bottom arm of the BC molecule, respectively, is tilted away from the vertical direction. It is important to note that in these 'leaning' phases the polarization is not necessarily parallel to the layers, but can have out-of-layer components, and they have $C_s$ symmetry.[15,16] Depending on the layer polarizations ('F' or 'A') in the leaning configurations we can again

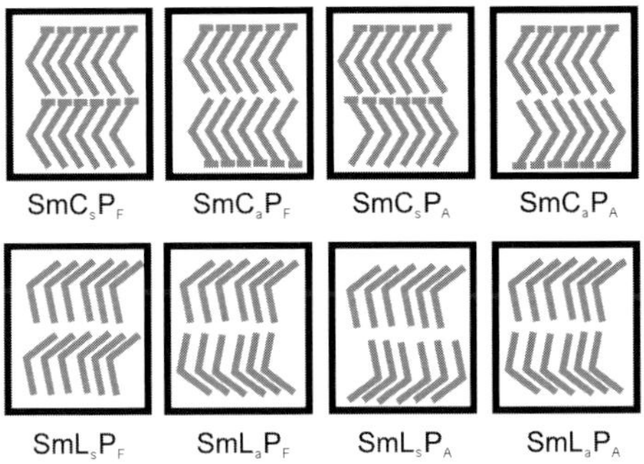


Fig. 3.3. All possible single-tilted bent-core smectic structures. Top: two-dimensional illustration of the four possible situations where only the molecular plane is tilted with respect to the layer normal. Bottom: two-dimensional illustration of the four possible situations where only the long axis is tilted (leaned) with respect to the layer normal.

have four subphases: $\mathrm{SmL_sP_A}$, $\mathrm{SmL_aP_F}$, $\mathrm{SmL_aP_A}$ and $\mathrm{SmL_sP_F}$, depending on the subsequent tilt and polarization direction combinations (see the bottom row of Fig. 3.3). Note that $\mathrm{SmC_aP_F}$ and $\mathrm{SmL_aP_F}$ look similar to $\mathrm{SmAP_F}$, while $\mathrm{SmC_aP_A}$ and $\mathrm{SmL_aP_A}$ are similar to $\mathrm{SmAP_A}$, as in all these cases the optic axis is parallel to the layer normal.

On the same grounds one can also imagine that both tilt and leaning occur together.[15,16] The possibility of those 'double-tilted' situations was mentioned in the 1974 edition of P.G. de Gennes's book[17] and were called $\mathrm{SmC_G}$ (G stands for 'general'). Following our previous nomenclature, we label them with C, L and P (SmCLP). Taking into account that the layer polarizations can either be parallel or antiparallel, we have altogether eight different subphases as illustrated in Fig. 3.4a.

Although in all situations the individual layer polarizations are not parallel to the smectic layers, the out-of-layer components average out and their macroscopic symmetry would be the same as that for SmAP (for the anticlinic cases such as $\mathrm{SmC_aL_aP_F}$ and $\mathrm{SmC_aL_sP_F}$) or as that for SmCP (for synclinic situations such as $\mathrm{SmC_sL_aP_F}$ and $\mathrm{SmC_sL_aP_F}$). It is important to note that the SmCLP structures can be equivalently described by a tilt of the molecular plane and a rotation of the layer polarization $\mathbf{P}$ around the long axis $\mathbf{n}$ by an angle $\phi$ (Fig. 3.4b). If $\phi = 0$ or $\pi$, we have the SmCP case; $\phi = \pi/2$ or $3\pi/2$ corresponds to the SmLP situations, otherwise we have the SmCLP phase. Double-tilted layers have triclinic symmetry; i.e., they are symmetric only with respect to a 360° rotation around the polar axis (in Schoenflies notation[18] $\mathrm{C_1}$). The detailed relations between the

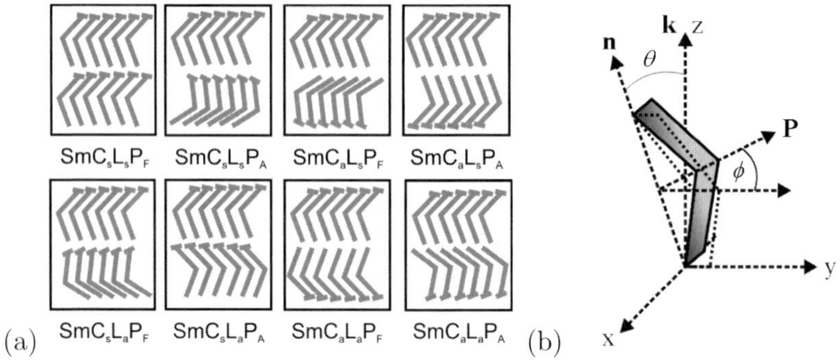


Fig. 3.4. (a) Two-dimensional illustrations of the eight possible double-tilted bent-core smectic structures. (b) Graphical explanation of the SmCLP structures in terms of two rotation angles $\theta$ and $\phi$.

physical and chemical structures and the physical properties of the smectic or columnar structures have been summarized recently in several review papers.[19–21] Instead of discussing these we have only noted those layered structures which are relevant for understanding the properties of bent-core nematics and especially the flexoelectricity of these materials.

### 3.1.2. *Bent-core nematics*

As described above, the kink in the molecular shape and the requirement to fill the space as effectively as possible are not compatible with a three-dimensional fluid order. In other words, when translating a bent-core molecule in the melt of the neighbouring bent-core molecules, it experiences a periodic potential with its periodicity determined by the length $l$ of the molecules. To allow for fluidity at the macroscopic level, one needs to frustrate the bent-core structure so that they do not lock into smectic layers easily. Such a frustration can be introduced by some steric or electrostatic disturbance of the bare bent-core (or 'peeled banana') shape, which has been seen in some modulated smectic phases.[22] As discussed by Bailey and Jákli,[23] a steric or electrostatic inclusion in the core of the molecules leads to layer modulation, an $\mathrm{SmC_G}$ structure and broken smectic layers that effectively correspond to a columnar phase, as the inclusions increase. Following this picture, here we postulate that bent-core nematics are probably more frustrated than the electrically unswitchable B7 (columnar) phases, in which the broken smectic ribbons are separated by melted fluid nematic regions. Such over-frustrated B7 materials are characterized by a macroscopical nematic order, though composed of floating smectic clusters (Fig. 3.5). This view has been supported by recent small-angle X-ray scattering (SAXS) measurements.[24–26] These smectic clusters are at least partially ferroelectric and they can explain the Goldstone mode-type low-frequency dielectric relaxations observed by Salamon *et al.*[27] and the giant flexoelectric effects observed first by Harden *et al.*[28,29]

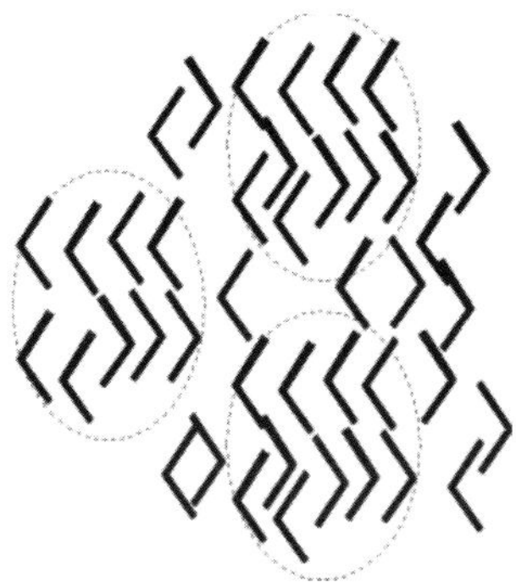

Fig. 3.5. Sketch of the cluster nanostructure of the nematic phase of bent-core molecules.

In the following sections of this chapter we will summarize the direct, as well as the converse, flexoelectric measurements in fluid and elastomeric (dry or swollen) bent-core nematic liquid crystals, and try to explain these observations using the structural model outlined above.

## 3.2. Flexoelectricity in Bent-core Liquid Crystals

### 3.2.1. *The flexoelectric coefficients*

Although most of the scientists working on liquid crystals may think that flexoelectricity is a special property of liquid crystals, it was actually first discussed in 1964 for crystals[30] as a response to strain (stress) gradients. The direct and converse flexoelectric coupling constants $e_{ijkl}$ were described with a fourth-rank tensor, as

$$P_i^{flexo} = e_{ijkl}\frac{\partial S_{jk}}{\partial x_l}, \quad A\frac{\partial T_{ij}}{\partial x_k} = e_{ijkl}E_l. \tag{3.1}$$

In these expressions $S_{jk}$ and $T_{ij}$ are the elements of the strain and the stress tensors, respectively; $A$ is the area of the sample. Flexoelectricity as defined by Meyer[31] for nematic liquid crystals is a special case of the general definition above, where the gradient of the strain is replaced by the gradient of the director $\mathbf{n}$.

Flexoelectricity in bent-core liquid crystals was discussed by Meyer.[31] In addition to the pear-shaped molecules, bent-shaped molecules were used to illustrate the dipolar origin of the flexoelectric effects in nematic liquid crystals. It was assumed that the constituent molecules of the nematic liquid crystals are free to rotate around their axes, and in the absence of electric fields, their dipole moments average out; so the net polarization of the material is zero. However, when liquid crystals made from polar pear- or banana-shaped molecules are subjected to splay or bend deformations, respectively, they can become macroscopically polar, because the polar structures correspond to a more efficient packing of the molecules. It follows from symmetry considerations that the deformation-induced flexoelectric polarization $\mathbf{P}_{\mathbf{fl}}$ can be written as:[17,31]

$$\mathbf{P}_{\mathbf{fl}} = e_1\mathbf{n}(\nabla \cdot \mathbf{n}) + e_3(\nabla \times \mathbf{n}) \times \mathbf{n}\,. \tag{3.2}$$

The magnitude of the effect is characterized by two flexoelectric coefficients, $e_1$ and $e_3$, for splay and bend, respectively. As far as the microscopic origin of these phenomenological flexocoefficients is concerned, two different mechanisms – dipolar and quadrupolar – have been identified; they are discussed in detail in Chapter 1.[32]

The dipolar mechanism is sensitive to the molecular shape. By dimensional considerations one can estimate[17] the flexocoefficients due to dipolar mechanism as $e_1, e_3 \leq \mu_e/a^2$, where $\mu_e \sim$ 1–5 debye (1 D $= 3.3\times10^{-30}$ Cm) is the molecular dipole moment and $a \sim$ 2–4 nm is the typical molecular dimension for a low molecular weight liquid crystal. This means that $e_1$ and $e_3$ are expected to be of the order of pCm$^{-1}$. Assuming a random three-dimensional distribution of the centre of masses of the constituent bent-core (banana-shaped) molecules, Helfrich[33,34] and Derzhanski and Petrov[35] derived a more precise expression for the macroscopically testable bend flexoelectric coefficient:

$$e_3 = \frac{\mu_\perp K_3}{2k_BT}\theta_0 \left(\frac{b}{a}\right)^{2/3} N^{1/3}. \tag{3.3}$$

Here $\mu_\perp$ is the dipole moment perpendicular to the long axis, $K_3$ is the bend elastic constant, $\theta_0$ is the kink angle (see Fig. 3.6), $a$ is the length, $b$ is the width, $N$ is the number density of the molecules and $k_BT$ is the thermal energy. The first measurements of $e_3$ in calamitic nematics yielded higher values than expected,[17] indicating that not only the dipolar mech-

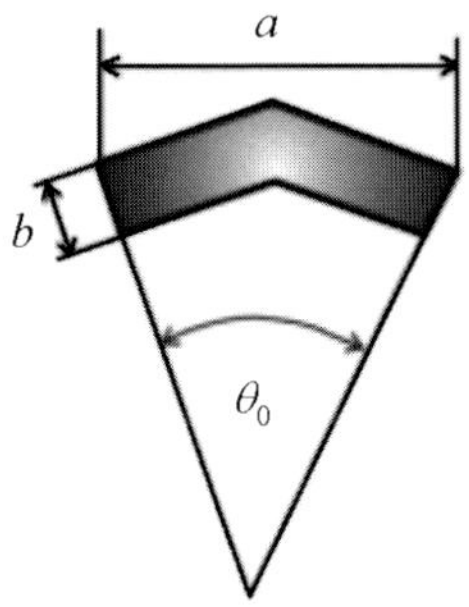


Fig. 3.6. Geometrical parameters of a bent-core molecule.

anism contributes to flexoelectricity. Indeed, it turned out that flexoelectricity can exist even in materials with non-polar and axially symmetric molecules. In this case it is caused by the interaction between the electric field gradient and the molecular quadrupole moments.[36] The flexoelectric coefficients of calamitic liquid crystals with a quadrupolar mechanism are actually typically in the range for non-liquid crystalline elastomers subjected to a deformation gradient.[37] Although these effects are very fundamental and interesting, the flexoelectric coefficients (being of the order of 1–10 $\mathrm{pCm}^{-1}$) are too small for practical applications.

For lyotropic systems, such as black lipid membranes (BLM), flexoelectric coefficients much larger than those estimated for thermotropic liquid crystals were observed as early as 1973,[38] and were called 'a peculiar kind of piezoelectric effect'. Later Petrov *et al.*[39,40] studied this effect in detail and found that the molecular basis of the flexoelectricity of lipid bilayers is due to the asymmetric redistribution of charges (C), dipoles (D) and the splayed uniaxial orientation of their quadruples (Q).[41] Converse flexoelectric measurements[42] found that the splay flexoelectric coefficient $e_1 \sim 100$ $\mathrm{pCm}^{-1}$, which is an order of magnitude larger than typical for calamitic thermotropic liquid crystals.

Converse flexoelectric studies of lyotropic liquid crystals, such as vesicles, is still an active subject.[43] Notably, the sensory mechanism of outer hair cell composite membranes[44–46] can be understood by the flexoelectric properties of the lipid bilayer.[39] The converse of this effect, i.e., a voltage-generated curvature, has also been observed and was discussed by Todorov *et al.*[47] Another related phenomenon is the ferroelectricity which results from the tilted layered structures of chiral molecules, which has been discussed extensively since the 1980s.[48–50] Ferroelectric phases are called either $L_\beta^*$ or SmC* phases, where L or Sm stands for lamellar or smectic; $\beta$ or C mean that the average molecular orientation (director) is tilted with respect to the layers and * indicates that the constituent molecules are chiral.

Molecules of BC nematics exhibit a large deviation from the rod shape; they are characterized by a kink angle $\theta_0 \approx 60°$, which is much larger than $\theta_0$ of calamitic nematics. Therefore in these banana-shaped compounds the dipolar contribution to $e_3$ is expected to grow at least an order of magnitude, while $e_1$ might remain at the usual order. In the following we will show that this anticipation has been met; moreover, BC nematics exhibit a giant flexoelectric effect with $e_3$ much larger than expected.

### 3.2.2. *A direct flexing method for measuring flexoelectric coefficients*

One might think that the most plausible way of measuring the flexoelectric coefficients should be based directly on their definition in Eq. (3.2), i.e. one should try to detect the charges arising due to the flexoelectric polarization in a splay-bend deformed nematic. While such a static deformation can easily be induced by surface treatments (as in a hybrid cell) or by external fields, mobile free charges can easily screen the resulting flexoelectric polarization, since neither liquid crystals nor the air surrounding the measuring cell are perfect insulators. This problem might be overcome, however, if oscillating deformations are utilized. The resulting oscillating flexoelectric polarization then yields an alternating flexoelectric current. The efficiency of screening is reduced by increasing the frequency of the oscillation, and the screening may become negligible typically above a few hertz. This method was first employed for lyotropic liquid crystals and biological membranes,[38,39,51] where a periodic bending deformation of the quasi two-dimensional object (a lipid bilayer) could be induced by pressure oscillations; this technique is described in more detail in Chapter 6.[52] This oscillating-drop method provides a very elegant way of performing qualitative measurements, such as checking the frequency and temperature dependences, but has not been used to measure the value of the flexoelectric constant, itself.

The set-ups developed for studying membranes cannot be directly applied to thermotropic liquid crystals, as most of them – especially the nematics – are usually not able to form self-supporting thin films; they require solid containers. The technique developed by the authors copes with this requirement, while still being able to utilize the direct flexoelectric effect to determine the flexoelectric coefficients.[28,29] The LC is filled into a sandwich cell made of flexible electrode plates, which are subjected to an oscillatory mechanical flexing deformation (Fig. 3.7). If the frequency of oscillations is low the orientation of the LC inside the cell follows that provided by the strong anchoring at the surfaces. The mechanical flexing of the plates thus produces an oscillatory bend deformation for planar, or a splay one for homeotropic, boundary conditions. The resulting flexoelectric polarization also varies in time; it induces an alternating electric current proportional to the flexoelectric coefficients, having a frequency identical to that of the flexing, which can be detected by a lock-in amplifier.

Let us now calculate the flexoelectric current occurring in the deformed cell. The edges of the cell are clamped so that the deformable length of the cell is $L$, and the surfaces are positioned parallel to the $xy$ plane at $z = 0$ and $z = d$ (thus $d$ is the thickness of the LC layer). The nematic fills the electrode area $X \times Y$ in the centre of the cell (Fig. 3.7a). Flexing is realized by translating the centre of the cell (at $x = 0$) with respect to the edges of the substrates by $S$ along $\mathbf{z}$ (Fig. 3.7b). This translation causes a deformation of the substrate, which on the one hand can be described by the local displacement $w(x)$ from the non-deformed state (plane plates).

On the other hand, the displacement is accompanied by a rotation of the local surface tangent by a bending angle $\psi$ (Fig. 3.7c). Trivially, $\tan(\psi) = dw/dx$. The calculation of the actual deformation profile $w(x)$ is a common problem in elasticity, known as the bending of a beam without a distributed load. Mathematically, for weak deformation, which is assumed as usual, $w(x)$ has to satisfy the differential equation[53]

$$\frac{d^4 w}{dx^4} = 0, \tag{3.4}$$

which has to be supplemented by proper boundary conditions. Due to the construction and the symmetry of the set-up (the $yz$ plane is a mirror plane), the centre of the cell is forced to displace by $S$ while the bending angle remains zero. Thus we have

$$w|_{x=0} = S, \quad \frac{dw}{dx}|_{x=0} = 0. \tag{3.5}$$

As the centre is displaced with respect to the edges, the positions of the edges have to be fixed. There are two ways to fulfil this requirement:

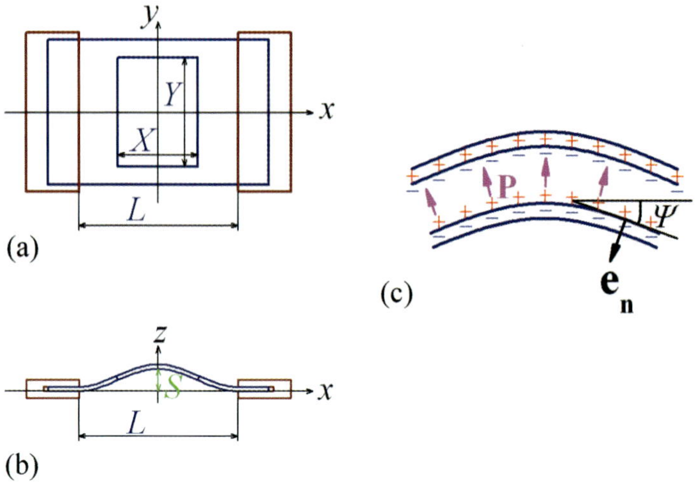


Fig. 3.7. (a) Schematic illustration of the measuring cell. The cell is parallel to the $xy$ plane. The distance between clamped edges is $L$. The nematic fills the area $X \times Y$ in the centre. (b) Deformation of the flexed cell. The centre of the cell is forced to swing by $S$ in the direction of the $z$ axis. As a result the excursion profile of the cell from the $xy$ plane is $w(x)$. (c) Schematic illustration of the polarization $\mathbf{P}$ and the charges (+ and −) in the cell and at the electrodes. The outward normal of the surface is $\mathbf{e_n}$ and the angle between the surface and the $x$ axis is $\psi$.

additional bending is blocked (case A) or allowed (case B) at the edges. Thus the boundary conditions at the edges are for case A (no displacement, no bending):

$$w|_{x=\pm L/2} = 0, \quad \frac{dw}{dx}|_{x=\pm L/2} = 0; \tag{3.6}$$

and for case B (no displacement, bending allowed, but no bending moment):

$$w|_{x=\pm L/2} = 0, \quad \frac{d^2w}{dx^2}|_{x=\pm L/2} = 0. \tag{3.7}$$

Taking into account the mirror symmetry of the set-up, after straightforward calculations we obtain the solution for case A as[a]

[a]Equation (3) in Harden *et al.*,[28] which gives the deformation profile of a cell moved at the edges, has unfortunately been mistyped; in its printed form it does not fulfil the requirement for cell symmetry, and it does not satisfy the boundary condition at the edges $[Z(\pm L/2) = S]$. Equation (3) should correctly read $\beta(x) = 3\left(\frac{2|x|}{L}\right)^2 - 2\left(\frac{2|x|}{L}\right)^3$.

$$
\begin{aligned}
w(x) &= S\left[2\left(\frac{2|x|}{L}\right)^{3} - 3\left(\frac{2|x|}{L}^{2}\right) + 1\right],\\
\frac{dw(x)}{dx} &= \tan\psi = 12\frac{S}{L}\left[\left(\frac{2|x|}{L}^{2}\right) - \left(\frac{2|x|}{L}\right)\right]\operatorname{sgn}(x),\\
\frac{d^2w(x)}{dx^2} &= 24\frac{S}{L^2}\left[2\left(\frac{2|x|}{L}\right) - 1\right].
\end{aligned}
\tag{3.8}
$$

Similarly for case B

$$
\begin{aligned}
w(x) &= S\left[\frac{1}{2}\left(\frac{2|x|}{L}\right)^{3} - \frac{3}{2}\left(\frac{2|x|}{L}\right)^{2} + 1\right],\\
\frac{dw(x)}{dx} &= \tan\psi = 3\frac{S}{L}\left[\left(\frac{2|x|}{L}\right)^{2} - 2\left(\frac{2|x|}{L}\right)\right]\operatorname{sgn}(x),\\
\frac{d^2w(x)}{dx^2} &= 12\frac{S}{L^2}\left[\left(\frac{2|x|}{L}\right) - 1\right].
\end{aligned}
\tag{3.9}
$$

Figures 3.8a and 3.8b illustrate that the deformation profile $w(x)$ and the surface tangent $dw/dx$, respectively, are both sensitive to the mechanical boundary conditions at the edges.

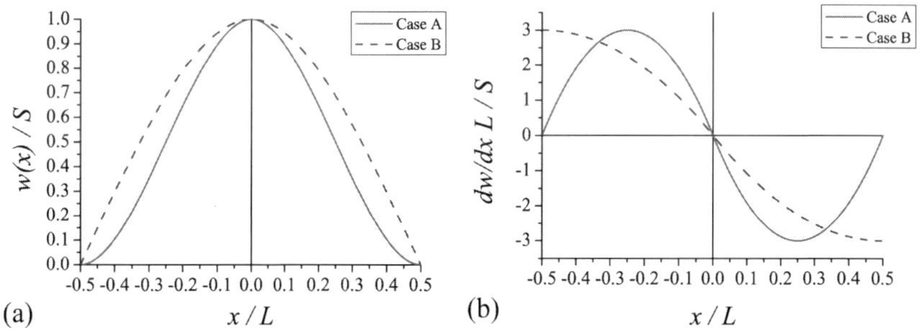


Fig. 3.8. Dependence of the substrate deformation on the mechanical boundary conditions. (a) The deformation profile $w(x)$. (b) The tangent $dw/dx$ of the surface. The solid lines correspond to case A (no bending at the edges) while the dashed lines are for case B (bending is allowed at the edges).

For a general orientation of the nematic director:

$$\mathbf{n} = (n_x; n_y; n_z) = (\Theta \sin\vartheta; n_y; \Theta \cos\vartheta) \text{ with } \Theta^2 = 1 - n_y^2. \tag{3.10}$$

We assume that the director is strongly anchored at the substrates. Therefore in the absence of deformation $\vartheta = \vartheta_0$, while in the deformed (flexed) state the director rotates with the substrates:

$$\vartheta = \vartheta_0 + \psi. \tag{3.11}$$

According to Eq. (3.2) the flexoelectric polarization $\mathbf{P_{fl}} = (P_{\mathrm{fl}x}; 0; P_{\mathrm{fl}z})$ in this geometry is given by

$$\begin{aligned} P_{\mathrm{fl}x} &= (e_1 + e_3)\Theta^2 \sin\vartheta \cos\vartheta \frac{\partial\vartheta}{\partial x} \\ &\quad - \Theta^2(e_1 \sin^2\vartheta - e_3 \cos^2\vartheta)\frac{\partial\vartheta}{\partial z}, \end{aligned} \tag{3.12}$$

$$\begin{aligned} P_{\mathrm{fl}z} &= \Theta^2(e_1 \cos^2\vartheta - e_3 \sin^2\vartheta)\frac{\partial\vartheta}{\partial x} \\ &\quad - (e_1 + e_3)\Theta^2 \sin\vartheta \cos\vartheta \frac{\partial\vartheta}{\partial z}. \end{aligned} \tag{3.13}$$

The presence of flexoelectric polarization induces a surface charge density of $\sigma_s = \mathbf{P_{fl}} \cdot \mathbf{e_n}$ in the electrodes; here $\mathbf{e_n}$ corresponds to the outside normal of the substrate. Therefore the total charge on the bottom electrode

$$Q_{fl} = \int\int_{\Sigma} (\mathbf{P_{fl}} \cdot \mathbf{e_n}) dA = \int_{-Y/2}^{Y/2} \int_{-X/2}^{X/2} (\mathbf{P_{fl}} \cdot \mathbf{e_n}) dx dy. \tag{3.14}$$

Here $dA$ is the surface element and the integration should extend to the active area of the cell (i.e. the area of electrodes filled with the nematic, see Fig. 3.7a) whose size is defined by $X$ and $Y$. The temporal variation of these surface charges of flexoelectric origin induces a flexoelectric current

$$I_{fl} = \frac{d}{dt} Q_{fl} = Y \frac{d}{dt} \int_{-X/2}^{X/2} (\mathbf{P_{fl}} \cdot \mathbf{e_n}) dx. \tag{3.15}$$

Using Eqs (3.12) and (3.13) we obtain

$$I_{fl} = \frac{d}{dt}\left\{Y\int_{-X/2}^{X/2}\left(\left[(e_1+e_3)\Theta^2\sin(\vartheta_0+\psi)\cos(\vartheta_0+\psi)\frac{\partial\psi}{\partial x}\right.\right.\right.$$

$$\left.\left.\left. -\Theta^2[e_1\sin^2(\vartheta_0+\psi)-e_3\cos^2(\vartheta_0+\psi)]\frac{\partial\psi}{\partial z}\right]\sin\psi\right)dx\right\}$$

$$-\frac{d}{dt}\left\{Y\int_{-X/2}^{X/2}\left(\left[\Theta^2[e_1\cos^2(\vartheta_0+\psi)-e_3\sin^2(\vartheta_0+\psi)]\frac{\partial\psi}{\partial x}\right.\right.\right.$$

$$\left.\left.\left. -(e_1+e_3)\Theta^2\sin(\vartheta_0+\psi)\cos(\vartheta_0+\psi)\frac{\partial\psi}{\partial z}\right]\cos\psi\right)dx\right\}. \qquad (3.16)$$

If the displacement $S$ is small compared to the length of the deformable region $L$, $\psi$ remains small and Eq. (3.16) can be expanded in powers of $\psi$. Keeping the lowest-order (linear) term in $\psi$ gives:

$$I_{fl} \approx Y\Theta^2\frac{d}{dt}\left\{\int_{-X/2}^{X/2}\left[(e_3\sin^2\vartheta_0 - e_1\cos^2\vartheta_0)\frac{\partial\psi}{\partial x}\right.\right.$$

$$\left.\left. +(e_1+e_3)\sin\vartheta_0\cos\vartheta_0\frac{\partial\psi}{\partial z}\right]dx\right\}$$

$$= Y\Theta^2\frac{d}{dt}(e_3\sin^2\vartheta_0 - e_1\cos^2\vartheta_0)\left(\psi|_{X/2}-\psi|_{-X/2}\right)$$

$$+Y\Theta^2\frac{d}{dt}\left[(e_1+e_3)\sin\vartheta_0\cos\vartheta_0\int_{-X/2}^{X/2}\frac{\partial\psi}{\partial z}dx\right]. \qquad (3.17)$$

Taking into account the mirror symmetry of the cell implies that $w(x)$ is an even function, while $dw/dx$ is an odd function of $x$; moreover for $\psi \ll 1$, which is assumed here, we can use the approximation $\psi \approx dw/dx$. The second term in Eq. (3.17) thus cancels out, and finally we obtain:

$$I_{fl} \approx 2Y\Theta^2\frac{d}{dt}\left(e_3\sin^2\vartheta_0 - e_1\cos^2\vartheta_0\right)\left.\frac{dw}{dx}\right|_{X/2}. \qquad (3.18)$$

Taking into account the deformation profile calculated above in Eqs (3.8) and (3.9) we can transform Eq. (3.18) into

$$I_{fl} \approx Y\Theta^2\left(e_3\sin^2\vartheta_0 - e_1\cos^2\vartheta_0\right)\beta\frac{d}{dt}\frac{S}{L}, \qquad (3.19)$$

where $\beta$ is a factor determined by the cell geometry (i.e. by the $X/L$ ratio) and by the mechanical boundary conditions at the cell edges (case A or B).

For case A:

$$\beta_A = 24\left[\left(\frac{X}{L}\right)^2 - \left(\frac{X}{L}\right)\right], \tag{3.20}$$

while for case B:

$$\beta_B = 6\left[\left(\frac{X}{L}\right)^2 - 2\left(\frac{X}{L}\right)\right]. \tag{3.21}$$

In particular for $X = L/2$ we obtain $\beta_A = -6$ and $\beta_B = -4.5$.

Flexing of the cell corresponds to a harmonic displacement of the cell centre, i.e. $S = S_0 \sin(\omega t)$. Thus the current generated by flexoelectricity is

$$I_{fl} \approx Y\Theta^2\left(e_3 \sin^2 \vartheta_0 - e_1 \cos^2 \vartheta_0\right)\beta\omega\frac{S_0}{L}\cos(\omega t). \tag{3.22}$$

It can be seen immediately from Eq. (3.22) that for a homeotropic orientation ($\Theta = 1$, $\vartheta_0 = 0$) the current depends on the splay flexoelectric coefficient $e_1$ only. For a planar orientation with the easy axis (the direction the molecules prefer to align on the substrate) along $\mathbf{x}$ ($\Theta = 1$, $\vartheta_0 = \pi/2$) the bend flexoelectric coefficient $e_3$ governs the response. For a planar orientation with the easy axis along $\mathbf{y}$ ($\Theta = 0$) no flexoelectric current is generated, since in that geometry flexing of the substrates does not induce director gradients.

### 3.2.3. *Giant flexoelectricity of bent-core nematics studied by the flexing method*

The direct method introduced in the previous section was first employed for studying the flexoelectric response of a bent-core nematic liquid crystal.[28,29] In the prototype experimental set-up thin brass plates ($L = 32$ mm) were used as flexible substrates and a 25-$\mu$m-thick Teflon spacer filled the gap between the electrodes except for the active area of $X \times Y = 1 \times 1$ cm$^2$ in the centre, provided for the LC. The flexible LC cell was inserted between two vertical slots in the movable side walls of a temperature-regulated box (controlled within 0.1 °C). Two vertical cylindrical posts standing on the fixed bottom plate of the box prevented the displacement of the cell centre, while the side walls were translated periodically by an audio speaker driven via a power amplifier by the built-in oscillator of a Perkin Elmer 7265 lock-in amplifier. Thus in this set-up the edges of the cell were displaced periodically by $S(t)$ with respect to the centre of the cell, while realizing the boundary conditions of case A. Displacement amplitudes up to 3 mm were achieved in the frequency range of 1–10 Hz. [Note that the flexing deformation for this set-up is equivalent to that induced by the displacement of the centre by $-S(t)$ as described in Section 3.2.2.] In later experiments the speaker was replaced by alternative mechanical driving technique: the rotation of an eccentric disc attached to the shaft of a DC motor was transformed into translational motion using either a Scotch–yoke[29] or a slide (as in a steam engine).

The flexoelectric current generated by the periodic flexing of the cell substrates can be measured by the lock-in amplifier using its built-in current-to-voltage (I/U) converter. Figure 3.9 shows an equivalent electric circuit of the set-up. The sample is represented by a current generator $I_{fl}$ according to Eq. (3.22) and by a parallel resistance $R_s$ and capacitance $C_s$. At the frequency range of the mechanical excitation ($f$ is a few hertz) the input impedance of the measuring device ($R_i \parallel C_i$) is much lower than the impedance of the sample or that of the cable ($R_c \parallel C_c$),[b] the leakage currents through the sample and the cable ($I_s$ and $I_c$, respectively) are negligible. Therefore the measured current $I = I_{fl} - I_s - I_c \approx I_{fl}$.

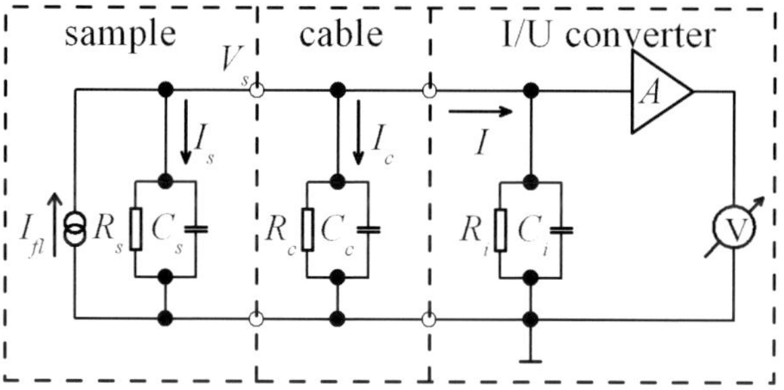


Fig. 3.9. Equivalent electrical circuit of the measuring set-up.

The first measurements were carried out using the banana nematic 4-chloro-1,3-phenylene bis 4-[4′-(9-decenyloxy) benzoyloxy] benzoate (**ClPbis10BB**).[28] The electrode surfaces were rubbed to obtain a planar alignment, which allows the determination of the bend flexocoefficient $e_3$. Measuring the root mean square (rms) value of the current $I$ and using Eq. (3.22) the magnitude of the flexocoefficient was determined as:

[b]Typical values: $R_i < 400\ \Omega$, $C_i \approx 30$ pF, $R_c > 10$ G$\Omega$, $C_c \approx 150$ pF, $R_s = O(1\ \text{M}\Omega)$, $C_s = O(100\ \text{pF})$ and $V_s < 1$ mV.

$$|e_3| = \sqrt{2}I\frac{L}{2\pi f\beta Y S_0}. \tag{3.23}$$

Figure 3.10a shows the temperature dependence of the magnitude of the flexoelectric coefficient $|e_3|$ of **ClPbis10BB** calculated from the flexoelectric current $I$ measured at $S_0 = 2$ mm, $f = 3.2$ Hz and corrected by subtracting a small background corresponding to the signal in the isotropic (I) phase. A big jump can be seen in the response at the isotropic-nematic (I-N) phase transition temperature $T_{\mathrm{NI}}$ indicating that this transition is of the first order. The maximum response, $|e_3| \approx 35$ nCm$^{-1}$, was detected a few degrees below $T_{\mathrm{NI}}$, then it fell back to almost zero when crystallization occurred. This is truly a giant value; it exceeds all flexocoefficients measured so far for various calamitic nematics by three orders of magnitude! In order to demonstrate this and at the same time to test the reliability and resolution of the new direct method, similar measurements were carried out on a well-known calamitic nematic, 4-pentyl-4′-cyanobiphenyl (**5CB**). Data obtained for **5CB** at $S_0 = 1.5$ mm and $f = 5$ Hz are plotted in Fig. 3.10b. The graph shows a rather gradual increase when passing $T_{\mathrm{NI}}$. Note, however, the different vertical scale! The flexoelectric response for the calamitic nematic **5CB** using the same cell geometry and at the same relative temperatures is about 1,000 times smaller, $|e_3| \approx 40$ pCm$^{-1}$, than for the BC compound.[28] This value for **5CB** matches roughly with data obtained by electro-optical methods.[54]

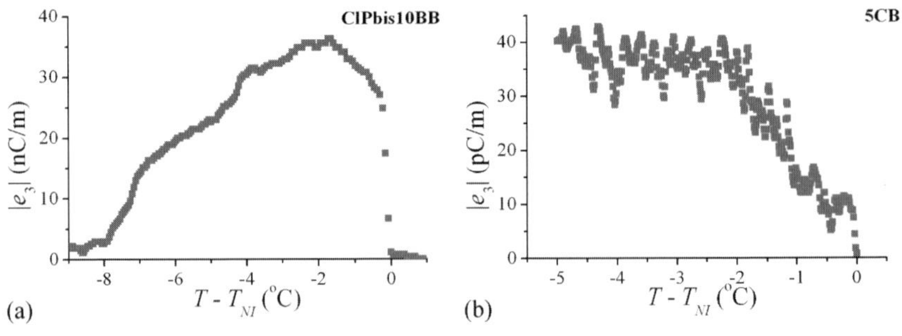


Fig. 3.10. Magnitude of the flexoelectric coefficient $|e_3|$ versus the temperature deviation from the nematic-isotropic phase transition for (a) **ClPbis10BB** and for (b) **5CB**. Data taken from Harden *et al.*[28] Note the three orders of magnitude difference in the vertical scales!

We would like to emphasize that for an imperfect orientation ($\Theta < 1$, $\vartheta_0 < \pi/2$) the flexoelectric current depends both on $e_3$ and $e_1$ [see

Eq. (3.22)]. Since for BC compounds $|e_3| \gg |e_1|$ is expected, misalignment may reduce the response considerably. Though the substrates were treated for planar anchoring, the quality of the actual orientation could not be checked by microscopy as the brass cells are not transparent. Therefore, the $|e_3|$ values shown in Fig. 3.10a for **ClPbis10BB** should be regarded as lower limits. In **5CB**, on the other hand, $|e_3|$ and $|e_1|$ are of the same order of magnitude[54] (mostly due to the quadrupolar contribution), so defects in the alignment do not have such a great influence.

Since the flexoelectric response is determined by the surface alignment angles, one might expect that transparent plastic sheets with indium-tin-oxide-(ITO)-coated surfaces would make better flexible electrodes since they would allow monitoring of the alignment. Indeed, earlier measurements on well-oriented cells made of transparent plastic sheets with ITO-coated surfaces have indicated $|e_3| \approx 60$ nCm$^{-1}$ for **ClPbis10BB**,[28] i.e. $e_3$ could be about twice as big as that shown in Fig. 3.10a. In contrast, unoriented samples provided an order of magnitude weaker flexoelectric response at small displacement amplitudes. Interestingly, however, increasing $S$ resulted in a non-linear increase of the flexocurrent and at larger displacements ($S \approx 2$ mm) the response approached that of the well-aligned sample. This observation implies that large amplitude flexing efficiently improves the orientation, presumably via a flexing-induced flow. Though using ITO-coated plastic sheets would seem to be a preferable choice for the measurements, unfortunately, the plastic cells had a substantially higher electric current background (with a large response in the isotropic phase or even from an empty cell) and the resulting lower resolution made them inappropriate for detecting the small flexoelectric response of **5CB**.

**ClPbis10BB** is not the only bent-core nematic that has a giant $e_3$ flexocoefficient; for a few other BC compounds (whose chemical structures and phase sequences are shown in Fig. 3.11) and mixtures a similar magnitude of $|e_3|$ has been obtained (see Table 3.1).

Mixtures of BC and calamitic nematics have also been investigated using the flexing technique. Due to the huge difference in the magnitude of the flexocoefficients of the components we would expect a substantial reduction of $|e_3|$ by increasing the calamitic content of the mixtures. Measurements have confirmed the reduction of $|e_3|$, however, strongly non-linear behaviour has been found.[29] At low as well as at high calamitic content the concentration dependence of $e_3$ is weak, while at intermediate concentrations (at about 40–50 wt%) there is a sudden fall in the value of the bend flexocoefficient.

**I**

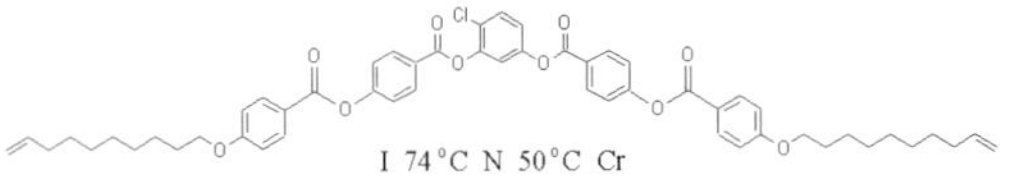


**II**

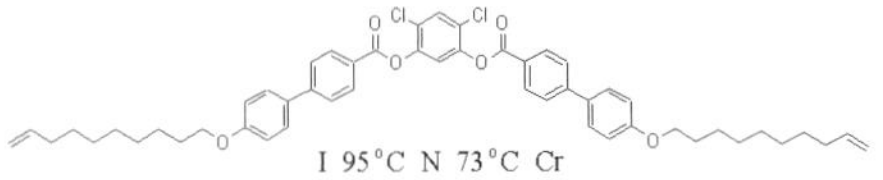


**III**

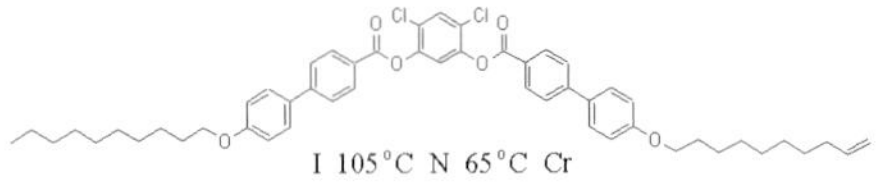


**IV**

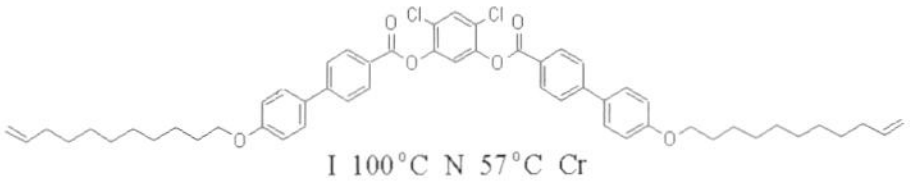


**V**

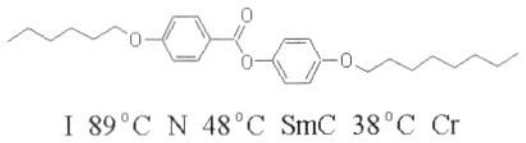


**VI**

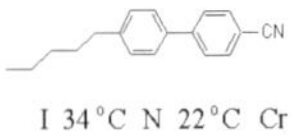


Fig. 3.11. Molecular structures and phase sequences of the studied BC and calamitic nematics. **I**: 4-chloro-1,3-phenylene bis 4-[4′-(9-decenyloxy)-benzoyloxy] benzoate (**ClPbis10BB**).[55] **II**: 4,6-dichloro-1,3-phenylene bis[4′-(9-decen-1-yloxy)-1,1′-biphenyl-4-carboxylate] (**10DClPBBC**).[56] **III**: 2,4-dichloro-5-[(4-decyloxy-1,1′-biphenyl)carbonyloxy]phenyl-4′-[(9-decen-1-yloxy)-1,1′-biphenyl]-4-carboxylate (**10DClPBCP**).[57] **IV**: 4,6-dichloro-1,3-phenylene bis[4′-(10-undecen-1-yloxy)-1,1′-biphenyl-4-carboxylate] (**11DClPBBC**).[56] **V**: 4-n-octyloxy-phenyl-4′-n-hexyloxybenzoate (**6OO8**).[58] **VI**: 4-pentyl-4′-cyanobiphenyl (**5CB**).[59] The first batches of compounds **I**, **II**, **IV** and **V** were obtained from K. Fodor-Csorba (Institute for Solid State Physics and Optics, Wigner Research Centre for Physics, Hungarian Academy of Sciences, Budapest). Several new batches of compound **I** were provided by R. Twieg at Kent State University, Kent, OH. Material **III** was synthesized by R. Verduzco at Rice University, Houston, TX. Compound **VI** was obtained from BDH.

Finally, we would like to emphasize that the flexing technique presented above has a greater potential for applications than techniques used so far. Namely, if the phase of the flexoelectric current is correctly traced with respect to that of the mechanical deformation, in addition to its magnitude, the sign of $e_3$ could also be obtained. Moreover, using homeotropically aligned cells in the same set-up one can measure the splay flexoelectric coefficient $e_1$ as well. Obtaining a good homeotropic alignment for BC nematics may, however, be a difficult task.

Table 3.1. The measured maximal bend flexoelectric constants of the BC nematics shown in Fig. 3.11 and their several mixtures. The range of the $e_3$ coefficients represents results obtained from several samples of the same material, however, different samples had different thicknesses and might contain compounds from different batches.

| Material | $e_3$ (nCm$^{-1}$) |
|---|---|
| **I** | 70–90 |
| **II** | 40–50 |
| **III** | 50–60 |
| **IV** | 35–40 |
| **V** | 0.07 |
| **VI** | 0.04 |
| 50 wt% **I** + 50 wt% **IV** | 45–50 |
| 60 wt% **I** + 40 wt% **V** | 60 |
| 40 wt% **I** + 60 wt% **V** | 50 |
| 30 wt% **I** + 70 wt% **V** | 5 |

## 3.3. The Inverse (Converse) Flexoelectric Effect

The direct flexoelectric effect means, as discussed earlier, that an electric field is produced when an LC film is cylindrically flexed. Symmetry predicts that the converse flexoelectric effect should also exist; i.e. the application of an electric field may induce mechanical distortions. One has to note here that converse flexoelectricity is just one (linear) coupling mechanism of several which can be responsible for field-induced deformations. It has to be distinguished from other linear mechanisms as ferroelectric coupling in ferroelectric LCs,[60] electromechanical coupling in cholesterics[61–64] or in chiral smectics C*[65–67] or piezoelectricity,[68–70] as well as from the more common quadratic mechanisms as the dielectric coupling responsible for most electro-optic effects in LCs[17] or electrostriction.[69,70] Converse flexoelectricity can manifest itself in two forms. Mostly, the applied electric field produces an orientational distortion in the LC, which can be analysed via some optical method. Typically that occurs if the LC film is bounded by stiff solid substrates. In fact the majority of techniques used to measure the flexoelectric coefficients rely on this converse flexoelectric effect; these methods are discussed in more detail in Chapter 2 by Madhusudana.[71]

Another type of converse flexoelectric effect can be observed using flexible substrates where orientational distortions may be accompanied by a flexing of the LC film (and the substrates). The curvature induced by flexo-electricity was first observed by Todorov *et al.*[47] in lipid membranes (lipids are akin to pear-shaped molecules), which are self-supporting and thus do not need to be constrained by substrates. The black lipid bilayer membranes showed a 20 $m^{-1}$ spherical curvature [47,72,73] at an applied field of 55 V/$\mu$m (see Chapter 6 for more detail[52]). This curvature corresponds to a splay flexoelectric coefficient[42] of $e_1 \sim 100$ pCm$^{-1}$, which is an order of magnitude larger than typical for calamitic thermotropic liquid crystals, but still two orders of magnitude smaller than the bend flexoelectric coefficient ($e_3 \sim 60$ nCm$^{-1}$) observed in the BC nematic **ClPbis10BB**.[28] In the following we will give an overview for how both types of converse flexoelectricity can be studied in BC nematics.

### 3.3.1. *Converse giant flexoelectric effect*

The small flexoelectric coefficients of calamitic nematics do not allow the observation of electric field-induced curvatures of the substrates of an LC film. The discovery of giant flexoelectricity made the observation of the converse flexoelectric effect – which could produce flexing large enough to be measured directly – possible. The Jákli group[29,74] used a cell made of flexible Mylar substrates of 100 $\mu$m thickness filled with **ClPbis10BB** and the resulting displacement (flexing) of the substrates was measured using a Mirau interferometer. The application of a voltage across the sample can cause two types of deflections, upward (closer to the interferometer) or downward (away from the interferometer). Flexoelectric flexing induces an upward deflection for one voltage polarity and a downward deflection for the opposite polarity. In addition, a polarity-independent (downward) deflection may also be present due to the Coulomb attraction between the conducting substrates. This attraction results from both the induced charge on the conducting coatings at a non-zero applied voltage $V$, and the static charge in the Mylar. The downward deflection due only to the induced charge can be determined by averaging the deflection at reversed polarities deep in the isotropic phase; the deflection due to the static charge is then half the difference between the deflections at reversed polarities.

The data after correcting for static charge, correspond to measurements of the voltage-dependent deflection under both polarities: $D(+|V|)$ and $D(-|V|)$. The asymmetric and symmetric parts of the deflections, $D_a$ and $D_s$, respectively, are defined as

$$D_a = \frac{D(+|V|) - D(-|V|)}{2}, \quad D_s = \frac{D(+|V|) + D(-|V|)}{2}. \tag{3.24}$$

The polarity-dependent $D_a$ describes a flexing of the plate due to converse flexoelectricity. Following the original definition of flexoelectricity by

Meyer,[31] the flexoelectric coefficients are positive if $D_a > 0$ and negative if $D_a < 0$. Measurements on **ClPbis10BB**[74] gave $e_3 > 0$ and showed that there is an effective flexoelectric effect even in the isotropic phase up to about 4 °C above the nematic phase. The converse flexoelectric coefficient was highest at about 3 °C below the I-N transition and disappeared in the crystal phase. The (quadratic) squeezing effect is obviously largest in the least viscous, isotropic phase and is basically zero in the crystal phase. Using the above measurements, the bend flexoelectric coefficient $e_3$ was calculated by equating the converse flexoelectric energy $E_{flex}$ with the bending energy $E_{bend}$ of the Mylar substrates. The energy of a cylindrically bending plate of thickness $h$ and Young's modulus $Y$ can be written as

$$E_{bend} = \frac{A_{sub}Y}{2R^2(1-\sigma^2)} \int_{-h/2}^{h/2} z^2 dz = \frac{Yh^3}{12} \frac{A_{sub}}{2R^2(1-\sigma^2)}, \qquad (3.25)$$

where $A_{sub} = 296$ mm$^2$ is the area of the substrates, $\sigma \sim 0.4$ is Poisson's ratio of Mylar,[75] $Y = 3.3$ GPa,[76] $R$ is the radius of curvature and $h = 100$ $\mu$m. It is important to note that the bending energy is proportional to the third power of the thickness of the bounding substrates, i.e., a decrease in the thickness of the bounding plates will lead to a substantial reduction of the energy needed to bend the substrates.

On the other hand, the flexoelectric energy for bend distortion of the director is:

$$E_{flex} = \frac{e_3}{R} V A_{LC}, \qquad (3.26)$$

where $V$ is the applied electric potential and $A_{LC} = 216$ mm$^2$ is the area of the liquid crystal between the electrodes. Equating Eqs (3.25) and (3.26) we can express $e_3$ as:

$$e_3 = \frac{A_{sub}}{A_{LC}} \frac{Yh^3}{24RV(1-\sigma^2)}. \qquad (3.27)$$

The $e_3$ values obtained from the measurements on **ClPbis10BB** using Eq. (3.27) are plotted in Fig. 3.12. It can be seen that $e_3$ deduced from these measurements is in satisfactory agreement with that obtained from the direct flexoelectric effect based on polarization current detection under an applied curvature strain (also plotted in Fig. 3.12).

We note here that, though the substrate curvature due to the converse giant flexoelectric effect could be observed, detection of the converse flexoelectric effect of calamitic nematics using the same geometry cannot be

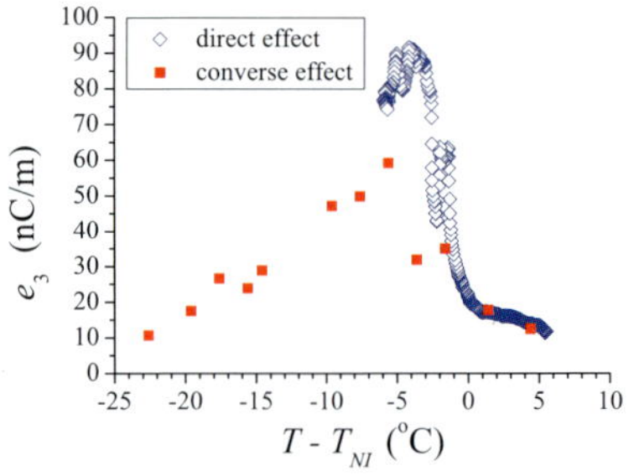


Fig. 3.12. Comparison of the temperature dependences of the direct and inverse flexoelectric effects for **ClPbis10BB**.

expected. While for the BC nematic $R \approx 1300$ m was measured as the radius of curvature,[74] for a calamitic molecule whose flexoelectric coefficient is 1,000 times smaller, the resulting radius of curvature would be about 1000 km. In other words, a calamitic would deflect the substrate by about one atomic length only (or we could just say there is no deflection).

### 3.3.2. *Flexoelectricity of bent-core molecules studied by indirect methods*

As discussed in detail in Chapter 2 by Madhusudana,[71] traditional methods of measuring the flexoelectric coefficients rely on the converse flexoelectric effect; the flexocoefficients can be determined indirectly from electric field-induced director distortions using mostly optical methods. Often the HAN (hybrid-aligned nematic) geometry is utilized where one substrate is planar and the other is homeotropic, providing an initial splay-bend distortion and thus flexoelectric polarization in the cell. The combination $e_1 + e_3$ can be obtained from the voltage dependence of the transmitted intensity in electric fields normal to the substrates.[77] In fields parallel to the substrates $e_1 - e_3$ can be determined from the rotation of the light polarization.[78] These techniques have been successfully applied for calamitic nematics and guest-host systems, i.e. for calamitic nematics doped by BC molecules [79,80] in less than 10 wt% concentrations. However, pure bent-core materials with negative dielectric anisotropy instead form a planar alignment in HAN cells and no intensity modulation was detected for applied AC voltages of up to 100 V.[28] On the other hand, HAN cells filled with mixtures of 4-n-octyloxy-4′-cyanobiphenyl (**8OCB**) and 1,3-phenylene bis[4-(3-methylbenzoyloxy)] 4′- n-dodecylbiphenyl 4′-carboxylate (**BC12**), which have positive dielectric anisotropy, form a homeotropic alignment, thus preventing measurement of the rotation of polarization.[80] The failure of the methods in both cases has a common origin. The flexoelectric polarization arising due to the initial splay-bend deformation yields an internal electric field, corresponding to a flexoelectric offset voltage[28]

$$V_o = \frac{e_1 + e_3}{2\epsilon_0 \epsilon_a} \ln\left(\frac{\epsilon_\perp}{\epsilon_\parallel}\right). \tag{3.28}$$

Consequently, in the absence of an external applied voltage, the cell actually behaves as if a DC voltage of $V_o$ were applied, distorting the director distribution toward the planar or the homeotropic state depending on the sign of $\epsilon_a$. While in calamitic nematics $V_o$ is a few volts, in BC nematics with giant flexoelectricity the offset voltage is orders of magnitude larger, thus saturating the distortion. The failure of the HAN techniques therefore indicates huge values of $e_1 + e_3$.

Homeotropic cells offer another way to detect the flexoelectric response via observing the bend director distortions induced by an electric field parallel to the substrates (the Helfrich method[81]). Takezoe's group[82] applied it to **ClPbis10BB** and found $e_3 \approx 20$ pCm$^{-1}$ (which is the order of flexocoefficients for calamitic nematics). As this value was obtained without taking into account the non-vanishing surface anchoring and the dielectric torques (both reduce the deformation), it should be regarded as a lower limit for $e_3$. Using the same method with a more complete interpretation Kumar *et al.*[83] obtained $e_3 < 10$ pCm$^{-1}$ for another BC nematic, 4-cyanoresorcinol bis[4-(4-n-dodecyloxybenzoyloxy) benzoate] (**CNRbis12OBB**).

Flexocoefficients have also been obtained from the threshold voltages and wave numbers of electric field-induced flexoelectric instabilities (which are discussed in more detail in Chapter 4 by Buka *et al.*[84]) for **ClPbis10BB**[82] and for the molecule 4-((3-(4-(4-(decyloxy) benzoyloxy)benzoyloxy) phenylimino) methyl)-3-hydroxyphenyl 4-(6-(4′-cyanobiphenyl-4-yloxy)hexyloxy) benzoate (**BCCB**), which is a dimer composed of a calamitic and a BC molecule.[83] In both cases the flexocoefficients were found to have the same order as that of calamitics.

The flexoelectro-optic effect in cholesterics (which is described in more detail in Chapter 7 by Rudquist and Lagerwall[85]) offers another way to measure $e_1 - e_3$. A tight-pitch helical structure can easily be induced in nematics by adding a few per cent of chiral dopant. Recently this technique has been adopted to test the flexoelectric response of another BC nematic, 4-cyano-1,3-phenylene bis [4-[4′-(hexyl) benzoyloxy] benzoate] (**C6ban**).[86]

The obtained $|e_1 - e_3| = 17.4$ pCm$^{-1}$ is not much bigger than the similar value for calamitic cyanobiphenyls.

Many bent-core molecules do not have nematic phases, but may be added to calamitic nematics to tune the flexoelectric behaviour. Several groups have reported measurement results in such guest-host systems[79,80,87] using either HAN cells or the flexoelectro-optic effect mentioned above.[88–91] It was found that some BC compounds used as dopants in small ($< 10\%$) concentration may effectively increase $e_1 - e_3$ of the host.[79,80] Other, mainly asymmetric, BC compounds had hardly any influence on $e_1 + e_3$ of the host nematic.[87]

The measurements reviewed here clearly demonstrate that all techniques based on the optical detection of converse flexoelectricity in cells with rigid substrates gives $e_3$ of the order of a few pCm$^{-1}$ for BC compounds, being even lower than the range expected from Helfrich's[34] and Derzhanski and Petrov's[35] statistical estimation for $\theta_0 \sim 60°$. In contrast, by using electromechanical techniques with flexible substrates a giant flexoelectric response could be detected, both in geometries utilizing direct and converse flexoelectricity. These discrepancies will further be addressed in the next section.

### 3.4. Physical Origin of Giant Flexoelectricity

Meyer's formula, Eq. (3.2), provides the most general form of the polarization induced by director gradients in a homogeneous, anisotropic, non-polar dielectric fluid (e.g. in a nematic liquid crystal). The microscopic statistical theories we are aware of – which relate the structure of individual molecules to the phenomenological flexocoefficients and are discussed in Chapter 1 by Osipov[32] – cannot provide $e_i$ values above a few tens of pCm$^{-1}$, independent of whether they are based on the dipolar[33–35] or on the quadrupolar[36] mechanism; therefore they are not able to explain giant flexoelectricity as summarized in Sections 3.2.3 and 3.3.1. Moreover, from energy conservation considerations (the electrical energy induced by the mechanical excitation cannot exceed the pumped in mechanical energy) it has been concluded recently[92] that a few tens of pCm$^{-1}$ represents a principal upper limit for the flexocoefficients in a homogeneous, non-polar nematic. These imply that the giant flexoelectricity of BC nematics should be related to the collective behaviour of a group of molecules called clusters, as proposed by Harden *et al.*[28]

We mentioned in Section 3.1 that bent-core molecules have a tendency to form polar smectic (SmCP) structures and in BC nematics this tendency is manifested by the presence of smectic-like clusters with polar layers floating in the nematic matrix. Thus BC nematics are neither homogeneous, nor non-polar; hence the limitations obtained from energy conversion considerations[92] do not apply directly.

The presence of clusters in BC nematics is now well established from various measurements. Recent studies[24–26] have in fact indicated a ferroelectric or an antiferroelectric response to an applied electric field, and an unusual low-frequency (presumably collective) mode has been detected in the dielectric spectra of bent-core nematics,[27] which might also be related to clusters. In spite of the intense studies, however, the exact structure and the physical properties of the clusters are still unknown. Therefore, not surprisingly, a precise physical model for the role of polar clusters in the flexoelectric response of BC nematics and a quantitative estimation of the resulting increment of the flexocoefficients has not yet been worked out.

Kumar *et al.*[83] proposed that a chain formed by intercalated bent-core molecules with alternating dipoles may have a considerable quadrupolar moment and they assumed that this idea is transferable to antiferroelectric layers. They estimated an increment in the quadrupole moment proportional to the number of molecules ($m$) in such clusters. As the typical cluster size indicated by X-ray measurements[25,26] is about 4–5 layers of 20–30 nm, $m$ may be of the order of 1,000. Therefore, on the one hand, the presence of clusters with an antiferroelectric SmCP structure may cause a huge increase in the quadrupolar contribution to the flexoelectric response compared to that of individual molecules.

On the other hand, if the clusters were composed of ferroelectric layers, one might expect that the huge net dipole moment of the clusters would lead to an enhanced dipolar contribution to the flexoelectric response (assuming no change in the shape factor, the association of molecules into ferroelectric clusters might yield a gain in $e_3$ scaling with $m^{2/3}$ according to Helfrich's formula Eq. (3.3)). Although so far there has been no direct experimental observation of the shape of the clusters, there are calculations indicating that a polarization splay (i.e. a configuration with $\nabla \cdot \mathbf{P} \neq 0$) leads to a bend of the layers, simply due to the steric inclusions[23] of the molecules (such as the presence of one or two chlorine atoms in the central benzene rings), which in fact are present in all bent-core nematics where giant flexoelectricity has been observed so far. In other words: when we bend the film, the ferroelectric clusters – and consequently their polarization – flip to the direction that is sterically favoured by the direction of the bend. Similarly, when we switch the direction of the polarization using an electric field, there is an enhanced mechanical effect. Transmission electron microscopic (TEM) studies are being used to verify and determine the shape of the clusters.[93]

Clusters can also account for the strongly non-linear concentration dependence of $e_3$ found in mixtures of bent-core and calamitic molecules. For a small calamitic content the cluster formation is still undisturbed, therefore giant flexoelectricity is still observable with a small reduction of $e_3$ due to the dilution. At around 40–50 wt%, however, the increasing number of calamitic molecules breaks the ability of BC molecules to form polar layers; therefore the cluster contribution to flexoelectricity fades away in a narrow concentration range resulting in the fall of $e_3$ back to a normal (calamitic) value. At the high calamitic concentration end the BC molecules just behave as dopants, increasing slightly the flexoelectric response of the calamitic host in proportion to their concentration. We still have to address the question why giant flexoelectricity is observed in some experiments and not in others. It was shown in Sections 3.2.3 and 3.3.1 that both direct and converse giant flexoelectricity is detectable in samples with flexible boundary plates. The insensitivity of HAN cells made of BC nematics to applied voltages (see Section 3.3.2) is also interpreted as the consequence of giant flexoelectricity (owing to the deformation induced by the internal electric field arising due to the huge flexoelectric polarization). In samples with rigid boundaries and where the deformation of the initial homogeneous state is expected to occur exclusively by converse flexoelectricity (as with the Helfrich geometry,[33,81] see Section 3.3.2), the measurements give flexocoefficients only for the typical order of calamitics.

This discrepancy clearly indicates that the BC nematics characterized by the presence of polar clusters cannot be regarded – at least from the point of view of the flexoelectric response – as a homogeneous continuum; instead the behaviour of the clusters in and their interaction with the surrounding nematic matrix should be handled separately. This challenging problem still waits for a solution.

A possible clue for resolving the problem of giant versus normal flexoelectricity of BC nematics may lie in recent observations of converse giant flexoelectricity.[74] As flexoelectricity is a linear phenomenon, the deformation of the substrates is expected to be proportional to the applied voltage. In this experiment, however, no flexing was detected below a critical voltage corresponding to a critical electric field of $1\mathrm{V}\mu\mathrm{m}^{-1}$. This may be interpreted that the cluster contribution to flexoelectricity requires a threshold field, below which only the normal flexoelectricity of the nematic matrix is active. Experiments[82,83] using the Helfrich geometry remained much below this threshold field, which may explain why no giant flexoelectricity was observed.

Summarizing, experimental observations suggest that the giant (direct or converse) flexoelectricity of bent-core nematics is related to the polar smectic clusters occurring in them. In order to explore the exact mechanism for how clusters contribute to the flexoelectric response, further experimental and theoretical studies are needed.

Finally, it is worth mentioning that a phenomenon analogous to the difference between the 'normal' and 'giant' flexoelectricity of calamitic and bent-core nematics, respectively, exists in crystals, ceramics and polymers too.[69] The flexoelectric response (defined in Eq. (3.1)) of perovskite-type ferroelectrics,[94] of relaxor ferroelectric ceramics[95–97] and polyvinylidene fluoride (PVDF) films[98] are four orders of magnitude larger than the flexoelectricity of dielectric crystals. In those solid ferroelectric materials the polarization induced by flexing is evidently of piezoelectric origin.

## 3.5. Giant Flexoelectric Effect in Liquid Crystalline Elastomers

The observation of giant flexoelectricity in BC nematics opened new horizons for their application in the conversion of mechanical energy into electrical energy (direct flexoelectricity) or vice versa (converse flexoelectricity). So far devices performing this energy conversion such as sensors or actuators have mainly been fabricated using perovskite-type ferroelectrics and relaxor ferroelectric ceramics.[95–97] For example, in a 1.8-mm-thick slab of barium strontium titanate one can induce a polarization of about 1 $\mathrm{Cm}^{-2}$ with a strain gradient of 0.01 $\mathrm{m}^{-1}$ (resulting in an effective $e \geq 40\ \mu\mathrm{Cm}^{-1}$).[94]

Efficient energy conversion has been demonstrated under dynamic flexing conditions (using a nanoscale cantilever beam[95,96]). Disadvantageously, however, ceramics are fragile and difficult to stress,[99] cannot maintain a large bend and suffer from fatigue in a dynamic environment. Furthermore, flexoelectric crystals must have a suitable shape,[100] which makes the fabrication process for crystalline or ceramic solid materials much more difficult.

In contrast to ceramics, BC nematics are fluids; therefore they require bulky elastic support or encapsulation to prevent cell leakage and electrical shorts. As noted above, the bending modulus of the substrate is cubic in its thickness; choosing thinner substrates would drastically reduce the mechanical resistance and thus could substantially increase the amplitude of the bending allowing more efficient energy conversion, however, at the expense of making shorts more likely.

The need for solid substrates may be lifted by using self-supporting liquid crystalline elastomers (LCEs). It has been found that conventional LCEs can be soaked in BC nematics producing a swollen elastomer,[57] which also shows giant flexoelectricity.

Unfortunately swollen LCEs are still not robust enough for long-term operation. However, if the rod-like molecules of a conventional LCE are substituted by BC nematics, a bent-core liquid crystalline elastomer (BCLCE) can be created. BCLCEs might combine excellent flexoelectric properties with rubber elasticity.[101]

BCLCEs have been synthesized only recently[102] and it has been proven that they possess the giant flexoelectric effect too.[103] Figure 3.13 shows the temperature dependence of $e_3$ for a BC monomer, a BCLCE and an LCE swollen with a BC nematic. It can be seen that the flexoelectric response reduces in this sequence, but $e_3$ still remains of the same order of magnitude for all three systems. Note that due to a domain structure with varying transition temperatures, the nematic to isotropic transition is gradual in the elastomers; unlike the pure BC nematic which has a weakly first-order phase transition.[104] The control groups, a calamitic LCE and a plain piece of rubber, did not show the giant flexoelectric effect; nor did the BCLCE when heated above the isotropic temperature.

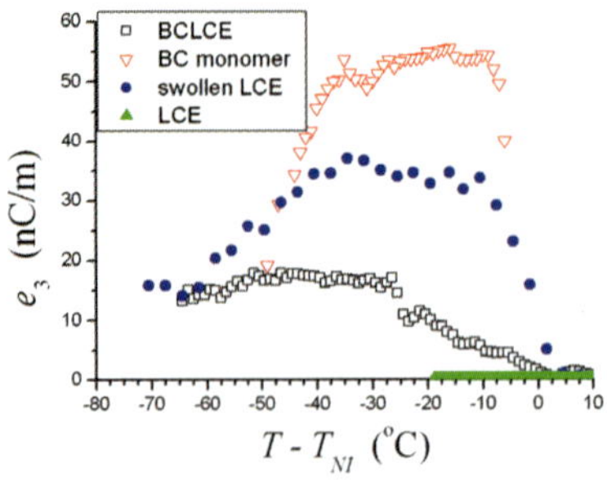


Fig. 3.13. Relative temperature dependences for the BC nematic **10DClPBCP** fluid monomer, for a BC nematic swollen in a calamitic liquid crystal elastomer (BCN-LCE) and a BC nematic elastomer (BCLCE).[103] The flexoelectric coefficient of an LCE is also shown (note that it is not distinguishable from the horizontal axis at the present scale due to the three orders of magnitude difference).

The frequency dependences of the bend flexoelectric coefficients were also measured for the same BC nematic fluid monomer, BC nematic swollen in a calamitic liquid crystal elastomer (BCN-LCE) and for the bent-core nematic elastomer (BCLCE) as shown in Fig. 3.14. One can see that for each material the flexoelectric effect was found to be zero below 1 Hz, then the response increases abruptly up to 2 Hz and then decreases slightly. The apparent absence of the response below 1 Hz is probably due to screening by free ions. The slow decrease of the flexoelectric coefficient at higher $f$ is not yet clear. We assume, however, that it is not a measurement error, because **5CB** showed a constant value in this frequency range.

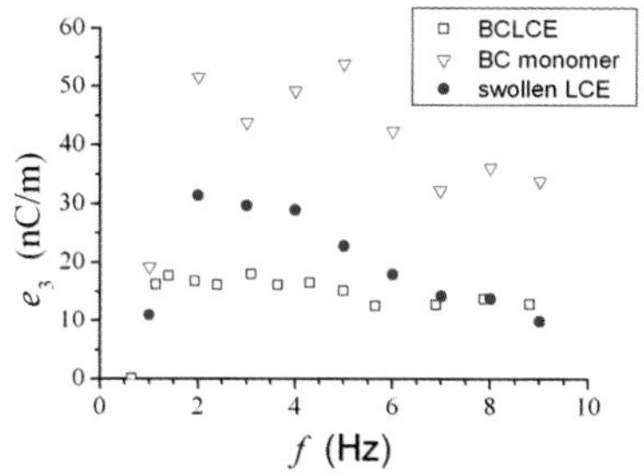


Fig. 3.14. Frequency dependences for the BC nematic (**10DClPBCP**) fluid monomer, the BC nematic swollen in a calamitic liquid crystal elastomer (BCN-LCE) and the BC nematic elastomer (BCLCE).

The BCLCEs could be flexed over ten thousand times with no sign of fatigue. Leakage of the material out of the substrates cannot occur since the BC units are chemically bound to the polymer backbone. Shorting was also not a problem since the elastomer acts as its own spacer, thus preventing the two substrates from coming into contact. Unlike the pure bent-core material, the nematic phase of the elastomer was enantiotropic and could therefore be reached by either heating from the glassy phase or cooling from the isotropic phase. In contrast, the pure bent-core material always had to be cooled from the isotropic phase in order to see any effect.

Some usability problems did, however, occur with the BCLCE too. Excessive heat cycles above the isotropic transitions deteriorated the alignment of the BCLCE; this, however, had to be risked since the isotropic phase was one of the control groups of the experiment. Flexing at or near room temperature (in the glassy phase of the elastomer) created tears and cracks in the system as well. In other words, the BCLCE has to operate in the nematic temperature range; not only because this is where the giant flexoelectric effect is seen, but also to prevent structural failures of the elastomer.

Another interesting aspect of the BCLCE is that it could manually be flexed to a 1 cm radius, which allows the output voltage to be seen on a simple oscilloscope instead of a costly lock-in amplifier. Of course, one needs to allow the elastomer to warm up into the nematic phase before flexing to prevent breakage. By flexing a 20 $mm^2$ piece of elastomer a voltage of 30 mV was produced. This means a current density of 2 $nAcm^{-2}$ at 3 Hz with a 0.1 $m^{-1}$ strain gradient, which corresponds to 1 $\mu Cm^{-2}$ polarization, i.e. the same charge per area but with a ten times larger strain gradient and four times smaller film thickness compared to the solid ferroelectric barium strontium titanate.[94]

To increase the induced current per unit volume one may need much thinner films that can be stacked together and connected in parallel circuitry. Hence, BCLCEs have a realistic potential for inexpensive and highly processable electromechanical energy conversion devices.

## Acknowledgments

This work was supported by the Hungarian Research Fund OTKA under grant No. K-81250 and by NSF grant DMR-0964765.

## References

1. D. Vorländer and A. Apel, Die Richtung der Kohlenstoff-Valenzen in Benzolabkömmlingen (II.), *Berichte der Deutschen Chemischen Gesellschaft* **65**(7), 1101–1109, (1932). doi: 10.1002/cber.19320650710
2. Y. Matsunaga and S. Miyamoto, Mesomorphic behavior of 2,4-bis-(4-alkoxybenzylidene)cyclopentanones and related compounds, *Mol. Cryst. Liq. Cryst.* **237**(1), 311–317, (1993). doi: 10.1080/10587259308030145
3. H. Matsuzaki and Y. Matsunaga, New mesogenic compounds with unconventional molecular structures 1,2-phenylene and 2,3-naphthylene bis[4-(4-alkoxyphenyliminomethyl)benzoates] and related compounds, *Liq. Cryst.* **14**(1), 105–120, (1993). doi: 10.1080/02678299308027306
4. T. Niori, T. Sekine, J. Watanabe, T. Furukawa and H. Takezoe, Distinct ferroelectric smectic liquid crystals consisting of banana shaped achiral molecules, *J. Mater. Chem.* **6**(7), 1231–1233, (1996). doi: 10.1039/JM9960601231
5. T. Sekine, T. Niori, M. Sone, J. Watanabe, S.W. Choi, Y. Takanishi and H. Takezoe, Origin of helix in achiral banana-shaped molecular systems, *Jpn. J. Appl. Phys. Part 1.* **36**(10), 6455–6463 (1997). doi: 10.1143/JJAP.36.6455

6. D.R. Link, G. Natale, R. Shao, J.E. Maclennan, N.A. Clark, E. Körblova and D.M. Walba, Spontaneous formation of macroscopic chiral domains in a fluid smectic phase of achiral molecules, *Science* **278**(5345), 1924–1927 (1997). doi: 10.1126/science.278.5345.1924
7. G. Heppke, A. Jákli, D. Krüerke, C. Löhning, D. Lötzsch, S. Paus, R. Rauch and K. Sharma, Polymorphism and electro-optical properties of banana shaped molecules. In *European Conference on Liquid Crystals (ECLC'97), Zakopane, March 1997*, p. 34.
8. R. Macdonald, F. Kentischer, P. Warnick and G. Heppke, Antiferroelectricity and chiral order in new liquid crystals of nonchiral molecules studied by optical second harmonic generation, *Phys. Rev. Lett.* **81**(20), 4408–4411, (1998). doi: 10.1103/PhysRevLett.81.4408
9. G. Heppke, A. Jákli, S. Rauch and H. Sawade, Electric-field-induced chiral separation in liquid crystals, *Phys. Rev. E* **60**(5), 5575–5579 (1999). doi: 10.1103/PhysRevE.60.5575
10. J. Szydlowska, J. Mieczkowski, J. Matraszek, D.W. Bruce, E. Gorecka, D. Pociecha and D. Guillon, Bent-core liquid crystals forming two- and three-dimensional modulated structures, *Phys. Rev. E* **67**(3), 31702/1–5,(2003). doi: 10.1103/PhysRevE.67.031702
11. M. Nakata, R.F. Shao, J.E. Maclennan, W. Weissflog and N.A. Clark, Electric-field-induced chirality flipping in smectic liquid crystals: The role of anisotropic viscosity, *Phys. Rev. Lett.* **96**(6), 067802/1–4, (2006). doi: 10.1103/PhysRevLett.96.067802
12. M.W. Schröder, S. Diele, G. Pelzl and W. Weissflog, Field-induced switching of the layer chirality in SmCP phases of novel achiral bent-core liquid crystals and their unusual large increase in clearing temperature under electric field application, *Chem. Phys. Chem.* **5**(1), 99–103, (2004). doi: 10.1002/cphc.200300904
13. J.P. Bedel, J.C. Rouillon, J.P. Marcerou, H.T. Nguyen and M.F. Achard, Evidence for different polymorphisms with and without an external electric field in a series of bent-shaped molecules, *Phys. Rev. E* **69**(6), 61702/1–9, (2004). doi: 10.1103/PhysRevE.69.061702
14. R. Amaranatha Reddy, M.W. Schröder, M. Bodyagin, H. Kresse, S. Diele, G. Pelzl and W. Weissflog, Field-induced switching of chirality in undulated ferroelectric and antiferroelectric SmCP phases formed by bent-core mesogens, *Angew. Chem.* **117**(5), 784–788, (2005). doi: 10.1002/ange.200461490
15. H.R. Brand, P.E. Cladis and H. Pleiner, Macroscopic properties of smectic $C_G$ liquid crystals, *Eur. Phys. J. B* **6**(3), 347–353, (1998). doi: 10.1007/s100510050560
16. P.E. Cladis, H.R. Brand and H. Pleiner, Erratum: Macroscopic properties of smectic CG liquid crystals, *Eur. Phys. J. B* **31**(1), 147, (2003). doi: 10.1140/epjb/e2003-00018-6
17. P.G. de Gennes, *The Physics of Liquid Crystals*, Clarendon Press, Oxford, 1974.
18. R.L. Flurry, *Symmetry Groups: Theory and Chemical Applications*, Prentice-Hall, 1980.

19. H. Takezoe and Y. Takanishi, Bent-core liquid crystals: Their mysterious and attractive world, *Jpn. J. Appl. Phys.* **45**(2A), 597–625, (2006). doi: 10.1143/JJAP.45.597
20. J. Etxebarria and M.B. Ros, Bent-core liquid crystals in the route to functional materials, *J. Mat. Chem.* **18**(25), 2919–2926, (2008). doi: 10.1039/B803507E
21. A. Jákli, C. Bailey and J. Harden, Chapter 2, Physical properties of banana liquid crystals, In ed. A. Ramamoorthy, *Thermotropic Liquid Crystals: Recent Advances*, Springer Publishers, 2007.
22. D.A. Coleman, J. Fernsler, N. Chattham, M. Nakata, Y. Takanishi, E. Körblova, D.R. Link, R.-F. Shao, W.G. Jang, J.E. Maclennan, O. Mondainn-Monval, C. Boyer, W. Weissflog, G. Pelzl, L.-C. Chien, J. Zasadzinski, J. Watanabe, D.M. Walba, H. Takezoe and N.A. Clark, Polarization-modulated smectic liquid crystal phases, *Science* **301**(568), 1204–1211, (2003). doi: 10.1126/science.1084956
23. C. Bailey and A. Jákli, Role of molecular shape on bent-core liquid-crystal structures, *Phys. Rev. Lett.* **99**(20), 207801/1–4, (2007). doi: 10.1103/PhysRevLett.99.207801
24. O. Francescangeli, V. Stanic, S.I. Torgova, A. Strigazzi, N. Scaramuzza, C. Ferrero, S. Orlandi and C. Zannoni, Ferroelectric response and induced biaxiality in the nematic phase of bent-core mesogens, *Adv. Func. Mater.* **19**(16), 2592-2600, (2009). doi: 10.1002/adfm.200801865
25. S.H. Hong, R. Verduzco, J.C. Williams, R.J. Twieg, E. DiMasi, R. Pindak, A. Jákli, J.T. Gleeson and S. Sprunt, Short-range smectic order in bent-core nematic liquid crystals, *Soft Matter* **6**(19), 4819–4827, (2010). doi: 10.1039/C000362J
26. S.H. Hong, R. Verduzco, J.T. Gleeson, S. Sprunt and A. Jákli, Nanostructures of liquid crystal phases in mixtures of bent-core and rod-shaped molecules, *Phys. Rev. E* **83**(6), 061702/1–5, (2011). doi: 10.1103/PhysRevE.83.061702
27. P. Salamon, N. Éber, Á. Buka, J.T. Gleeson, S. Sprunt and A. Jákli, Dielectric properties of mixtures of a bent-core and a calamitic liquid crystal, *Phys. Rev. E* **81**(3), 031711/1–11, (2010). doi: 10.1103/PhysRevE.81.031711
28. J. Harden, B. Mbanga, N. Éber, K. Fodor-Csorba, S. Sprunt, J.T. Gleeson and A. Jákli, Giant flexoelectricity of bent-core nematic liquid crystals, *Phys. Rev. Lett.* **97**(15), 157802/1–4, (2006). doi: 10.1103/PhysRevLett.97.157802
29. A. Jákli, M. Chambers, J. Harden, M. Madhabi, R. Teeling, J. Kim, Q. Li, G.G. Nair, N. Éber, K. Fodor-Csorba, J.T. Gleeson and S. Sprunt, Extraordinary properties of nematic phases of bent-core liquid crystals, (Emerging Liquid Crystal Technologies III, San Jose, January 20–24, 2008). *Proc. SPIE* **6911**, 691105/1–10, (2008). doi: 10.1117/12.768866
30. Sh.M. Kogan, Piezoelectric effect on non-uniform deformation and acoustical scattering of carriers in crystals, *Fiz. Tverd. Tela*, **5**(10), 2829–2831 (1963). [*Sov. Phys. Solid State* **5**, 2069–2071, (1964).]

31. R.B. Meyer, Piezoelectric effects in liquid crystals, *Phys. Rev. Lett.* **22**(18), 918–921, (1969). doi: 10.1103/PhysRevLett.22.918
32. M.A. Osipov, Chapter 1. Molecular theory of flexoelectricity in nematic liquid crystals. In eds. Á. Buka and N. Éber, *Flexoelectricity in Liquid Crystals. Theory, Experiments and Applications*, Imperial College Press, London, 2012. pp. 9–32.
33. W. Helfrich, A simple method to observe the piezoelectricity of liquid crystals, *Phys. Lett.* **35A**(6), 393–394, (1971). doi: 10.1016/0375-9601(71)90160-5
34. W. Helfrich, The strength of piezoelectricity in liquid crystals, *Z. Naturforsch.* **26A**(5), 833–835, (1971).
35. A. Derzhanski and A.G. Petrov, A molecular-statistical approach to the piezoelectric properties of nematic liquid crystals, *Phys. Lett.* **36A**(6), 483–484, (1971). doi: 10.1016/0375-9601(71)90657-8
36. J. Prost and J.P. Marcerou, On the microscopic interpretation of flexoelectricity, *J. Phys. France* **38**(3), 315–324, (1977). doi: 10.1051/jphys:019770038030031500
37. M. Marvan and A. Havranek, Flexoelectric effect in elastomers, In ed. I. Chudácek, *Relationships of Polymeric Structure and Properties, Progress in Colloid and Polymer Science.* Vol. **78**, Springer-Verlag, New York, 1988. pp. 33–36. doi: 10.1007/BFb0114342
38. V.I. Passechnik and V.S. Sokolov, Permeability change of modified bimolecular phospholipid membranes accompanying periodical expansion, *Biofizika (Moscow)* **18**, 655–660, (1973).
39. A.G. Petrov and A. Derzhanski, Membranes and lyotropic phases on some problems in the theory of elastic and flexoelectric effects in bilayer lipid membranes and biomembranes, *J. Phys. Colloques* **37**(C3), C3-155–C3-160, (1976). doi: 10.1051/jphyscol:1976330
40. A.G. Petrov, Flexoelectricity of lyotropics and biomembranes, *Il Nuovo Cimento* **3D**(1), 174–192, (1984). doi: 10.1007/BF02452210
41. A.G. Petrov and V.S. Sokolov, Curvature-electric effect in black lipid membranes, *Eur. Biophys. J.* **13**(3), 139–155, (1986). doi: 10.1007/BF00542559
42. A. Derzhanski, A.G. Petrov, A.T. Todorov and K. Hristova, Flexoelectricity of lipid bilayers, *Liq. Cryst.* **7**(3), 439–449, (1990). doi: 10.1080/02678299008033820
43. L-T. Gao, X-Q. Feng, Y-J. Yin and H. Gao, An electromechanical liquid crystal model of vesicles, *J. Mech. Phys. Solids* **56**(9), 2844–2862, (2008). doi: 10.1016/j.jmps.2008.04.006
44. R.M. Raphael, A.S. Popel and W.E. Brownell, A membrane bending model of outer hair cell electromotility, *Biophys. J.* **78**(6), 2844–2862, (2000). doi: 10.1016/S0006-3495(00)76827-5
45. A.A. Spector, N. Deo, K. Grosh, J.T. Ratnanather and R.M. Raphael, Electromechanical models of the outer hair cell composite membrane, *J. Membrane Biol.* **209**(2-3), 135–152, (2006). doi: 10.1007/s00232-005-0843-7

46. A.G. Petrov, Flexoelectric model for active transport, In ed. J. Vassileva, *Physical and Chemical Bases of Biological Information Transfer*, Plenum Press, New York, 1975. pp. 167.
47. A.T. Todorov, A.G. Petrov and J.H. Fendler, First observation of the converse flexoelectric effect in bilayer lipid membranes, *J. Phys. Chem.* **98**(12), 3076–3099, (1994). doi: 10.1021/j100063a004
48. L. Beresnev and L.M. Blinov, Ferroelectric liquid crystals: molecular aspects, *Mendeleev J. All-Union Chem. Soc.* **28**, 149–155, (1982).
49. H.R. Leuchtag and V.S. Bystrov, Theoretical models of conformational transitions and ion conduction in voltage-dependent ion channels: Bioferroelectricity and superionic conduction, *Ferroelectrics* **220**(1), 157–204, (1999). doi: 10.1080/00150199908216214
50. B. Hille, *Ionic Channels of Excitable Membranes*, Sinauer, Sunderland, 1992.
51. A. Derzhanski, A.G. Petrov and Y.V. Pavloff, Curvature induced conductive and displacement currents through lipid bilayers, *J. Physique Lett.* **42**(6), L-119–L-122, (1981). doi: 10.1051/jphyslet:01981004206011900
52. A.G. Petrov, Chapter 6. Flexoelectricity in lyotropics and in living liquid crystals. In eds. Á. Buka, and N. Éber, *Flexoelectricity in Liquid Crystals. Theory, Experiments and Applications*, Imperial College Press, London, 2012. pp. 177–210.
53. L.D. Landau and E.M. Lifshitz, *Theory of Elasticity, 3rd edition*, Butterworth Heinemann, Oxford, England, 1986.
54. P.R. Maheswara Murthy, V.A. Ragunathan and N.V. Madhusudana, Experimental determination of the flexoelectric coefficients of some nematic liquid crystals, *Liq. Cryst.* **14**(2), 483–496, (1993). doi: 10.1080/02678299308027664
55. K. Fodor-Csorba, A. Vajda, G. Galli, A. Jákli, D. Demus, S. Holly and E. Gács-Baitz, Ester-type banana-shaped monomers and investigations of their electro-optical properties, *Macromol. Chem. Phys.* **203**(10-11), 1556–1563, (2002). doi: 10.1002/1521-3935(200207)
56. K. Fodor-Csorba, A. Vajda, A. Jákli, Ch. Slugovc, G. Trimmel, D. Demus, E. Gács-Baitz, S. Holly and G. Galli, Ester type banana-shaped liquid crystalline monomers: synthesis and physical properties, *J. Mater. Chem.* **14**(16), 2499–2506, (2004). doi: 10.1039/B400246F
57. M. Chambers, R. Verduzco, J.T. Gleeson, S. Sprunt and A. Jákli, Calamitic liquid-crystalline elastomers swollen in bent-core liquid-crystal solvents, *Adv. Mater.* **21**(16), 1622–1626, (2009). doi: 10.1002/adma.200802739
58. J.P. van Meter and B.H. Klanderman, Mesomorphic properties of some phenyl benzoate derivatives, *Mol. Cryst. Liq. Cryst.* **22**(3-4), 271–284, (1973). doi: 10.1080/15421407308083350
59. G.W. Gray, K.J. Harrison and J.A. Nash, New family of nematic liquid crystals for displays, *Electronics Letters* **9**(6), 130–131, (1973). doi: 10.1049/el:19730096
60. S.T. Lagerwall, *Ferroelectric and Antiferroelectric Liquid Crystals*, Wiley-VCH, Weinheim, 1999.

61. N.V. Madhusudana and R. Pratibha, Electromechanical coupling in cholesteric liquid crystals, *Mol. Cryst. Liq. Cryst. Lett.* **5**(2), 43–51, (1987).
62. N.V. Madhusudana and R. Pratibha, An experimental investigation of electromechanical coupling in cholesteric liquid crystals, *Liq. Cryst.* **5**(6), 1827–1840, (1989). doi: 10.1080/02678298908045691
63. N.V. Madhusudana, R. Pratibha and H.P. Padmini, Electromechanical effect in cholesteric liquid crystals with fixed boundary conditions, *Mol. Cryst. Liq. Cryst.* **202**(1), 35–49, (1991). doi: 10.1080/00268949108035658
64. H.P. Padmini and N.V. Madhusudana, Electromechanical effect in cholesteric mixtures with a compensation temperature, *Liq. Cryst.* **14**(2), 497–511, (1993). doi: 10.1080/02678299308027665
65. A. Jákli, L. Bata, Á. Buka, N. Éber and I. Jánossy, New electromechanical effect in chiral smectic C* liquid crystals, *J. Physique Lett.* **46**(16), L-759–L-761, (1985). doi: 10.1051/jphyslet:019850046016075900
66. A. Jákli, L. Bata, Á. Buka and N. Éber, Electromechanical effect in $S_C^*$ liquid crystals, *Ferroelectrics* **69**(1), 153–163, (1986). doi: 10.1080/00150198608008189
67. A. Jákli, N. Éber and L. Bata, Electromechanical effects in ferroelectric liquid crystalline polymers, *Polym. Adv. Technol.* **3**(5), 269–274, (1992). doi: 10.1002/pat.1992.220030512
68. A. Jákli and N. Éber, Piezoelectric effects in liquid crystals, In A. Buka ed., *Modern Topics in Liquid Crystals*. World Scientific, Singapore, 1993. pp. 235–256.
69. A. Jákli, Electro-mechanical effects in liquid crystals, *Liq. Cryst.* **37**(6–7), 825–837, (2010). doi: 10.1080/02678291003784081
70. W.G. Cady, *Piezoelectricity*, McGraw-Hill, London, 1946.
71. N.V. Madhusudana, Chapter 2. Flexoelectro-optics and measurements of flexocoefficients. In eds. Á. Buka and N. Éber, *Flexoelectricity in Liquid Crystals. Theory, Experiments and Applications*, Imperial College Press, London, 2012. pp. 33–60.
72. A.G. Petrov, M. Spassova and J.H. Fendler, Flexoelectricity and photoflexoelectricity in model and biomembranes, *Thin Solid Films* **284–285**, 845–848, (1996). doi: 10.1016/S0040-6090(95)08461-4
73. A.G. Petrov, Electricity and mechanics of biomembrane systems: Flexoelectricity in living membranes, *Anal. Chim. Acta* **568**(1–2), 70–83, (2006). doi: 10.1016/j.aca.2006.01.108
74. J. Harden, R. Teeling, J.T. Gleeson, S. Sprunt and A. Jákli, Converse flexoelectric effect in a bent-core nematic liquid crystal, *Phys. Rev. E* **78**(3), 031702/1–5, (2008). doi: 10.1103/PhysRevE.78.031702
75. T.J. Ma, B. Bhushan, H. Murooka, I. Kobayashi and T. Osawa, A novel technique to measure the Poissons ratio and submicron lateral dimensional changes of ultrathin polymeric films, *Rev. Sci. Instrum.* **73**(4), 1813–1820, (2002). doi: 10.1063/1.1458062
76. K. Matsuzawa, Measurements of Youngs modulus of 'Mylar' polyester film, *J. Acoust. Soc. Jpn.* **15**(3), 147–150, (1959).

77. N.V. Madhusudana and G. Durand, Linear flexo-electro-optic effect in a hybrid aligned nematic liquid crystal cell, *J. Physique Lett.* **46**(5), L-195–L-200, (1985). doi: 10.1051/jphyslet:019850046050019500
78. I. Dozov, Ph. Martinot Lagarde and G. Durand, Flexoelectrically controlled twist of texture in a nematic liquid crystal, *J. Physique Lett.* **43**(10), L-365–L-369, (1982). doi: 10.1051/jphyslet:019820043010036500
79. J.H. Wild, K. Bartle, N.T. Kirkman, S.M. Kelly, M. O'Neill, T. Stirner and R.P. Tuffin, Synthesis and investigation of nematic liquid crystals with flexoelectric properties, *Chem. Mater.* **17**(25), 6354–6360, (2005). doi: 10.1021/cm051682y
80. B. Kundu, A. Roy, R. Pratibha and N.V. Madhusudana, Flexoelectric studies on mixtures of compounds made of rodlike and bent-core molecules, *Appl. Phys. Lett.* **95**(8), 081902/1–3, (2009). doi: 10.1063/1.3213544
81. D. Schmidt, M. Schadt and W. Helfrich, Liquid-crystalline curvature electricity: The bending mode of MBBA, *Z. Naturforsch.* **27A**(2), 277–280, (1971).
82. K.V. Le, F. Araoka, K. Fodor-Csorba, K. Ishikawa and H. Takezoe, Flexoelectric effect in a bent-core mesogen, *Liq. Cryst.* **36**(10–11), 1119–1124, (2009). doi: 10.1080/02678290902854086
83. P. Kumar, Y.G. Marinov, H.P. Hinov, U.S. Hiremath, C.V. Yelamaggad, K.S. Krishnamurthy and A.G. Petrov, Converse flexoelectric effect in bent-core nematic liquid crystals, *J. Phys. Chem. B* **113**(27), 9168–9174, (2009). doi: 10.1021/jp903241z
84. Á. Buka, T. Tóth-Katona, N. Éber, A. Krekhov and W. Pesch, Chapter 4. The role of flexoelectricity in pattern formation. In eds. Á. Buka and N. Éber, *Flexoelectricity in Liquid Crystals. Theory, Experiments and Applications*, Imperial College Press, London, 2012. pp. 101–135.
85. P. Rudquist and S.T. Lagerwall, Chapter 7. Applications of flexoelectricity. In eds. Á. Buka and N. Éber, *Flexoelectricity in Liquid Crystals. Theory, Experiments and Applications*, Imperial College Press, London, 2012. pp. 211–247.
86. P.S. Salter, C. Tschierske, S.J. Elston and E.P. Raynes, Flexoelectric measurements of a bent-core nematic liquid crystal, *Phys. Rev. E* **84**(3), 031708/1–5, (2011). doi: 10.1103/PhysRevE.84.031708
87. N. Aziz, S.M. Kelly, W. Duffy and M. Goulding, Banana-shaped dopants for flexoelectric nematic mixtures, *Liq. Cryst.* **35**(11), 1279–1292 (2008). doi: 10.1080/02678290802522114
88. J.S. Patel and R.B. Meyer, Flexoelectric electro-optics of a cholesteric liquid crystal, *Phys. Rev. Lett.* **58**(15), 1538–1540, (1987). doi: 10.1103/PhysRevLett.58.1538
89. S.-D. Lee, J.S. Patel and R.B. Meyer, Effect of flexoelectric coupling on helix distortions in cholesteric liquid crystals, *J. Appl. Phys.* **67**(3), 1293–1297, (1990). doi: 10.1063/1.345679
90. P. Rudquist, L. Komitov and S.T. Lagerwall, Linear electro-optic effect in a cholesteric liquid crystal, *Phys. Rev. E* **50**(6), 4735–4743, (1994). doi: 10.1103/PhysRevE.50.4735

91. B. Musgrave, P. Lehmann and H.J. Coles, A new series of chiral nematic bimesogens for the flexoelectro-optic effect, *Liq. Cryst.* **26**(8), 1235-1249, (1999). doi: 10.1080/026782999204255
92. F. Castles, S.M. Morris and H.J. Coles, The limits of flexoelectricity in liquid crystals, *AIP Advances* **1**(3), 032120/1–4, (2011). doi: 10.1063/1.3624725
93. C. Zhang, N. Diorio, M. Gao, S. Sprunt, J.T. Gleeson and A. Jákli, Direct observation of the smectic clusters in the nematic phase of bent core liquid crystals, to be published.
94. W. Ma and L.E. Cross, Flexoelectric polarization of barium strontium titanate in the paraelectric state, *Appl. Phys. Lett.* **81**(18), 3440–3442, (2002). doi: 10.1063/1.1518559
95. W. Ma and L.E. Cross, Observation of the flexoelectric effect in relaxor $Pb(Mg_{1/3}Nb_{2/3})O_3$ ceramics, *Appl. Phys. Lett.* **78**(19), 2920–2921, (2001). doi: 10.1063/1.1356444
96. W. Ma, Flexoelectricity: strain gradient effects in ferroelectrics, *Physica Scripta* **T129**, 180–183, (2007). doi: 10.1088/0031-8949/2007/T129/041
97. W. Ma, L.E. Cross and A. Large, Large flexoelectric polarization in ceramic lead magnesium niobate, *Appl. Phys. Lett.* **79**(26), 4420–4422, (2001). doi: 10.1063/1.1426690
98. S. Baskaran, N. Ramachandran, X. He, S. Thiruvannamalai, H.J. Lee, H. Heo, Q. Chen and J.Y. Fu, Giant flexoelectricity in polyvinylidene fluoride films, *Phys. Lett. A* 375, 2082–2084, (2011). doi: 10.1016/j.physleta.2011.04.011
99. J.Y. Fu, Z. Zhu and L.E. Cross, Experimental studies of the converse flexoelectric effect induced by inhomogeneous electric field in a barium strontium titanate composition, *J. Appl. Phys.* **100**(2), 024112/1–6, (2006). doi: 10.1063/1.2219990
100. J.W. Fu and L.E. Cross, Separate control of direct and converse piezoelectric effects in flexoelectric piezoelectric composites, *Appl. Phys. Lett.* **91**(16), 162903/1–3, (2007). doi: 10.1063/1.2790476
101. M. Warner and T.M. Terentjev, *Liquid Crystal Elastomers*, Oxford University Press, Oxford, 2003.
102. R. Verduzco, P. Luchette, S.H. Hong, J. Harden, E. DiMasi, P. Palffy-Muhoray, S.M. Kilbey II, S. Sprunt, J.T. Gleeson and A. Jákli, Bent-core liquid crystal elastomers, *J. Mater. Chem.* **20**(39), 8488–8495, (2010). doi: 10.1039/C0JM01920H
103. J. Harden, M. Chambers, R. Verduzco, P. Luchette, J. Gleeson and S. Sprunt, Giant flexoelectricity in bent-core nematic liquid crystal elastomers, *Appl. Phys. Lett.* **96**(10), 102907/1–3, (2010). doi: 10.1063/1.3358391
104. J.V. Selinger, H.G. Jeon and B.R. Ratna, Isotropic-nematic transition in liquid-crystalline elastomers, *Phys. Rev. Lett.* **89**(22), 225701/1–4, (2002). doi: 10.1103/PhysRevLett.89.225701

# Chapter 4

# The Role of Flexoelectricity in Pattern Formation

Ágnes Buka*, Tibor Tóth-Katona*, Nándor Éber*,
Alexei Krekhov† and Werner Pesch†

* *Institute for Solid State Physics and Optics*
*Wigner Research Centre for Physics*
*Hungarian Academy of Sciences*
*H-1525 Budapest, P.O.B. 49, Hungary*
*e-mail: buka.agnes@wigner.mta.hu, tothkatona.tibor@wigner.mta.hu,*
*eber.nandor@wigner.mta.hu*

† *Physikalisches Institut, Universität Bayreuth*
*D-95440 Bayreuth, Germany*
*e-mail: alexei.krekhov@uni-bayreuth.de, werner.pesch@uni-bayreuth.de*

In this chapter the influence of flexoelectricity on pattern formation induced by an electric field in nematics will be summarized. Two types of patterns will be discussed in the linear regime, the equilibrium structure of flexoelectric domains and the dissipative electroconvection (EC) rolls. In a separate section, recent experimental and theoretical results on the competition and crossover between the flexoelectric domains and EC patterns will be described.

## Contents

## 4.1. Introduction

Patterns, i.e. regular spatiotemporal structures, can easily be generated in liquid crystals via a large variety of external stresses, e.g., by mechanical shear, temperature or pressure gradients, electric or magnetic fields, etc.; representative examples can be found in Buka and Kramer.[1] Here we concentrate on patterns induced by electric fields in nematics and in particular on the implications of flexoelectricity.

Nematics are uniaxial fluids; the preferred axis is defined by a unit vector $\boldsymbol{n}$, the director.[2] We consider a thin nematic layer of thickness $d$ confined between two plates (parallel to the $xy$ plane), which impose the initial direction $\boldsymbol{n}_0$ of $\boldsymbol{n}$ in the basic state. The plates also serve as electrodes for the application of an electric field $\boldsymbol{E}$ along the $z$ axis.

As a consequence of the uniaxial symmetry all material properties of nematics have to be represented by tensors. For instance, the dielectric displacement $\boldsymbol{D}$ and $\boldsymbol{E}$ are connected by the dielectric susceptibility tensor $\boldsymbol{\epsilon}$ as $\boldsymbol{D} = \epsilon_0 \boldsymbol{\epsilon} \boldsymbol{E} \equiv \epsilon_0[\epsilon_\perp \boldsymbol{E} + (\epsilon_\| - \epsilon_\perp)(\boldsymbol{n}\cdot\boldsymbol{E})\boldsymbol{n}]$. Thus $\boldsymbol{\epsilon}$ depends in general on the local director orientation and is specified by two dielectric constants, $\epsilon_\|$ and $\epsilon_\perp$ (for $\boldsymbol{E}$ parallel and perpendicular to $\boldsymbol{n}$, respectively). An analogous representation applies to the electric conductivity tensor $\boldsymbol{\sigma}$.

Any spatial distortion of $\boldsymbol{n}$ leads to elastic restoring torques, which are determined in the standard continuum description of nematics (exclusively used in this review) by three elastic constants $K_1$ (splay), $K_2$ (twist) and $K_3$ (bend).[2] In addition, the electric field $\boldsymbol{E}$ gives rise to an electric torque on the director. The balance of these torques, reflected in the resulting equilibrium director configuration, corresponds to the minimum of the orientational free energy $\mathcal{F}(\boldsymbol{n})$. For positive dielectric anisotropy ($\epsilon_a = \epsilon_\| - \epsilon_\perp$) the dielectric torque ($\propto |\boldsymbol{E}|^2$) is destabilizing in the planar director configuration ($\boldsymbol{n}_0 \parallel \boldsymbol{x}$). When $|\boldsymbol{E}|$ increases above a certain threshold $E_F$ the electric torque is obviously larger than the stabilizing elastic torque (determined by the elastic constants). Thus an initial planar director configuration will experience a splay distortion in the form of an out-of-plane rotation of $\boldsymbol{n}$. This process, the *Freedericksz transition*,[2] is a prominent example of an orientational transition in nematics under the influence of an applied electric field. Since $\mathcal{F}(\boldsymbol{n})$ of the distorted state is lower than that of the basic state at $E_F$, the notion of an *equilibrium phase transition* is common in analogy to standard thermodynamics.

In most cases the stationary director configuration resulting from the

Freedericksz transition in planar geometry is uniform in the plane of the layer and varies only in the $z$ direction. However, in some exceptional cases, when the splay elastic constant $K_1$ is much larger than the twist elastic constant $K_2$ (e.g., in liquid crystal polymers), a spatially periodic out-of-plane director distortion becomes energetically favourable. The resulting splay-twist (ST) Freedericksz state is manifested in experiments in the form of a 'longitudinal' stripe pattern[3] running parallel to the initial director alignment $\boldsymbol{n}_0 \parallel \boldsymbol{x}$.

Besides the elastic and the electric torques the so-called *flexoelectric* (or *flexo*) torques on the director play an important role as well. Their effect on pattern-forming instabilities in nematics is the main issue of this chapter. Flexotorques originate from the fact that typically (in some loose analogy to piezoelectricity) any director distortion is accompanied by an electric flexopolarization $\boldsymbol{P}_{\text{fl}}$ (characterized by the two flexocoefficients $e_1$, $e_3$).[2,4] From a microscopic point of view, finite $e_1$ and $e_3$ naturally arise when the nematic molecules have a permanent dipole moment.[4] But also for molecules with a quadrupolar moment, finite $e_1$ and $e_3$ are possible[5] (see also Chapter 1 in this book[6]). Flexopolarization has to be incorporated into the free energy $\mathcal{F}(\boldsymbol{n})$ for finite $\boldsymbol{E}$. It is not surprising that this leads to quantitative modifications of phenomena, which exist also for $e_1 = e_3 = 0$. Though, for example, the Freedericksz threshold field $E_F$ is not modified, the presence of flexoelectricity leads to considerable modifications of the Freedericksz distorted state for $|\boldsymbol{E}| > E_F$.[7]

Much more exciting is the possibility of qualitatively new phenomena, which are generically related to flexopolarization. A prominent example is provided by the so-called flexodomains. They appear as the result of an equilibrium transition from the basic planar state if the applied electric field strength exceeds a certain threshold, $E_{\text{fl}}$. Flexodomains are stripe patterns parallel to the imposed preferred direction $\boldsymbol{n}_0 \parallel \boldsymbol{x}$, i.e. with a wave vector $\boldsymbol{q}_c \perp \boldsymbol{x}$.[8–12] In contrast to the standard Freedericksz transition, the sign of $\epsilon_a$ plays no role, but the difference $|e_1 - e_3|$ has to be large enough.

More frequently, instead of the equilibrium pattern sketched so far, one observes electroconvection (EC) patterns in nematics, which present dissipative structures characterized by director distortions, space charges and material flow. A necessary requirement for their existence is the presence of charge carriers in the nematic. In a distorted nematic, where $\boldsymbol{n}$ is neither parallel nor perpendicular to $\boldsymbol{E}$, the generation of a non-zero space charge, $\rho_{el}$, by charge separation is then inevitable. The resulting Coulomb force in the flow equations (generalized Navier–Stokes equations) drives a flow, which in turn exerts a destabilizing *viscous torque* on the director. Under favourable conditions the stabilizing elastic and electric torques may be overcompensated leading to a *non-equilibrium phase transition* from the basic quiescent state. As a result a periodic array of convection rolls appears if the strength of $\boldsymbol{E}$ is above the EC threshold.[13–16] To understand the majority of EC convection patterns investigated it has been sufficient to analyse the *standard* nemato-electrohydrodynamic model,[15] where flexoelectricity is not included. More recently a specific class of nematic materials has been studied,[17,18] where the interpretation of the observed EC patterns ('non-standard EC') definitively requires the inclusion of flexoelectricity in the theoretical description.[19]

Patterns in nematics are easily observed by optical means where the anisotropy of the refractive index is exploited. In this way the stripe patterns in electroconvection in the planar geometry are easily discriminated from flexodomains: the angle $\alpha$ between the wave vector $\boldsymbol{q}$ of the EC stripes and the preferred direction $\boldsymbol{n}_0 \parallel \boldsymbol{x}$ is small (normal or oblique rolls) in contrast to $\alpha = 90°$ (longitudinal stripes) in flexodomains.

In the following we will exclusively concentrate on the final states, which are reached by the system in an applied field when all transients have died out. To resolve these transients in experiments and to analyse them in theory is a highly demanding task, but certainly gives important additional insights. For instance in the development of the homogeneous splay Freedericksz state one observes two kinds of *transient patterns*.[20] For nematics with small (positive) $\epsilon_a$ the stripes are oriented almost perpendicular to the initial director alignment $\boldsymbol{n}_0$ ($\alpha \approx 0$) while for large $\epsilon_a$ the stripes are parallel to $\boldsymbol{n}_0$ ($\alpha \approx 90°$). In these stripe patterns flow is present until the final equilibrium structure is established; thus the theoretical analysis has to be based on the full electrohydrodynamic equations. The stripes perpendicular to $\boldsymbol{n}_0$ have been explained using a linear stability analysis of the basic planar state.[20] The appearance of parallel stripes has been explained by a non-linear analysis taking into account the time evolution of the Freedericksz state.[21] Here the effect of field inhomogeneity due to the large $\epsilon_a$ plays an important role, which favours stripes parallel to $\boldsymbol{n}_0$. Another example of transients is the decay of an EC pattern when turning off the applied electric field.[22,23] In these cases one has indeed obtained deeper insights into the complex interplay of the various mechanisms responsible for patterns in nematics, which also opens additional routes to determine some material parameters. As far as we know, the effect of flexoelectricity has not yet been studied for these transient patterns. It would be certainly rewarding to study other situations where flexoelectricity is more important during the transients than in the final state.

In this review we focus in particular on recent theoretical and experimental investigations of the planar geometry, which are mostly restricted to the linear regime, i.e. to applied electric field strengths $|\boldsymbol{E}|$ slightly above the respective threshold fields. We put some emphasis on the unique aspect of nematics: they allow us to study in the same experimental set-up the competition between the (equilibrium) flexodomains and the dissipative EC patterns by changing the circular frequency $\omega$ of the applied AC electric field.

The chapter is organized as follows: Section 4.2 describes flexodomains in the planar geometry. Particular emphasis is placed on the most recent theoretical results, where for the first time arbitrary ratios of the elastic constants $K_1, K_2$ are considered as well as driving by an AC electric field. Section 4.3 deals with the effects of flexoelectricity on dissipative EC patterns. The focus is on qualitatively new phenomena that are not covered by the standard model of EC. In Section 4.4 we analyse the competition between flexodomains and EC patterns at low AC driving frequencies. The chapter ends with a discussion and some concluding remarks in Section 4.5.

## 4.2. Equilibrium Structures: Flexodomains

As noted in the introduction, flexotorques in the presence of an electric field in nematics can lead to spatially periodic, equilibrium director distortions, the flexodomains. The characteristic static, 'longitudinal stripes' of flexodomains oriented parallel to $\boldsymbol{n}_0 \parallel \boldsymbol{x}$ were first observed more than 40 years ago under the action of a DC electric field[8,9] in the planar geometry (for an example, see Fig. 4.1). As first described by Bobylev and Pikin,[10] *flexoelectricity* indeed provides a natural mechanism (independent of the sign of $\epsilon_a$) to explain these stripes using the well-known fact that splay or bend director distortions are generically associated with the flexopolarization

$$\boldsymbol{P}_{\text{fl}} = e_1 \boldsymbol{n}(\nabla \cdot \boldsymbol{n}) + e_3(\boldsymbol{n} \cdot \nabla)\boldsymbol{n}, \tag{4.1}$$

where $e_1$ and $e_3$ are the splay and bend flexoelectric coefficients, respectively.[4] Thus in the presence of an electric field $\boldsymbol{E}$ the free energy density of nematics contains, in addition to the elastic and electric contributions, the flexo term

$$\mathcal{F}_{\text{fl}} = -\boldsymbol{P}_{\text{fl}} \cdot \boldsymbol{E}, \tag{4.2}$$

which results in the flexoelectric torque $\boldsymbol{n} \times (\delta\mathcal{F}_{\rm fl}/\delta\boldsymbol{n})$. The flexoelectric torque depends only on the difference $(e_1 - e_3)$ and favours, as will be shown below, flexodomains for sufficiently large $|e_1 - e_3|$ if the strength of $\boldsymbol{E}$ is above a certain threshold $|\boldsymbol{E}| = E_{\rm fl}$.

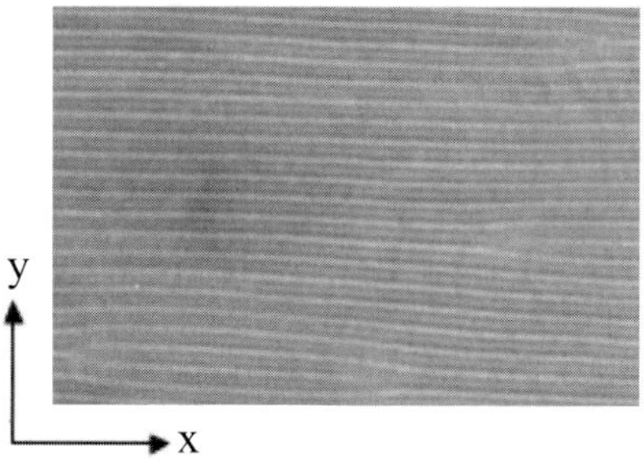


Fig. 4.1. Photograph of the flexodomains in a planar nematic **Phase 4** with the preferred axis parallel to $\boldsymbol{n}_0 \parallel \boldsymbol{x}$.

The theoretical analysis of flexodomains in planar nematics, which is briefly reviewed in the following, exploits the balance of torques by minimizing the total free energy $\mathcal{F}(\boldsymbol{n})$. If not otherwise stated, an assumption of *strong anchoring* of the director $\boldsymbol{n}$ at the confining plates is used; i.e. the director at the boundaries remains parallel to $\boldsymbol{n}_0 \parallel \boldsymbol{x}$, irrespective of director distortions in the bulk of the nematic layer.

Let us start with the case of DC driving. In the one-elastic-constant approximation (*isotropic elasticity*, i.e. $K_1 = K_2 = K_3 = K_{av}$) the director equations originating from the torque balance can be solved analytically in the linear regime of small distortions of the planar basic state. This gives a closed threshold formula for the critical voltage $U_c = E_{\rm fl} d$ and the critical wave number $q_c$ of the longitudinal rolls:[10,11,24,25]

$$U_c = \frac{2\pi K_{av}}{|e_1 - e_3|(1+\mu)}, \quad q_c = \frac{\pi}{d}\left(\frac{1-\mu}{1+\mu}\right)^{1/2}, \tag{4.3}$$

where

$$\mu = \frac{\epsilon_0 \epsilon_a K_{av}}{(e_1 - e_3)^2}. \tag{4.4}$$

According to Eq. (4.3) the flexodomains exist only ($q_c$ has to be finite!) if $|\mu| < 1$, i.e.,

$$|\epsilon_a| < \frac{(e_1 - e_3)^2}{\epsilon_0 K_{av}}. \tag{4.5}$$

As a function of $\mu$ both $U_c$ and $q_c$ rise monotonically from $\mu = 1$ until they diverge in the limit $\mu \to -1$.

The distortions of the basic director orientation $\boldsymbol{n}_0 \parallel \boldsymbol{x}$ in the flexodomains are characterized by an out-of-plane component ($n_z \neq 0$) and an in-plane rotation ($n_y \neq 0$), which are periodic in the $y$ direction and depend on $z$.

The analysis of flexodomains in the general case of anisotropic orientational elasticity ($K_1 \neq K_2$) is in principle straightforward.[26] It is quite easy to derive a transcendental equation for the 'neutral curve' $U_0(q)$ (first given by Bobylev *et al.*[24]) at which the bifurcation of flexodomains with wave number $q$ from the basic planar state takes place. Solving numerically for $U_0(q)$ and subsequently minimizing $U_0(q)$ with respect to $q$ yields the critical wave number $q_c$ and the critical voltage $U_c \equiv U_0(q_c)$.

One of the central results of the analysis of anisotropic elasticity at DC voltage driving is given in Fig. 4.2. It shows the range for flexodomains (marked as the grey region) in the $(\mu, \delta k)$ plane, where $\mu$ is defined in Eq. (4.4). The elastic constants $K_1$ and $K_2$ are parameterized in terms of the average value $K_{av} = (K_1 + K_2)/2$ of the elastic constants, and by the relative deviation $\delta k$ from $K_{av}$ in the following manner:

$$K_1 = K_{av}(1 + \delta k), \quad K_2 = K_{av}(1 - \delta k). \tag{4.6}$$

Obviously $|\delta k| < 1$ is required. The case $\delta k > 0$ (i.e., $K_2 < K_1$) corresponds to rod-like nematics like **MBBA** [N-(4-metoxybenzylidene)-4′-butylaniline] while $\delta k < 0$ (i.e., $K_2 > K_1$) holds for discotic nematics.

The $(\mu, \delta k)$ range of flexodomains is limited from above by the upper limiting curve, $\mu_{max}(\delta k) > 0$ (dashed line in Fig. 4.2). At larger $\mu$, e.g. at larger $\epsilon_a$, there is the homogeneous Freedericksz state. Note that $\mu_{max}(\delta k)$ diverges at $\delta k \approx 0.53$. This divergence is closely related to the existence of the spatially periodic splay-twist (ST) Freedericksz stripes for $\delta k \gtrsim 0.53$ in the absence of the flexotorque ($e_1 - e_3 = 0$) for $\epsilon_a > 0$.[3,27,28] The ST director state (known for more than two decades and shown as ST in the upper right corner of Fig. 4.2) expands into the range for flexodomains for non-zero flexoelectricity and is discussed in more detail in Krekhov *et al.*[26]

On the other hand, the existence regime for flexodomains is limited from below by the lower limit curve $\mu_{min}(\delta k) < 0$ (the solid line in Fig. 4.2). Since the electric torque is stabilizing for $\epsilon_a < 0$ ($\mu < 0$) in the planar geometry such a line must exist, at which the destabilization of the basic state by the flexotorque becomes impossible. Since with decreasing $\delta k$ the relative strength of the flexotorque increases,[26] $\mu_{min}(\delta k)$ decreases leading to $\mu_{min}(\delta k) \to -\infty$ for $\delta k \to -1$. Figure 4.2 also demonstrates that in the one-elastic-constant approximation ($\delta k = 0$, $K_1 = K_2$) the limits $\mu_{min}(0) = -1$ and $\mu_{max}(0) = 1$ in Eq. (4.3) are recovered.

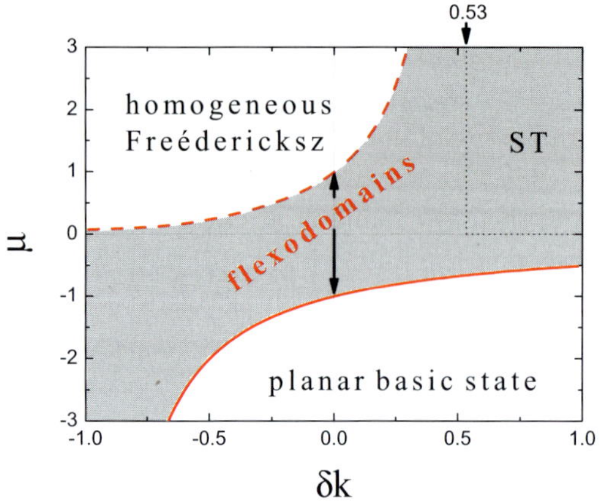


Fig. 4.2. Upper and lower limit curves, $\mu_{max}(\delta k)$ (dashed line) and $\mu_{min}(\delta k)$ (solid line), respectively, determining the range for flexodomains (grey region) in the $(\delta k, \mu)$ plane. The rectangular region marked as ST shows the range for the spatially periodic splay-twist Freedericksz structure for $\delta k \gtrsim 0.53$, $e_1 - e_3 = 0$ and $\epsilon_a > 0$. The double arrow indicates in the one-elastic-constant approximation ($\delta k = 0$) the range for flexodomains ($|\mu| < 1$).

Rigorous closed expressions for $U_c$ and $q_c$, as given in Eq. (4.3) for the special case $\delta k = 0$, do not exist for the general case of anisotropic orientational elasticity ($K_1 \neq K_2$). Some characteristic curves for typical material parameter sets for nematics are given in Krekhov *et al.*[26] Qualitatively the $U_c$, $q_c$ curves as functions of $\mu$ look very similar to the ones in Eq. (4.3) for $\delta k = 0$, except that they are supported on a larger interval $\mu_{min}(\delta k) < \mu < \mu_{max}(\delta k)$. As long as $|\delta k| < 0.3$ and $|\mu| < 0.3$ the deviations of $U_c(\mu)$ and $q_c(\mu)$ from the values for $\delta k = 0$ [Eq. (4.5)] are quite small. This applies, for instance, to the nematic **Phase 4** where $\delta k = 0.21$ and $\mu = -0.3$, as follows from May *et al.*[29]

For completeness we would like to mention that a recent approximate numerical analysis[30] of the flexodomains in the DC case taking into account anisotropic elasticity reproduces the general features of the rigorous calculations.[26] Note, however, that an analytical treatment of the problem[31,32] suffers from a methodological error as has been pointed out recently.[26,33] Therefore the expressions for $U_c$ and $q_c$ given by Marinov and Hinov[31,32] for $\delta k \neq 0$ are not correct.

For an AC driving voltage flexodomains have been analysed using numerical solutions of the linear director (torque balance) equations.[26] In general, the existence of flexodomains for a specific material parameter set in the DC case seems to be a necessary prerequisite for their existence when an AC voltage is applied. The analysis has been restricted to the low-frequency range $\omega\tau_\mathrm{d} < 20$, which corresponds to frequencies $f = \omega/2\pi$ up to 20 Hz for a 10-$\mu$m-thick **MBBA** layer.[26] (The director relaxation time $\tau_\mathrm{d}$ is defined in the following section.) There seems to be no need to study flexodomains for larger $\omega$, since in all situations studied so far they are replaced either by the equilibrium Freedericksz state or by an EC pattern.

The solutions of the linear equations for the director distortions $n_z(y,z,t)$ and $n_y(y,z,t)$ are periodic in the $y$ direction as in the DC case, but depend explicitly on time. Since the linear equations are invariant against a time shift by half a period, $\mathcal{T} : t \to t + \pi/\omega$, we have two classes of solutions characterized by $\mathcal{T} n_z(t) = \mathfrak{p} n_z(t)$ with $\mathfrak{p} = \pm 1$. Which symmetry class is realized for flexodomains depends on $\omega$. The case $\mathfrak{p} = 1$ defines solutions with the so-called 'conductive symmetry', where the time average of $n_z$ over one period is finite; the complementary case $\mathfrak{p} = -1$ corresponds to solutions with 'dielectric symmetry' where the time average of $n_z(t)$ vanishes. The time symmetry of the in-plane director component is opposite, i.e. $\mathcal{T} n_y(t) = -\mathfrak{p} n_y(t)$.

As a consequence the limit $\omega \to 0$ is not smooth, since in the DC case ($\omega = 0$) the time average of both $n_z$ and $n_y$ are finite. A closer look at the director dynamics in the AC case shows that non-zero values of $n_z(t)$ and $n_y(t)$ appear only during a very small fraction of the period $T = 2\pi/\omega$. This means that the patterns appear in the experiments only for very short time intervals as a flash. This phenomenon has indeed been observed in the experiments discussed in Section 4.4.

The dependence of $U_c$ and $q_c$ for flexodomains on the strength of the flexocoefficients, on the frequency $\omega$ and on the elastic constants was discussed by Krekhov *et al.*[26] In general both $U_c$ (see Fig. 4.5 below) and $q_c$ rise very steeply as functions of $\omega$. Thus for $\mu > 0$ ($\epsilon_a > 0$) the bifurcation to the homogeneous Freedericksz state (with an $\omega$ independent critical voltage) will prevail for very small $\omega$. The competition with standard EC will be discussed in Section 4.4. In view of the discontinuities of $n_z(t)$ and

$n_y(t)$ at $\omega \to 0$ discussed before it is not surprising that $U_c(\omega)$ and $q_c(\omega)$ are also discontinuous at $\omega \to 0$.[26] Note, that this was not anticipated in earlier investigations. As a caveat, however, it is worth mentioning that these discontinuities are much less expressed when using a square waveform instead of a sinusoidal one.

There has been only one investigation of flexodomains in the weakly nonlinear regime for $U \gtrsim U_c$ in the DC case.[34] Based on a clever variational *ansatz* for the director distortion it has been demonstrated that the director amplitudes grow continuously as $(U - U_c)^{1/2}$ (forward bifurcation). As a by-product an approximate expression for the $\mu_{max}(\delta k)$ shown in Fig. 4.2 was obtained.

So far we have only considered strong anchoring of the director at the confining plates. The case of weak anchoring, where the director orientation at the plates is sensitive to the distortions in the bulk, has been considered[35,36] for $K_1 = K_2$. No qualitatively new scenarios only quantitative corrections of $U_c$ and $q_c$ were predicted. These depend on two additional material parameters to describe the 'surface potential' of the director, which in most cases have not been measured.

*Experimental observations* of the flexodomains have been made in *planar geometry* for various nematics either with $\epsilon_a < 0$ (where the Freedericksz transition is excluded)[11,12,29,37,38] or with $\epsilon_a > 0$.[11,12,34,39] Precise comparison of the theory of flexodomains with experiments requires a knowledge of the material parameters $\epsilon_a$, $K_1$ and $K_2$. These have been measured independently for some nematics like **MBBA**, **Phase 4**, **Phase 5**. However, since these parameters are typically not measured *in situ* (in the same cell), some scatter of their values in different experiments cannot be excluded: the compounds may come from different manufacturers, could be contaminated, etc. The situation with the flexocoefficients is much more unsatisfactory (see, e.g. Chapter 2 in this book[40]). First of all their direct measurements are quite complicated and are thus rare. In addition the results of different measurements deviate substantially in almost all cases. Thus the flexocoefficients are often used as fitting parameters in the comparison of theory and experiments. We will postpone a more detailed discussion of the flexocoefficients to Section 4.5, after having discussed EC patterns and their possible competition with flexodomains at low $\omega$.

Flexodomains have also been observed and analysed in twisted geometry (with a twist angle $\pi/2$) in **BMAOB** (4-n-butyl-4′-methoxyazoxybenzene) in DC electric fields.[41] The flexodomains are oriented almost parallel to the initial director orientation in the midplane of the cell (i.e. at an angle of $\pi/4$). The threshold voltage was found to be slightly higher than for the planar oriented sample. In addition the instabilities for twist angles in the range 0–5 $\pi/2$ were studied by adding a small amount of a cholesteric compound to **BMAOB**. The onset characteristics of the flexodomains in twisted geometry are well described by the theory.[41]

Flexoelectric patterns also exist for nematic layers with asymmetric boundary conditions, i.e. with homeotropic anchoring on one surface and planar anchoring on the other one (*hybrid-aligned nematics*).[42,43] The critical voltage and the critical wave number obtained with the one-elastic-constant approximation are in a good agreement with experimental results.[42]

Besides the flexodomains, Hinov and his co-workers demonstrated the existence of another flexoelectrically driven stripe pattern in nematics, which they described as *flexo-dielectric walls*.[44–46] Though these patterns, which have mostly been investigated in **MBBA** and in **BMAOB**, are also oriented parallel to the preferred director alignment, they seem to differ substantially in other respects from the flexodomains discussed above.[45] The flexo-dielectric walls have been observed under special boundary conditions, e.g. pretilt of the director and weak surface anchoring. A finite electrical conductivity and apparently a negative dielectric anisotropy of the nematics used in the experiments seems to be crucial. Moreover, it is reported that the flexo-dielectric walls are more concentrated near one of the electrodes depending on the polarity, in clear contrast to the standard flexodomains residing in the bulk of the nematic layer. Since the experimental data are so far insufficient to clarify the origin of the flexo-dielectric walls and convincing theoretical models are still missing, flexo-dielectric walls will not be discussed in detail.

## 4.3. Dissipative Structures: Electroconvection

Electroconvection in nematics is certainly a prominent paradigm for non-equilibrium pattern-forming instabilities in anisotropic systems. As mentioned in the introduction, the viscous torques induced by a flow field are decisive. The flow field is caused by an induced charge density $\rho_{el}$ when the director varies in space. The electric properties of nematics with their quite low electric conductivity [$\sim 10^{-8}$ $(\Omega\ \mathrm{m})^{-1}$] are well described within the electric quasi-static approximation, i.e. by charge conservation and Poisson's law. Thus $\rho_{el}$ is determined by the equations:

$$\frac{d}{dt}\rho_{el} + \nabla \cdot \boldsymbol{j}_{el} = 0, \qquad \rho_{el} = \nabla \cdot (\epsilon_0 \boldsymbol{\epsilon} \boldsymbol{E} + \boldsymbol{P}_{\text{fl}}), \tag{4.7}$$

where the current density is given as $\boldsymbol{j}_{el} = \boldsymbol{\sigma}\boldsymbol{E}$. In analogy to $\boldsymbol{\epsilon}$ the electrical conductivity tensor $\boldsymbol{\sigma}$ is characterized by the conductivities $\sigma_{\|}$ and $\sigma_{\perp}$.

The physical phenomena involved in electroconvection are characterized by three different time scales: the director relaxation time $\tau_{\text{d}} = \frac{\gamma_1 d^2}{K_1 \pi^2}$, the charge relaxation time $\tau_{\text{q}} = \frac{\epsilon_0 \epsilon_{\perp}}{\sigma_{\perp}}$ and the viscous relaxation time $\tau_{\text{v}} = \frac{\rho d^2}{\alpha_4/2}$. Here $d$ is the layer thickness, $\rho$ is the mass density, $\gamma_1$ is the rotational viscosity and $\alpha_4/2$ is the isotropic viscosity. Typically $\tau_{\text{d}} \gg \tau_{\text{q}} \gg \tau_{\text{v}}$,[16] such that the flow dynamics follows adiabatically the dynamics of the director and of the electric charges. As in most studies in this field, we will focus on the EC instability driven by a sinusoidal AC electric field, where the circular frequency $\omega = 2\pi f$ serves as an important control parameter besides the voltage amplitude. We will exclusively concentrate on anisotropic EC in planar geometry, which has been discussed in a number of reviews in recent years (see for example Buka and Kramer,[1] Kramer and Pesch,[16,47] Buka *et al.*[48]). According to the seminal ideas of Carr[13] and Helfrich[14] the mechanism underlying EC is most transparent in the planar geometry when the electric anisotropies fulfil the conditions $\epsilon_a < 0$ and $\sigma_a = (\sigma_{\|} - \sigma_{\perp}) > 0$. In this case and for $\omega\tau_{\text{d}} \gg 1$ the 'standard model' without flexoelectricity is sufficient to understand the main features of EC. The roll patterns are usually oriented perpendicular (normal rolls, Fig. 4.3a), or nearly perpendicular (oblique rolls, Fig. 4.3b) to the preferred direction $\boldsymbol{n}_0 \parallel \boldsymbol{x}$. The latter appear usually at low frequencies below the so-called Lifshitz frequency $\omega_L$ and are replaced by the normal rolls above $\omega_L$. Oblique rolls have two symmetry degenerate species: *zig* and *zag*. They appear either in separated patches as in Fig. 4.3b or are superimposed leading to rhomboidal grid patterns[49] (not shown here). In analogy to the flexodomains, we find EC patterns both with conductive temporal symmetry (finite time average of $n_z$) and with dielectric symmetry (zero time average of $n_z$). Conductive symmetry is observed below a cut-off frequency $\omega_c$ above which dielectric symmetry prevails.

Regarding the impact of flexoelectricity, the initial theoretical investigations basically focused on the conductive regime[50,51] in planar electroconvection for $\epsilon_a < 0$ and $\sigma_a > 0$. The resulting changes arising in the AC threshold voltage $U_c$ are small, in contrast to the DC threshold voltage of EC, which changes by about 25% for the material parameters of the nematic

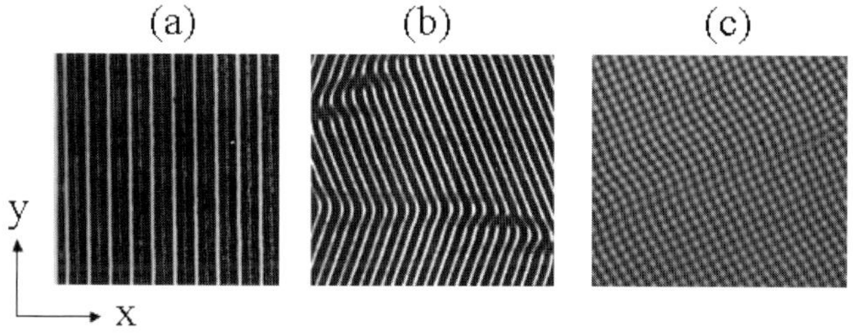


Fig. 4.3. Photographs of (a) normal rolls, (b) oblique rolls and (c) squares.

**MBBA**.[50] The wave number $|\boldsymbol{q}_c|$ of the patterns is not much influenced by the inclusion of flexoelectricity in contrast to considerable changes with respect to the direction of the wave vector.

In the high-frequency (dielectric) regime, however, the flexoeffect is much more important, as has been demonstrated recently. It considerably modifies the threshold and the critical wave vector.[26,52,53] In general the inclusion of the flexoeffect also breaks temporal symmetry: in the conductive regime contributions due to dielectric symmetry appear and vice versa in the dielectric regime. We will give more details in Section 4.3.1. In addition, we will discuss why flexoelectricity is important for understanding EC for very low frequencies ($\omega\tau_{\rm d} \lesssim 1$) and thin cells ($d < 10\ \mu$m).

For the parameter combination $\epsilon_a < 0$ and $\sigma_a < 0$, which can be found in some nematic compounds, electroconvection is definitely excluded within the standard model. Nevertheless, EC has surprisingly been observed in this case (for recent examples see, e.g. Kochowska *et al.*[17] and Tóth-Katona *et al.*[18]). The theoretical analysis has proved that flexoelectricity is crucial for understanding this *non-standard* EC[19] because in Eq. (4.7) the contribution $\nabla \cdot \boldsymbol{P}_{\rm fl}$ to $\rho_{el}$ is dominant. It is interesting that the flexotorque on the director is determined by the difference $(e_1 - e_3)$ of the flexocoefficients while the sum $(e_1 + e_3)$ governs the flexocharge and thus its contribution to the viscous torque. Further details will be sketched in Section 4.3.2.

So far we have discussed EC instabilities driven by a sinusoidal AC voltage. When the AC driving voltage $U(t)$ with period $T$ is 'asymmetric', i.e. $U(t + T/2) \neq -U(t)$, besides the conductive and the dielectric symmetries there is a 'subharmonic' pattern where the director dynamics is $2T$-periodic in time.[54] The effect of flexopolarization on standard and non-standard EC for an asymmetric driving voltage has been analysed.[55,56] One recovers in principle the scenarios with symmetric driving described above, except there are subharmonic patterns in the standard EC.

For completeness we would like to mention that the effects of flexoelectricity have not been analysed for all material parameter combinations, where EC has been observed in nematics. An overview of the various cases has been given in a recent review,[57] which contains a systematic discussion of the sensitive influence of the sign of the anisotropies $\epsilon_a$, $\sigma_a$, and of the basic director configuration on the patterns. For instance, in the homeotropic configuration with $\epsilon_a > 0$ and $\sigma_a < 0$, isotropic convection (where the roll axis is selected spontaneously in the EC bifurcation) has been experimentally observed and theoretically analysed.[58–60] In particular the appearance of square patterns (see Fig. 4.3c), which are typical for isotropic convection, is well understood. In general, one finds satisfactory agreement between theory and experiment for this interesting EC scenario without including flexoelectricity.

### 4.3.1. *Standard electroconvection*

This section deals with the influence of flexoelectricity on electroconvection with planar geometry and the most studied material parameter combination $\epsilon_a < 0$ and $\sigma_a > 0$. The analysis makes use of the common nemato-electrohydrodynamic equations,[2,15] where in addition the flexopolarization is included. This leads to modifications in the electric torques and influences also the charge density $\rho_{el}$ [Eq. (4.7)].[19,47]

In order to consider a specific example, we have mainly used in our analysis the material parameters of **Phase 5**.[52] However, in order to study specifically the effects of flexoelectricity, a 'theoretical' scaling factor $\xi$ has been introduced to tune the strength of the flexoelectricity, i.e. we use $\xi_-(e_1-e_3)$ instead of $(e_1-e_3)$ and $\xi_+(e_1+e_3)$ instead of $(e_1+e_3)$. In Fig. 4.4 the calculated critical voltage $U_c(f)$ as a function of the AC frequency $f$ is shown for three different values of $\xi = \xi_+ = \xi_-$. In line with the standard model, EC with conductive symmetry at low $\omega$ switches to 'dielectric' EC at a certain crossover frequency $\omega_c$ for $\xi = \xi_+ = \xi_- = 0$ (i.e. in the absence of flexoelectricity).

For a finite flexoelectric contribution of realistic magnitude ($\xi = \xi_+ = \xi_- = 1$), the dielectric threshold and the crossover frequency $\omega_c$ decrease, while the conductive threshold is not affected significantly. The results of the experiments on **Phase 5** shown in Fig. 4.4 as circles agree very well with the theory. Measuring frequencies in units of the director relaxation time $\tau_\mathrm{d}$ the range $0 < \omega\tau_\mathrm{d} < 140$ is covered in the figure. A closer look

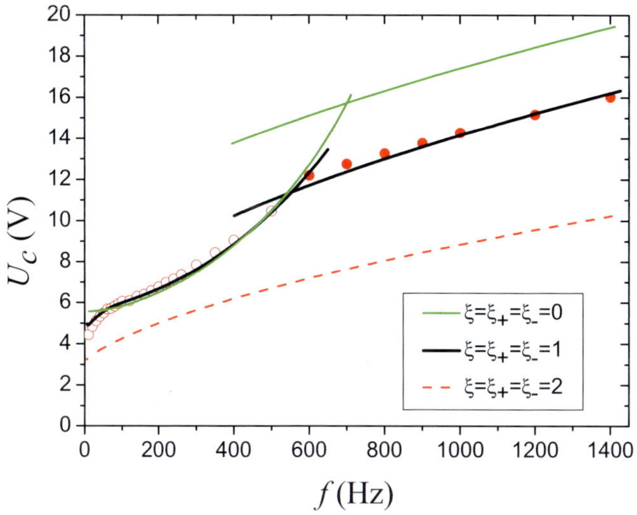


Fig. 4.4. Threshold voltage $U_c$ as a function of the frequency $f$ calculated using **Phase 5** material parameters and for different values of the flexoelectric strength $\xi$. Circles and bullets represent the experimental data for conductive and dielectric EC, respectively.

reveals a further consequence of flexoelectricity: the slope of $U_c(f)$ versus $f$ increases quite strongly at small $f < 100$ Hz.[52] If the flexostrength is further increased the dielectric $U_c$ branch expands further towards small frequencies, until the conductive range totally vanishes and EC with dielectric symmetry bifurcates in the whole frequency range. This is documented in Fig. 4.4 for $\xi = \xi_+ = \xi_- = 2$. Similar theoretical curves for $U_c$ as in Fig. 4.4 have also been obtained when using **MBBA** material parameters.[26] In contrast to **Phase 5** we find, however, oblique rolls in the dielectric regime (for further details see Krekhov *et al.*[26]).

The onset characteristics of planar EC with $\epsilon_a < 0$ and $\sigma_a > 0$ depend strongly on the magnitude of $\sigma_a$. In the absence of flexoelectricity the crossover frequency $\omega_c$ between EC with conductive and dielectric symmetry moves towards $\omega = 0$ when decreasing $\sigma_a$. At the same time the critical voltage $U_c$ diverges. For finite $e_1$, $e_3$ the shift of $\omega_c$ is observed as well but $U_c$ remains finite. This scenario is documented in the upper panel of Fig. 4.5 for **MBBA** material parameters[19] except that a larger flexostrength with $\xi = \xi_+ = \xi_- = 2$ is used and $\sigma_a$ is varied. Time is measured in units of the

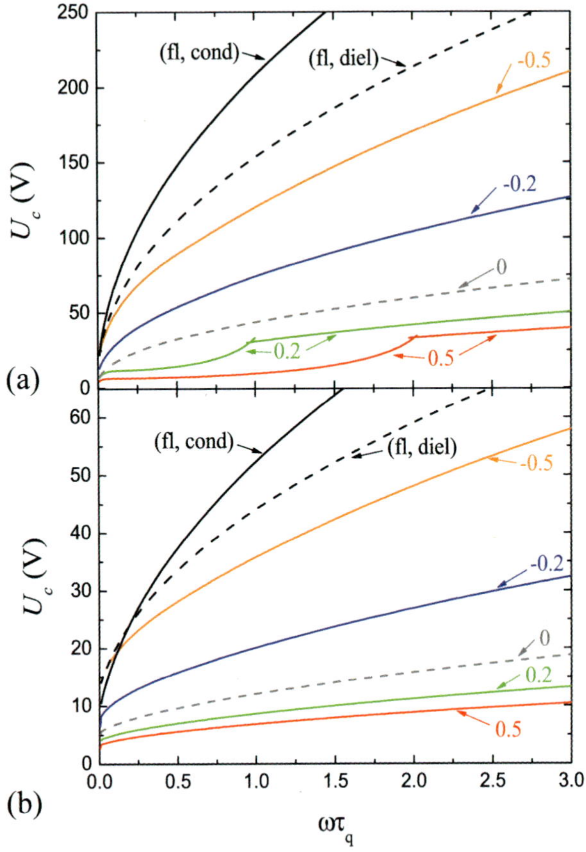


Fig. 4.5. Threshold voltage of EC as a function of the dimensionless frequency $\omega\tau_{\rm q}$ calculated with **MBBA** material parameters and $\xi = \xi_+ = \xi_- = 2$ for five different values of $\sigma_a/\sigma_\perp$ between 0.5 and −0.5. The corresponding critical voltages for the flexodomains with conductive symmetry (fl, cond) and with dielectric symmetry (fl, diel) are included as well. Sample thickness: (a) $d = 40$ $\mu$m; (b) $d = 10$ $\mu$m.

charge relaxation time $\tau_{\rm q} = 4.7 \times 10^{-3}$ s. For $\sigma_a = 0.5$ we have dielectric rolls for $\omega_c\tau_{\rm q} > 2$ and conductive rolls for $\omega_c\tau_{\rm q} < 2$, while $\omega_c\tau_{\rm q} \approx 1$ for $\sigma_a = 0.2$. For $\sigma_a = 0$ we find dielectric rolls in the whole frequency range. On the other hand $U_c$ monotonically increases with decreasing $\sigma_a$. The discussion of the cases $\sigma_a < 0$ is postponed to the following section.

Besides $\sigma_a$, the cell thickness $d$ of the nematic layer, which has almost no effect in the case $e_1 = e_3 = 0$, reveals a strong influence on EC for finite flexocoefficients. This is demonstrated in the lower panel of Fig. 4.5, where for $d = 10$ $\mu$m the conductive branch is totally absent. Then, as with the conductive regime, one can find a transition from oblique to normal dielectric rolls above a Lifshitz frequency $\omega_L$. In a recent experiment the oblique dielectric rolls at small $\omega$ have indeed been observed.[29] The threshold characteristics $U_c$ and $q_c$ and the obliqueness angle $\alpha$ could be well reproduced by a theoretical analysis of the nemato-electrohydrodynamic equations including flexopolarization.[29]

Finally we would like to point out that the magnitude $\sigma_0$ of the electrical conductivity in nematics plays an important role. For Fig. 4.5 we chose $\sigma_0 = 10^{-8}$ $(\Omega\,\text{m})^{-1}$, which is the typical scale of the electrical conductivities in nematics commonly used in experiments. A closer look at the linear nemato-electrohydrodynamic equations (see the Appendix in Krekhov *et al.*[19]) shows that the thickness $d$ as well as $\sigma_0$ appear only through the dimensionless parameter $Q \propto \tau_\text{d}/\tau_\text{q} \propto \sigma_0 d^2$. Thus decreasing $\sigma_0$ by a factor of 16 is equivalent to a reduction of $d$ by a factor of 4 (compare Fig. 4.5a and Fig. 4.5b).

In Fig. 4.5 we have also included the ($\sigma_a$-independent!) critical voltage curves for flexodomains of conductive and of dielectric symmetry. They exist for the standard **MBBA** material parameters[26] except that the flexostrength is increased by a factor of two ($\xi = \xi_+ = \xi_- = 2$). Obviously, for finite frequencies the flexodomains play no role compared to the EC rolls with their much lower $U_c$ values. However, in the limit $\omega \to 0$ both critical voltages decrease and approach each other, which will become important in Section 4.4.

### 4.3.2. *Non-standard electroconvection*

As already stated, electroconvection cannot be explained by the standard model for $\sigma_a < 0$ and $\epsilon_a < 0$. Surprisingly, EC has been observed also for this parameter combination in certain calamitic nematics.[17,18,61–64] These EC patterns differ clearly from the standard EC patterns: the rolls are dominantly parallel to the initial director alignment (see Fig. 4.6a). They are not observable using the common shadowgraph technique (single polarizer) but are by using crossed polarizers (plus sometimes an additional

$\lambda/4$ plate). Investigation of these 'non-standard EC' rolls has recently been intensified.[18,19,26,52,53,65] Of particular importance are materials for which the conductive anisotropy $\sigma_a$ changes from negative to positive as a function of temperature while the other material parameters do not change much and, in particular, the sign of $\epsilon_a$ does not change. Thus by simply increasing the temperature we have a continuous movement from non-standard EC (ns-EC) to standard EC (s-EC).

Figure 4.5 demonstrates that a finite threshold can be obtained for $\sigma_a < 0$ and $\epsilon_a < 0$, when the standard electrohydrodynamic description of nematics is extended by including flexoelectricity; thus non-standard EC can in fact be explained. According to the theory ns-EC is characterized by dielectric symmetry and the ns-EC rolls have an orientation almost parallel to the preferred direction $\boldsymbol{n}_0 \parallel \boldsymbol{x}$, in distinct contrast to s-EC rolls. The critical voltage is predicted to increase almost linearly with $\omega$ in contrast to a rise roughly $\propto \sqrt{\omega}$ in standard dielectric electroconvection. These general trends are in satisfactory agreement with the recent experimental findings. Even quantitatively, theory and experiments in the standard as well as in the non-standard regime match very well in some cases.[19] The material parameters are mostly measured in this case and the flexocoefficients $e_1$ and $e_3$ have been used as fit parameters. It is worth mentioning that in some cases a Hopf bifurcation to travelling ns-EC rolls has been observed.[18] To describe this scenario, flexoelectricity should be incorporated into the weak electrolyte model (WEM),[66] which can explain travelling s-EC rolls. This is certainly a demanding task, which would involve a detailed description of the electrical conductivity in terms of the participating ions, their mobility and their recombination rates.

Non-standard EC has also been observed in bent-core nematics,[67–70] where $\sigma_a < 0$, $\epsilon_a < 0$ is also realized and where flexoelectricity is strong. There is still no detailed theoretical description, which would have to include the strong frequency dependence of $\sigma_a$[67] as well as the unusual viscosity and elastic properties; these might indicate smectic cluster formations not only in the nematic but even in the isotropic phase.[70–74]

Convective patterns were observed some time ago for the $\epsilon_a > 0$ and $\sigma_a > 0$ parameter combination with an initial homeotropic director orientation.[37,75–78] They can be observed in a polarizing microscope (with crossed polarizers) as an arrangement of 'Maltese crosses'. The so-called isotropic mechanism has been proposed to explain this instability. Alternatively, 'Maltese crosses' have been interpreted using the so-called surface polarization mechanism.[79,80] However, the theoretical treatments of both

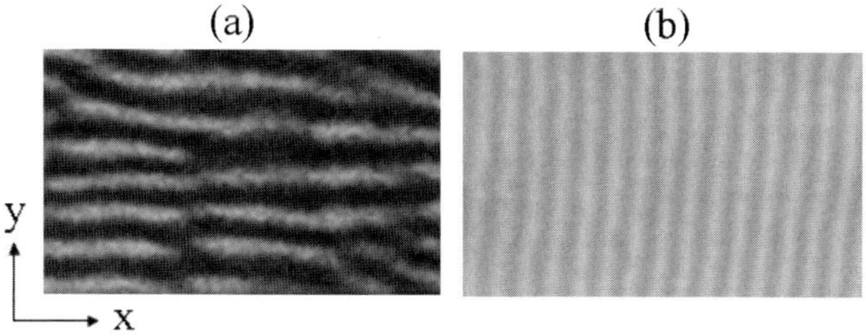


Fig. 4.6. Photographs of two types of non-standard EC patterns in planar nematics ($\boldsymbol{n}_0 \parallel \boldsymbol{x}$) with (a) $\epsilon_a < 0$, $\sigma_a < 0$ and (b) $\epsilon_a > 0$, $\sigma_a > 0$.

mechanisms have not been worked out in sufficient detail to allow for comparison with experiments.

It is unclear why instead of 'Maltese crosses' convective cellular and subsequently roll patterns resulting from a secondary instability were detected in a recent study.[81] Note that these patterns have been observed both in the homeotropic and in planar geometry (for the rolls, see Fig. 4.6b). There is still no theoretical analysis. For a homeotropic alignment the inclusion of flexopolarization in the standard model of EC does not provide the mechanisms needed to drive electroconvection. First, there is no direct flexoelectric torque on the director in this geometry. Second, the flexocharge contribution is too small and the originating viscous torque cannot overcome the strong stabilizing electric torque, since the substances are characterized by a large $\epsilon_a > 0$. In the planar case the electric torque leads at first to the Freedericksz transition from the planar to the homeotropic orientation of the director, except in a thin boundary layer. So we have essentially the homeotropic case where flexoelectricity does not support EC.

## 4.4. Crossover between Flexodomains and Electroconvection

It is very interesting to see that a crossover between the equilibrium flexodomains and the dissipative EC patterns can be observed in the same experiment (planar geometry) by merely increasing the AC frequency $\omega$. Inspection of Fig. 4.5 reveals that at very small $\omega$ and for suitable material parameter combinations of the nematics, the critical voltages of the equilibrium flexodomains and those of the dissipative electroconvection patterns might approach each other. In fact, very recent theoretical and experimental studies on the calamitic nematic **Phase 4**, have demonstrated the existence of a crossover between these two qualitatively different patterns at a very low transition frequency $f_t$.[29]

Before discussing these experiments we will present new measurements[82] using the nematic **Phase 5**, which show a similar crossover at $f_t \lesssim 0.1$ Hz. Thus for $f < f_t$ there are flexodomains as a first instability while for $f > f_t$ there are the usual EC roll patterns with conductive symmetry. Above the Lifshitz frequency $f_L \approx 40$ Hz there are normal rolls, which are replaced by oblique rolls in the region $f_t < f < f_L$. The cell thickness was $d = 11.3$ $\mu$m. Roughly speaking the frequency dependence of $U_c$ is analogous to the lowest curve in Fig. 4.5a. Detailed studies regarding the main characteristics [such as $U_c(f)$ and $\mathbf{q_c}(f)$] of both patterns are in progress.

Regarding the temporal evolution of the pattern, the contrast $C(t)$ is in general time periodic with a frequency twice the external AC frequency $f$. It was, however, not expected that the functional form of $C(t)$ would become increasingly spiky with decreasing $f$ (see Fig. 4.7d). In a recent theoretical study it was shown that this behaviour is generic for flexodomains and also for EC rolls at frequencies $\omega\tau_d < 1$.[26] Both patterns are indeed expressed in Fig. 4.7d, where $C(t)$ is shown over one period $T = 1/f = 10$ s. In the interval $T/2 < t < T$ the first maximum (a) of $C(t)$ corresponds to longitudinal flexodomains (Fig. 4.7a), then the contrast decays and remains on a flat plateau (b) value (Fig. 4.7b), i.e. the nematic is in the quiescent state. Later oblique EC rolls appear with a second steep maximum (c) of $C(t)$ (Fig. 4.7c). The whole sequence repeats itself periodically. At the moment it is unclear why in particular the flexodomain peak in the interval $0 < t < T/2$ is much less expressed than in the interval $T/2 < t < T$. Asymmetries in the boundary conditions (e.g. a pretilt) come immediately to mind. Moreover, even a small DC offset in the applied AC voltage, which practically would not change the threshold values of $U_c$ and $\boldsymbol{q}_c$, could lead to a strong asymmetry of the $C(t)$ peaks over one AC period in the spiky regime.

The transition between the flexodomains and the conductive oblique EC rolls within the period $T$ has also been confirmed in Fourier space by light diffraction experiments.[82] The incoming light beam had nearly normal incidence and the fringes were recorded on a screen at a distance of about 0.6 m from the sample. Flexodomains (Fig. 4.7a), which are oriented along $\boldsymbol{n}_0 \parallel \boldsymbol{x}$, produce an obvious set of fringes along the $y$ axis as shown in Fig. 4.8, left panel. The zig and zag oblique roll patches visible in Fig. 4.7c are directly responsible for the set of fringes along the two lines through the origin at an angle $2\alpha$; here $\alpha$ denotes the angle between the wave vector

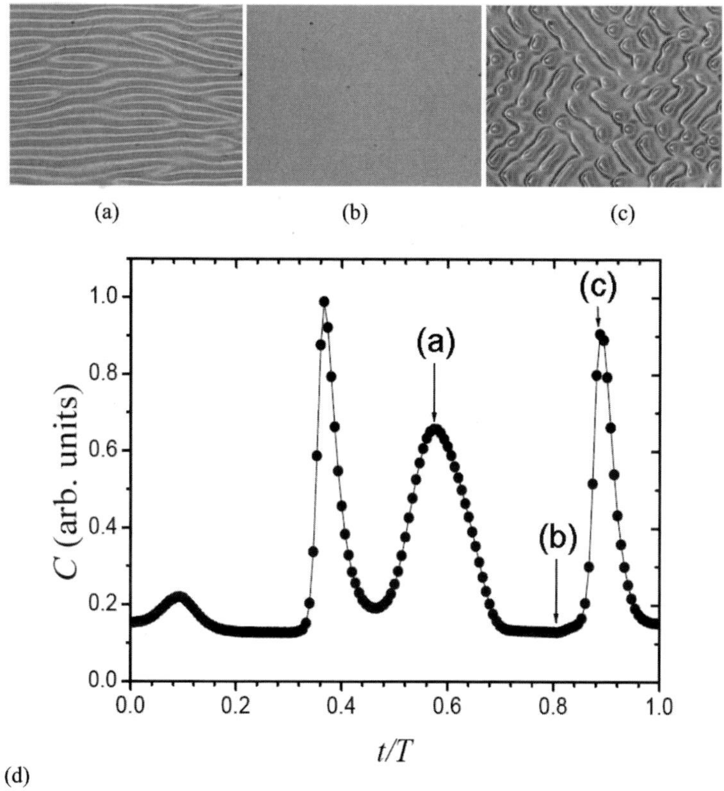


Fig. 4.7. Flexodomains (a), quiescent state (b) and conductive oblique EC rolls (c) in **Phase 5**. The initial director is horizontal. (d) Temporal evolution of the contrast $C$ within a period $T$ of the driving voltage above the onset of instabilities at $f = 0.1$ Hz, close to the transition frequency $f_t$. Arrows indicate the times when the photographs in the corresponding subfigures were taken.

of the rolls and the $x$ axis (Fig. 4.8, right panel). The fringes along the $x$ and $y$ axes in Fourier space (Fig. 4.8, right panel) are due to non-linear effects; they correspond to the sum and difference of the zig and zag wave vectors.[49] There is still no detailed theoretical analysis of the competition between flexodomains and the conductive EC rolls.

The competition between *dielectric* oblique EC rolls and flexodomains was recently studied using the nematic **Phase 4**,[29] which has a low electrical conductivity $\sigma_0$. The cell thickness was $d = 11.4$ $\mu$m. Obviously the constellation of the various material parameters is such that the scenario in Fig. 4.5b is realized, where only dielectric EC rolls exist near the onset

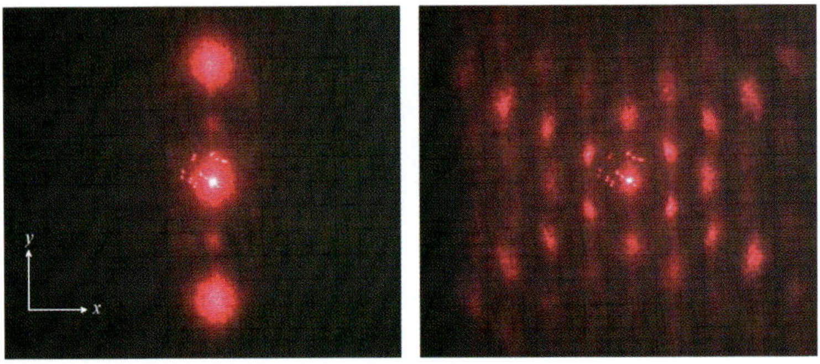


Fig. 4.8. Light diffraction patterns for flexodomains (left) and for conductive oblique EC rolls (right) at different instants within the same period of the driving voltage, slightly above the onset of the instabilities and at $f = 0.1$ Hz (close to the transition frequency $f_t$).

for positive values of $\sigma_a/\sigma_\perp$. For frequencies above the Lifshitz frequency, $f_L = 50$ Hz, there are normal dielectric rolls and for $f < f_L$ oblique ones. The transition to flexodomains is observed at $f_t \approx 0.1$ Hz. A recent theoretical analysis[26] has demonstrated that such a scenario is possible within the nemato-electrohydrodynamic equations including flexopolarization.

In Fig. 4.9 we show the critical voltage $U_c$, the critical wavelength $\lambda$ and the angle of obliqueness $\alpha$ of the rolls at onset. The latter quantity reflects in particular the transition from the longitudinal flexodomains ($f < f_t$) with $\alpha = 90°$ to the oblique EC rolls ($f > f_t$) with much smaller $\alpha$.

Since not all of the material parameters are known for **Phase 4**, fitting procedures were used in the theoretical analysis. In particular the flexocoefficients were chosen in such a way that the theoretical curves for the critical data ($U_c$, $\boldsymbol{q}_c$) at larger $f$, i.e. for 5 Hz $< f <$ 200 Hz agreed very well with the experimental ones (not shown here). It was very satisfactory to see an internal consistency: the experimental and the theoretical DC critical voltages for flexodomains that were calculated with the same parameters agreed very well.

It should be stressed that the negative slope of $U_c$ for the EC rolls in Fig. 4.9a when approaching $f_t$, cannot be described by the existing theories. It seems to be very plausible that at very small $f$ the measured $U_c$ has to be corrected by contributions due to alignment coatings on the confining electrode plates and from Debye screening layers at the boundaries. Some additional remarks are postponed to the next section.

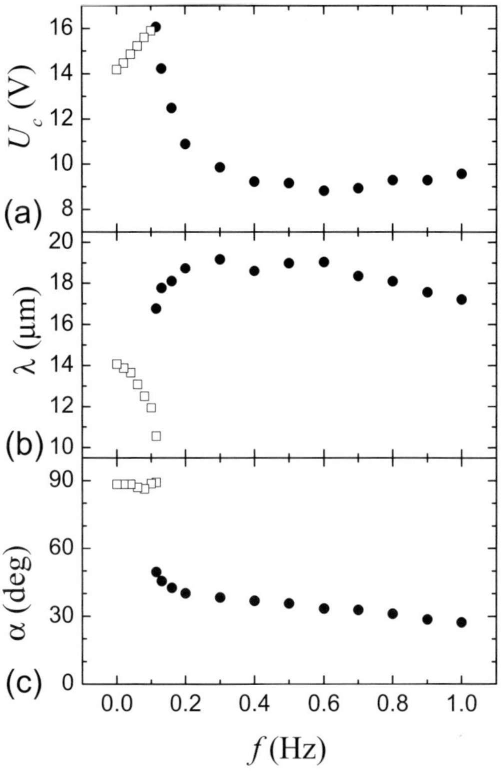


Fig. 4.9. The threshold voltage $U_c$ (a), the wavelength $\lambda$ (b) and the angle of obliqueness $\alpha$ (c) as a function of the driving frequency $f$ for flexodomains (□) and for oblique dielectric EC rolls (•) in **Phase 4**.

## 4.5. Discussions and Conclusions

Convection instabilities in simple isotropic fluids, like Rayleigh–Bénard convection (for a recent review see Bodenschatz *et al.*[83]), are completely understood near their onset because the number of relevant material parameters is fairly small. In contrast, pattern-forming instabilities in nematic liquid crystals are very rich as documented in this review; their detailed exploration, however, requires a demanding numerical analysis of the complex nemato-electrohydrodynamic equations and therefore values for many more material parameters. Some of these like the dielectric permittivity or the electrical conductivity are relatively easy to measure, but this is not the case, however, for the five viscosity coefficients $\alpha_i$ in nematics.[2] Most difficult is the determination of the flexocoefficients, which have a central role in this chapter. In general, the flexocoefficients have been obtained only for a few substances by purposefully directed measurements. Therefore we will discuss how the analysis of the pattern-forming instabilities described in this chapter can contribute to the determination of the flexocoefficients.

We note first that according to Eqs (4.1) and (4.2), simultaneously reversing the signs of $e_1$, $e_3$ and $\boldsymbol{E}$ does not change the flexoelectric torque on the director. The same holds for the total torque too, as the elastic and viscous contributions are independent of $\boldsymbol{E}$, while the dielectric one depends only on $\boldsymbol{E}^2$. Consequently, the resulting patterns are also invariant with respect to the same transformation. On the other hand, inspection of the general electrohydrodynamic equations shows that at least the threshold characteristics of the flexodomains (and also of the electroconvection patterns) are independent of the sign of $\boldsymbol{E}$ and hence these characteristics would also remain unaltered if the signs of both flexocoefficients are inverted together. Consequently only the relative signs of $e_1$ and $e_3$ can be extracted from an analysis of the flexoelectric or EC patterns.

Let us first concentrate on the flexodomains. With the help of the general theory[26] $|e_1 - e_3|$ can be determined from the measurements of $U_c$ and $q_c$ of the flexodomains if the elastic coefficients $K_1$ and $K_2$ and the anisotropy $\epsilon_a$ are known. In this way the value of $|e_1 - e_3| = 5.7$ pC/m was obtained[11,12] for **BMAOB**. Note that this value has been confirmed for twisted cell geometry.[41] In a later study **BMAOB** doped with a 'swallow-tail' compound in concentrations of up to 4.55 mole% was analysed.[84] The value of $|e_1 - e_3|$ for the mixture was found to slightly increase with increasing concentration of the 'swallow-tail' compound (see Fig. 6 in Marinov *et al.*[84]).

It is interesting that flexodomains have also been detected in a *metallo-organic* nematic liquid crystal, which was made by complexation with palladium.[85] The value of $|e_1 - e_3|$ was found to be of the same order of magnitude as in the usual nematics.[25,78]

As indicated in Chapter 2 of this book,[40] different measurements may show large discrepancies in the values of the flexocoefficients. This becomes evident when one considers for instance **MBBA**, one of the most used nematics in our context. According to the 'standard' data set[19,50] where $e_1 - e_3 = 4$ pC/m and $\delta k = 0.23$, we do not expect flexodomains since $\mu = -1.6 < \mu_{min}(\delta k) = -0.8$ as shown in Fig. 4.2. In fact flexodomains have not been reported in most experiments. In a few cases, however, flexodomains have been observed for **MBBA** both in the standard planar geometry[46,86] as well as in the hybrid (planar-homeotropic) configuration,[42] where they should be excluded as well. If we increase, however, the difference $|e_1 - e_3|$ by a factor of about two, flexodomains are immediately allowed ($\mu = -0.4 > \mu_{min}$ as shown in Fig. 4.2). A report comparing the various measurements of $e_1$ and $e_3$ in **MBBA**[87] as well as Table A.1 in Appendix A shows that the smallest values of $|e_1 - e_3|$ and the largest ones differ by a factor of about 4.5.

For completeness we would like to mention some studies on flexodomains in **BMAOB** under the *combined action of DC and high frequency AC voltages*.[32,88] It was found that the critical DC voltage and the critical wave number of the patterns increased when increasing the AC voltage. On the basis of the standard director equations these findings are very plausible. The quantitative analysis in Hinov and Marinov[32] and Marinov *et al.*,[88] however, suffers from the same mistake as reported recently for the pure DC case.[33] Thus the estimation of the flexocoefficients in these papers is not convincing.

Another complication is the finite, though very small electric conductivity of most nematics, which easily leads to material flow. This has been systematically measured for flexodomains in the hybrid geometry[42] above the onset by following the trajectories of tracer particles. In general we expect that the existence of flow will lead to perturbations of the ideal flexodomain patterns for instance by the generation of defects such as dislocations and disclinations.

As mentioned, measurements of the critical voltage $U_c$ and the critical wave number $q_c$ of flexodomains in the DC case allow us to obtain an estimate of the magnitude of $e_1 - e_3$ but not its sign. In principle much more information is contained in the EC pattern, since the frequency serves as an important secondary control parameter besides the voltage amplitude, and the sum $e_1 + e_3$ can also be tested. As documented in Section 4.3.1 for s-EC, the location of the Lifshitz frequency or the sensitive dependence of the $U_c$ and $\boldsymbol{q}_c$ curves on $\omega$ in the dielectric regime give valuable insights. Moreover, materials that have a transition between flexodomains and EC, are of particular interest, since both the difference and the sum of $e_1$ and $e_3$ can be extracted.

For instance from fits (based on the extended standard model of EC including flexoelectricity) to the experimental EC threshold voltages $U_c(\omega)$ for **Phase 4** we found $e_1 - e_3 = -4.7$ pC/m and $e_1 + e_3 = -31.5$ pC/m.[29]

The difference $e_1 - e_3$ corresponds to $\mu = -0.3$ ($\delta k = 0.21$), which is inside the existence regime of the flexodomains (see Fig. 4.2) and indeed flexodomains were seen. From independently measuring the $U_c$ and $q_c$ values, $|e_1 - e_3| = 4.7$ pC/m and $|e_1 - e_3| = 4.1$ pC/m were obtained, respectively; they match nicely with the EC fit value. From similar EC fits we found for **Phase 5** the values $e_1 - e_3 = -2.9$ pC/m and $e_1 + e_3 = -50.1$ pC/m, and for **MBBA** $e_1 - e_3 = 6.0$ pC/m and $e_1 + e_3 = -35.0$ pC/m.[52] In this latter case the difference $e_1 - e_3$ for **MBBA** corresponds to $\mu = -0.6$, which is slightly larger than $\mu_{min} = -0.8$; thus we would be inside the existence regime of the flexodomains (see Fig. 4.2). Note that $U_c$ and $\boldsymbol{q}_c$ of the EC pattern are insensitive against the sign inversion of both flexocoefficients, as mentioned earlier in this section. Thus the EC fits can provide only the relative signs of $e_1 - e_3$ and $e_1 + e_3$.

Finally, we mention that the longitudinal rolls observed in a *bent-core* nematic have been associated by Wiant *et al.*[67] with ns-EC patterns. In a recent paper,[89] however, similar patterns were interpreted as flexodomains and in this way the value of $|e_1 - e_3| \approx 6$ pC/m was estimated. This value is of the same order of magnitude as in calamitic nematics. Similarly, for the flexocoefficients of another bent-core nematic a value of $O(10$ pC/m$)$ was obtained by studying the instability due to the surface polarization mechanism.[90]

In any case, EC measurements should be used as a test bed and a consistency check for the various methods described in Chapter 2 of this book[40] where the flexocoefficients are directly measured by appropriate deformations of a homogeneous nematic layer.

While the linear description of flexodomains that yields $q_c$ and $U_c$ is quite well worked out in theory, the weakly non-linear regime and the secondary bifurcations from the flexodomains have not yet been explored. This is certainly a rewarding problem in view of the interesting patterns observed for instance in experiments in a DC electric field. In addition, the secondary bifurcations are much more sensitive to the values of the material parameters as the first instability. If the nematic is practically an insulator, the flexodomains have been reported as being stable for applied DC voltages $U$ considerably larger than $U_c$.[8] For higher electrical conductivities, however, the flexodomains were found to be destabilized by EC rolls in **MBBA** when increasing the DC voltage.[42,44,86] As expected, the EC rolls are oriented perpendicular to the preferred director orientation, i.e. perpendicular to the orientations of the flexodomains (see, e.g. Fig. 9 in Hinov and Vistin[44]). Without detailed calculations this scenario allows for a simple qualitative explanation. The optimal action of the Carr–Helfrich mechanism is prohibited by the distortions of the underlying director structure in the flexodomains where $n_z$ and $n_y$ vary periodically along the $y$ direction. The $n_y$ modulations are considered to be less prohibitive for EC as evident from the existence of the so-called *abnormal* EC rolls.[48,91] They are characterized by the common, spatially periodic $n_z$ distortions in the presence of a spatially homogeneous $n_y$ distortion. Thus it is not surprising that the transverse EC rolls are restricted to the regions where $|n_y| > |n_z|$ and thus would not run continuously through the flexodomains. Note that the same scenario has also been observed in the secondary bifurcations of the flexodomains in the hybrid geometry.[42]

The analysis of an alternation of flexodomains and EC patterns at low AC frequencies discussed in Section 4.4 is much more complicated. For frequencies $f \approx f_t$, where the threshold voltages of both patterns are near to each other, one expects them to flash up independently at the onset. With increasing voltage the flexodomains and rolls will start to interact, apart from the fact that each pattern type might develop its own secondary instabilities. To disentangle these processes is certainly a very demanding task both in theory and in experiment.

In the theoretical analysis of electrically driven pattern formation in nematics one deals only with the 'theoretical' AC voltage $U_{theo}$, which drops over the nematic layer. $U_{theo}$ differs, however, from the 'experimental' voltage $U_{exp}$ applied to the whole LC cell and recorded in experiments. Thus a quantitative comparison between experiments and theory is far from trivial as has been emphasized for instance by Krekhov *et al.*[26] Typical liquid crystal cells consist of a nematic layer confined between ITO- or $SnO_2$-coated glass plates covered with a thin film of an aligning polymer. As the polymer is a quite good insulator this sandwich has fairly complicated electric properties. In particular, at low frequencies the whole system has to be represented by a complex equivalent electric circuit model.[92,93]

Furthermore, the ionic character of the electrical conductivity of nematics, which in many cases is satisfactorily described through a simple ohmic conductivity, is certainly important at low $\omega$. Debye screening layers build up in the nematic near the electrodes, where part of the applied voltage drops as well. This happens only if the ions are able to follow the external driving. It is easy to see that a new characteristic time $\tau_{\text{mig}} \propto d^2$, which is needed to form the Debye layer, is relevant.[66] The Debye layer and the voltage drop across it are considered to be important only for frequencies $\omega\tau_{\text{mig}} < 1$. This additional voltage drop is most probably relevant in explaining the large negative slope of the measured $U_{exp}$ at low frequencies and small $d$.[29,52] Given that $U_{theo}$ mostly decays monotonically with decreasing $\omega$ (for an exception see the next paragraph below), a minimum of $U_{exp}$ at $\omega_{min} \sim \tau_{\mathrm{mig}}^{-1} \propto d^{-2}$ is expected. In fact a minimum and a subsequent large increase of $U_{exp}(\omega)$ has been observed in experiments.[29,52] For thin cells the proportionality $\omega_{min} \propto d^{-2}$ has been confirmed as well. The question remains whether this thickness dependence really originates from $\tau_{\mathrm{mig}}^{-1}$ or from the inverse director relaxation time ($\tau_{\mathrm{d}}^{-1} \propto d^{-2}$) as proposed by Tóth-Katona *et al.*[52] There the relation $\omega_{min} \approx 2\pi\tau_{\mathrm{d}}^{-1}$ was found for $2\ \mu\mathrm{m} \leq d \leq 10\ \mu\mathrm{m}$ in **Phase 5**. In May *et al.*,[29] however, a much smaller proportionality factor was measured in **Phase 4** since $\omega_{min} \approx 0.2\tau_{\mathrm{d}}^{-1}$ for $d = 11.4\ \mu\mathrm{m}$.

Note that in Tóth-Katona *et al.*[52] the increase of $U_{exp}$ was tentatively associated with increase of $U_{theo}$ at low $\omega$. Such a dependence of $U_{theo}$ has been found in simulations only for small ratios $\sigma_a/\sigma_\perp \lesssim 0.1$, however, the nematic used by Tóth-Katona *et al.*[52] had $\sigma_a/\sigma_\perp = 0.7$.

In summary, we have shown that flexopolarization has a strong impact on the pattern-forming instabilities in nematics subjected to the action of an electric field. This applies not only to flexodomains but also to EC patterns. At present the theoretical analysis of the nemato-electrohydrodynamic equations in the linear regime allows us to calculate the critical voltage and the critical wave vector of the patterns. For better quantitative comparison with experiment one would need a more precise knowledge of the various material parameters, in particular, of the crucial flexocoefficients $e_1$ and $e_3$. For that purpose more experiments with stable nematics, which show both flexodomains and EC patterns, are most welcome. In particular systematic studies with respect to the frequency, the thickness and the electrical conductivity dependencies of the patterns are of great importance. Certainly finding chemically stable nematics with small dielectric anisotropy and low electric conductivity would allow a big step forward.

## Acknowledgements

Financial support by the Hungarian Research Fund under contract No. OTKA K81250 is gratefully acknowledged.

## References


1. A. Buka and L. Kramer, eds., *Pattern Formation in Liquid Crystals*, Springer-Verlag, New York, 1996.
2. P.G de Gennes and J. Prost, *The Physics of Liquid Crystals*, Clarendon Press, Oxford, 1993.
3. F. Lonberg and R.B. Meyer, New ground state for the splay-Freedericksz transition in a polymer nematic liquid crystal, *Phys. Rev. Lett.* **55**(7), 718–721, (1985). doi: 10.1103/PhysRevLett.55.718
4. R.B. Meyer, Piezoelectric effects in liquid crystals, *Phys. Rev. Lett.* **22**(18), 918–921, (1969). doi: 10.1103/PhysRevLett.22.918
5. J. Prost and J.P. Marcerou, On the microscopic interpretation of flexoelectricity, *J. Phys. France* **38**(3), 315–324, (1977). doi: 10.1051/jphys:01977003803031500
6. M.A. Osipov, Chapter 1. Molecular theory of flexoelectricity in nematic liquid crystals. In eds. Á. Buka and N. Éber, *Flexoelectricity in Liquid Crystals. Theory, Experiments and Applications*, Imperial College Press, London, 2012. pp. 9–32.
7. C.V. Brown and N.J. Mottram, Influence of flexoelectricity above the nematic Freedericksz transition, *Phys. Rev. E* **68**(3), 031702/1–5, (2003). doi: 10.1103/PhysRevE.68.031702
8. L.K. Vistin', Electrostructural effect and optical properties of a certain class of liquid crystals and their binary mixtures, *Sov. Phys. Crystallogr.* **15**(3), 514–515, (1970) [*Kristallografiya* **15**(3), 594–595, (1970)].
9. L.K. Vistin', New electrostructural effect in liquid crystals of nematic type, *Sov. Phys. Dokl.* **15**(10), 908–910, (1971) [*Dokl. Akad. Nauk SSSR* **194**(6), 1318–1321, (1970)].
10. Yu.P. Bobylev and S.A. Pikin, Threshold piezoelectric instability in a liquid crystal, *Sov. Phys. JETP* **45**(1), 195–198, (1977) [*Zh. Eksp. Teor. Fiz.* **72**(1), 369–374, (1977)].
11. M.I. Barnik, L.M. Blinov, A.N. Trufanov and B.A. Umanski, Flexoelectric domains in nematic liquid crystals, *Sov. Phys. JETP* **46**(5), 1016–1019, (1977) [*Zh. Eksp. Teor. Fiz.* **73**(5), 1936–1943, (1977)].
12. M.I. Barnik, L.M. Blinov, A.N. Trufanov and B.A. Umanski, Flexo-electric domains in liquid crystals, *J. Phys. France* **39**(4), 417–422, (1978). doi: 10.1051/jphys:01978003904041700
13. E.F. Carr, Influence of electric fields on the molecular alignment in the liquid crystal p-(anisalamino)-phenyl acetate, *Mol. Cryst. Liq. Cryst.* **7**(1), 253–268, (1969). doi: 10.1080/15421406908084876
14. W. Helfrich, Conduction-induced alignment of nematic liquid crystals: basic model and stability considerations, *J. Chem. Phys.* **51**(9), 4092–4105, (1969). doi: 10.1063/1.1672632
15. E. Bodenschatz, W. Zimmermann and L. Kramer, On electrically driven pattern-forming instabilities in planar nematics, *J. Phys. France* **49**(11), 1875–1899, (1988). doi: 10.1051/jphys:0198800490110187500

16. L. Kramer and W. Pesch, Electrohydrodynamic instabilities in nematic liquid crystals. In eds. Á. Buka and L. Kramer, *Pattern Formation in Liquid Crystals*, Springer-Verlag, New York, 1996. pp. 221–256.
17. E. Kochowska, S. Németh, G. Pelzl and Á. Buka, Electroconvection with and without the Carr–Helfrich effect in a series of nematic liquid crystals, *Phys. Rev. E* **70**(1), 011711/1–9, (2004). doi: 10.1103/PhysRevE.70.011711
18. T. Tóth-Katona, A. Cauquil-Vergnes, N. Éber and Á. Buka, Nonstandard electroconvection with Hopf bifurcation in a nematic liquid crystal with negative electric anisotropies, *Phys. Rev. E* **75**(6), 066210/1–12, (2007). doi: 10.1103/PhysRevE.75.066210
19. A. Krekhov, W. Pesch, N. Éber, T. Tóth-Katona and Á. Buka, Nonstandard electroconvection and flexoelectricity in nematic liquid crystals, *Phys. Rev. E* **77**(2), 021705/1–11, (2008). doi: 10.1103/PhysRevE.77.021705
20. Á. Buka and L. Kramer, Linear and non-linear transient patterns in the splay Freedericksz transition of nematics, *J. Phys. II (France)* **2**(3), 315–326, (1992). doi: 10.1051/jp2:1992136
21. Á. Buka and L. Kramer, Theory of nonlinear transient patterns in the splay Freedericksz transition, *Phys. Rev. A* **45**(8), 5624–5631, (1992). doi: 10.1103/PhysRevA.45.5624
22. N. Éber, S.A. Rozanski, S. Németh, Á. Buka, W. Pesch and L. Kramer, Decay of spatially periodic patterns in a nematic liquid crystal, *Phys. Rev. E* **70**(6), 061706/1–8, (2004). doi: 10.1103/PhysRevE.70.061706
23. W. Pesch, L. Kramer, N. Éber and Á. Buka, The role of initial conditions in the decay of spatially periodic patterns in a nematic liquid crystal, *Phys. Rev. E* **73**(6), 061705/1–10, (2006). doi: 10.1103/PhysRevE.73.061705
24. Y.P. Bobylev, V.G. Chigrinov and S.A. Pikin, Threshold flexoelectric effect in nematic liquid crystal, *J. Phys. Coll. (France)* **40**(C3), C3-331–C3-333, (1979). doi: 10.1051/jphyscol:1979364
25. S.A. Pikin, *Structural Transformations in Liquid Crystals*, Gordon and Breach Science Publishers, New York, 1991.
26. A. Krekhov, W. Pesch and Á. Buka, Flexoelectricity and pattern formation in nematic liquid crystals, *Phys. Rev. E* **83**(5), 051706/1–13, (2011). doi: 10.1103/PhysRevE.83.051706
27. C. Oldano, Comment on 'New ground state for the splay-Freedericksz transition in a polymer nematic liquid crystal', *Phys. Rev. Lett.* **56**(10), 1098–1098, (1986). doi: 10.1103/PhysRevLett.56.1098
28. W. Zimmermann and L. Kramer, Periodic splay-twist Freedericksz transition in nematic liquid crystals, *Phys. Rev. Lett.* **56**(24), 2655–2655, (1986). doi: 10.1103/PhysRevLett.56.2655
29. M. May, W. Schöpf, I. Rehberg, A. Krekhov and Á. Buka, Transition from longitudinal to transversal patterns in an anisotropic system, *Phys. Rev. E* **78**(4), 046215/1–9, (2008). doi: 10.1103/PhysRevE.78.046215
30. G. Derfel and M. Buczkowska, Numerical study of flexoelectric longitudinal domains, *Mol. Cryst. Liq. Cryst.* **547**(1), 213–221, (2011). doi: 10.1080/15421406.2011.572787

31. Y.G. Marinov and H.P. Hinov, On the threshold characteristics of the flexoelectric domains arising in a homogeneous electric field: the case of anisotropic elasticity, *Eur. Phys. J. E* **31**(2), 179–189, (2010). doi: 10.1140/epje/i2010-10560-0
32. H.P. Hinov and Y. Marinov, Theoretical considerations and experimental illustration of the electro-optic behavior of longitudinal flexoelectric domains under the joint action of DC and AC voltages: the case of strong anchoring, *Mol. Cryst. Liq. Cryst.* **503**(1), 45–68, (2009). doi: 10.1080/15421400902841494
33. A. Krekhov, W. Pesch and Á. Buka, Comment on 'On the threshold characteristics of the flexoelectric domains arising in a homogeneous electric field: the case of anisotropic elasticity' by Y.G. Marinov and H.P. Hinov, *Eur. Phys. J. E* **34**(8), 80/1–2, (2011). doi: 10.1140/epje/i2011-11080-1
34. P. Schiller, G. Pelzl and D. Demus, Analytical theory for flexo-electric domains in nematic layer, *Cryst. Res. Technol.* **25**(1), 111–116, (1990). doi: 10.1002/crat.2170250121
35. M.F. Lednei and I.P. Pinkevich, Threshold spatially periodic structure of the director in a nematic flexoelectric cell with finite anchoring energy, *Crystallography Rep.* **50**(3), 471–477, (2005) [*Kristallografiya* **50**, 517–523, (2005)]. doi: 10.1134/1.1927612
36. M. Ledney and I. Pinkevych, Influence of anchoring at a nematic cell surface on threshold spatially periodic reorientation of a director, *Liq. Cryst.* **34**(5), 577–583, (2007). doi: 10.1080/02678290500248160
37. M.I. Barnik, L.M. Blinov, S.A. Pikin and A.N. Trufanov, Instability mechanism in the nematic and isotropic phases of liquid crystals with positive dielectric anisotropy, *Sov. Phys. JETP* **45**(2), 396–398, (1977) [*Zh. Eksp. Teor. Fiz.* **72**(2), 756–761, (1977)].
38. Kh. Khinov, L.K. Vistin' and Yu.G. Magakova, The flexoelectric character of longitudinal domains in liquid crystals, *Sov. Phys. Crystallogr.* **23**(3), 323–325, (1978) [*Kristallografiya* **23**(3), 583–587, (1978)].
39. P. Kumar and K.S. Krishnamurthy, Gradient flexoelectric switching response in a nematic phenyl benzoate, *Liq. Cryst.* **34**(2), 257–266, (2007). doi: 10.1080/02678290601111192
40. N.V. Madhusudana, Chapter 2. Flexoelectro-optics and measurements of flexocoefficients. In eds. Á. Buka and N. Éber, *Flexoelectricity in Liquid Crystals. Theory, Experiments and Applications*, Imperial College Press, London, 2012. pp. 33–60.
41. B.A. Umanskii, V.G. Chigrinov, L.M. Blinov and Y.B. Podyachev, Flexoelectric effect in twisted liquid-crystal structures, *Sov. Phys. JETP* **54**(4), 694–699, (1981) [*Zh. Eksp. Teor. Fiz.* **81**(4), 1307–1317, (1981)].
42. V.A. Delev, A.P. Krekhov and L. Kramer, Crossover between flexoelectric stripe patterns and electroconvection in hybrid aligned nematics, *Mol. Cryst. Liq. Cryst.* **366**(1), 849–856, (2001). doi: 10.1080/10587250108024026
43. S.P. Palto, N.J. Mottram and M.A. Osipov, Flexoelectric instability and spontaneous chiral-symmetry breaking in a nematic liquid crystal with asymmetric boundary conditions, *Phys. Rev. E* **75**(6), 061707/1–8, (2007). doi: 10.1103/PhysRevE.75.061707

44. H.P. Hinov and L.K. Vistin, Parallel and cross-like domains due to d.c. and low frequency ($< 2$ Hz) electric fields in nematic liquid crystal layers with negative dielectric anisotropy, *J. Phys. France* **40**(3), 269–292, (1979). doi: 10.1051/jphys:01979004003026900
45. H.P. Hinov, I. Bivas, M.D. Mitov, K. Shoumarov and Y. Marinov, A further experimental study of parallel surface-induced flexoelectric domains (PSIFED) (flexo-dielectric walls), *Liq. Cryst.* **30**(11), 1293–1317, (2003). doi: 10.1080/02678290310001607198
46. H.P. Hinov, On the coexistence of the flexo-dielectric walls–flexoelectric domains for the nematic MBBA – A new estimation of the modulus of the difference between the flexoelectric coefficients of splay and bend $|e_{1z} - e_{3x}|$, *Mol. Cryst. Liq. Cryst.* **524**(1), 26–35, (2010). doi: 10.1080/15421400903568161
47. L. Kramer and W. Pesch, Electrohydrodynamics in nematics. In eds. D.A. Dummur, A. Fukuda and G.R. Luckhurst, *Physical Properties of Nematic Liquid Crystals*, Inspec, London, 2001. pp. 441–454.
48. À. Buka, N. Éber, W. Pesch and L. Kramer, Isotropic and anisotropic electroconvection, *Phys. Rep.* **448**(5–6), 115–132, (2007). doi: 10.1016/j.physrep.2007.02.013
49. M. Dennin, D.S. Cannell and G. Ahlers, Patterns of electroconvection in a nematic liquid crystal, *Phys. Rev. E* **57**(1), 638–649, (1998). doi: 10.1103/PhysRevE.57.638
50. W. Thom, W. Zimmermann and L. Kramer, The influence of the flexoelectric effect on the electrohydrodynamic instability in nematics, *Liq. Cryst.* **4**(3), 309–316, (1989). doi: 10.1080/02678298908029184
51. N.V. Madhusudana and V.A. Raghunathan, Influence of flexoelectricity on electrohydrodynamic instabilities in nematics, *Liq. Cryst.* **5**(6), 1789–1812, (1989). doi: 10.1080/02678298908045689. *Mol. Cryst. Liq. Cryst. Lett.* **5**(6), 201–209, (1988).
52. T. Tóth-Katona, N. Éber, Á. Buka and A. Krekhov, Flexoelectricity and competition of time scales in electroconvection, *Phys. Rev. E* **78**(3), 036306/1–12, (2008). doi: 10.1103/PhysRevE.78.036306
53. T. Tóth-Katona, N. Éber and Á. Buka, Flexoelectricity in electroconvection, *Mol. Cryst. Liq. Cryst.* **511**(1), 11–24, (2009). doi: 10.1080/15421400903048461
54. T. John, J. Heuer and R. Stannarius, Influence of excitation wave forms and frequencies on the fundamental time symmetry of the system dynamics studied in nematic electroconvection, *Phys. Rev. E* **71**(5), 056307-10, (2005). doi: 10.1103/PhysRevE.71.056307
55. J. Low and S.J. Hogan, Subharmonic pattern formation in a nematic cell – a 1d flexoelectric model approach, *Mol. Cryst. Liq. Cryst.* **478**(1), 67–81, (2007). doi: 10.1080/15421400701735640
56. J. Low and S.J. Hogan, Standard and nonstandard nematic electrohydrodynamic convection in the presence of asymmetric ac electric fields, *Phys. Rev. E* **78**(4), 041706-15, (2008). doi: 10.1103/PhysRevE.78.041706

57. Á. Buka, N. Éber, W. Pesch and L. Kramer, Convective patterns in liquid crystals driven by electric field. In eds. A.A. Golovin and A.A. Nepomnyashchy, *Self Assembly, Pattern Formation and Growth Phenomena in Nano-Systems.* Springer, Dordrecht, 2006. pp. 55–82.
58. Á. Buka, B. Dressel, W. Otowski, K. Camara, T. Tóth-Katona, L. Kramer, J. Lindau, G. Pelzl and W. Pesch, Electroconvection in nematic liquid crystals with positive dielectric and negative conductivity anisotropy, *Phys. Rev. E* **66**(5), 051713/1–8, (2002). doi: 10.1103/PhysRevE.66.051713
59. Á. Buka, B. Dressel, L. Kramer and W. Pesch, Isotropic convection scenarios in an anisotropic fluid, *Phys. Rev. Lett.* **93**(4), 044502/1–4, (2004). doi: 10.1103/PhysRevLett.93.044502
60. Á. Buka, B. Dressel, L. Kramer and W. Pesch, Direct transition to electroconvection in a homeotropic nematic liquid crystal, *Chaos* **14**(3), 793–802, (2004). doi: 10.1063/1.1774412
61. M. Goscianski and L. Leger, Electrohydrodynamic instabilities above a nematic to smectic A (or C) transition, *J. Phys. Coll. (France)* **36**(C1), C1-231–C1-236, (1975). doi: 10.1051/jphyscol:1975141
62. N.A. Tikhomirova, A.V. Ginzberg, E.A. Kirsanov, Yu.P. Bobyshev, S.A. Pikin and P.V. Adomenas, Concerning a new type of instability in liquid crystals, *JETP Lett.* **24**(5), 269–272, (1976) [*Pis'ma Zh. Eksp. Teor. Fiz.* **24**(5), 301–304, (1976)].
63. L.M. Blinov, M.I. Barnik, V.T. Lazareva and A.N. Trufanov, Electrohydrodynamic instabilities in the liquid crystalline phases with smectic ordering, *J. Phys. Coll. (France)* **40**(C3), C3-263–C3-268, (1979). doi: 10.1051/jphyscol:1979350
64. H.R. Brand, C. Fradin, P.L. Finn, W. Pesch and P.E. Cladis, Electroconvection in nematic liquid crystals: comparison between experimental results and the hydrodynamic model, *Phys. Lett. A* **235**(5), 508–514, (1998). doi: 10.1016/S0375-9601(97)00680-4
65. T. Tóth-Katona, N. Éber and Á. Buka, Temporal response to harmonic driving in electroconvection, *Phys. Rev. E* **83**(6), 061704/1–8, (2011). doi: 10.1103/PhysRevE.83.061704
66. M. Treiber and L. Kramer, Bipolar electrodiffusion model for electroconvection in nematics, *Mol. Cryst. Liq. Cryst.* **261**(1), 311–326, (1995). doi: 10.1080/10587259508033478
67. D. Wiant, J.T. Gleeson, N. Éber, K. Fodor-Csorba, A. Jákli and T. Tóth-Katona, Nonstandard electroconvection in a bent-core nematic liquid crystal, *Phys. Rev. E* **72**(4), 041712/1–12, (2005). doi: 10.1103/PhysRevE.72.041712
68. S. Tanaka, S. Dhara, B.K. Sadashiva, Y. Shimbo, Y. Takanishi, F. Araoka, K. Ishikawa and H. Takezoe, Alternating twist structures formed by electroconvection in the nematic phase of an achiral bent-core molecule, *Phys. Rev. E* **77**(4), 041708/1–5, (2008). doi: 10.1103/PhysRevE.77.041708
69. S. Tanaka, H. Takezoe, N. Éber, K. Fodor-Csorba, A. Vajda and Á. Buka, Electroconvection in nematic mixtures of bent-core and calamitic molecules, *Phys. Rev. E* **80**(2), 021702/1–8, (2009). doi: 10.1103/PhysRevE.80.021702

70. P. Tadapatri, U.S. Hiremath, C.V. Yelamaggad and K.S. Krishnamurthy, Patterned electroconvective states in a bent-core nematic liquid crystal, *J. Phys. Chem. B* **114**(1), 10–21, (2010). doi: 10.1021/jp9058802
71. D. Wiant, K. Neupane, S. Sharma, J.T. Gleeson, S. Sprunt, A. Jákli, N. Pradhan and G. Iannacchione, Observation of a possible tetrahedric phase in a bent-core liquid crystal, *Phys. Rev. E* **77**(6), 061701/1–7, (2008). doi: 10.1103/PhysRevE.77.061701
72. E. Dorjgotov, K. Fodor-Csorba, J.T. Gleeson, S. Sprunt and A. Jákli, Viscosities of a bent-core nematic liquid crystal, *Liq. Cryst.* **35**(2), 149–155, (2008). doi: 10.1080/02678290701824199
73. C. Bailey, K. Fodor-Csorba, J.T. Gleeson, S. Sprunt and A. Jákli, Rheological properties of bent-core liquid crystals, *Soft Matter* **5**(19), 3618–3622, (2009). doi: 10.1039/B907261F
74. M. Majumdar, P. Salamon, A. Jákli, J.T. Gleeson and S. Sprunt, Elastic constants and orientational viscosities of a bent-core nematic liquid crystal, *Phys. Rev. E* **83**(3), 031701/1–8, (2011). doi: 10.1103/PhysRevE.83.031701
75. A.N. Trufanov, M.I. Barnik, L.M. Blinov and V.G. Chigrinov, Electrohydrodynamic instability in homeotropically oriented layers of nematic liquid crystals, *Sov. Phys. JETP* **53**(2), 355–361, (1981) [*Zh. Eksp. Teor. Fiz.* **80**(2), 704–715, (1981)].
76. D.K. Rout and R.N.P. Choudhary, Electrohydrodynamic instability in some nematic cyanobiphenyls in an AC electric field, *Liq. Cryst.* **4**(4), 393–398, (1989). doi: 10.1080/02678298908035485
77. E.I. Rjumtsev and S.G. Polushin, Electrohydrodynamic instabilities in nematic liquid crystals with large positive dielectric anisotropy, *Liq. Cryst.* **13**(5), 623–628, (1993). doi: 10.1080/02678299308026335
78. L.M. Blinov and V.G. Chigrinov, *Electro-optic Effects in Liquid Crystal Materials*, Springer, Berlin, 1996.
79. O.D. Lavrentovich, V.G. Nazarenko, V.M. Pergamenshchik, V.V. Sergan and V.M. Sorokin, Surface-polarization electrooptic effect in a nematic liquid crystal, *Sov. Phys. JETP* **72**(3), 431–444, (1991) [*Zh. Eksp. Teor. Fiz.* **99**(3), 777–802, (1991)].
80. O.D. Lavrentovich, V.G. Nazarenko, V.V. Sergan and G. Durand, Dielectric quenching of the electric polar surface instability in a nematic liquid crystal, Phys. Rev. A **45**(10), R6969–R6972, (1992). doi: 10.1103/PhysRevA.45.R6969
81. P. Kumar, J. Heuer, T. Tóth-Katona, N. Éber and Á. Buka, Convection-roll instability in spite of a large stabilizing torque, *Phys. Rev. E* **81**(2), 020702(R)/1–4, (2010). doi: 10.1103/PhysRevE.81.020702
82. N. Éber, L.O. Palomares, P. Salamon, A. Krekhov and Á. Buka, Temporal evolution and morphological transitions of electric field induced patterns at low frequency driving, (unpublished).
83. E. Bodenschatz, W. Pesch and G. Ahlers, Recent developments in Rayleigh–Bénard convection, *Annu. Rev. Fluid Mech.* **32**, 709–778, (2000). doi: 10.1146/annurev.fluid.32.1.709

84. Y. Marinov, J. Kosmopoulos, W. Weissflog, A.G. Petrov and D.J. Photinos, Flexoelectricity of wedge-like molecules in nematic mixtures, *Mol. Cryst. Liq. Cryst.* **357**(1), 221–228, (2001). doi: 10.1080/10587250108028255
85. N. Scaramuzza and M.C. Pagnotta, Evidence for flexoelectric effect in pallated metallorganic liquid crystals, *Mol. Cryst. Liq. Cryst.* **239**(1), 263–267, (1994). doi: 10.1080/10587259408047188
86. L.K. Vistin', I.G. Chistyakov, R.I. Zharenov and S.S. Yakovenko, Changes in the domain pattern of nematic liquid crystals in electric fields, *Sov. Phys. Crystallogr.* **21**(1), 91–93, (1976) [*Kristallografiya* **21**(1), 173–177, (1976)].
87. T. Takahashi, S. Hashidate and T. Akahane, Novel measurement method for flexoelectric coefficients of nematic liquid crystals, *Jpn. J. Appl. Phys.* **37**(4A), 1865–1869, (1998). doi: 10.1143/JJAP.37.1865
88. Y. Marinov, A.G. Petrov and H.P. Hinov, On a simple way for obtaining important material constants of a nematic liquid crystal: longitudinal flexoelectric domains under the joint action of DC and AC voltages, *Mol. Cryst. Liq. Cryst.* **449**(1), 33–45, (2006). doi: 10.1080/15421400600580188
89. K. Van Le, F. Araoka, K. Fodor-Csorba, K. Ishikawa and H. Takezoe, Flexoelectric effect in a bent-core mesogen, *Liq. Cryst.* **36**(10-11), 1119–1124, (2009). doi: 10.1080/02678290902854086
90. P. Kumar, Y.G. Marinov, H.P. Hinov, U.S. Hiremath, C.V. Yelamaggad, K.S. Krishnamurthy and A.G. Petrov, *J. Phys. Chem. B* **113**(27), 9168–9174, (2009). doi: 10.1021/jp903241z
91. E. Plaut, W. Decker, A.G. Rossberg, L. Kramer and W. Pesch, New symmetry breaking in nonlinear electroconvection of nematic liquid crystals, *Phys. Rev. Lett.* **79**(12), 2367–2370, (1997). doi: 10.1103/PhysRevLett.79.2367
92. M. Oh-e, K. Kondo and Y. Kando, Theoretical consideration of the drop in threshold voltage at low frequencies in nematic liquid crystals, *Liq. Cryst.* **17**(1), 95–107, (1994). doi: 10.1080/02678299408036551
93. H. Seiberle and M. Schadt, LC-conductivity and cell parameters; their influence on twisted nematic and supertwisted nematic liquid crystal displays, *Mol. Cryst. Liq. Cryst.* **239**(1), 229–244, (1994). doi: 10.1080/10587259408047185

# Chapter 5

# Flexoelectricity in Chiral Polar Smectics

Mojca Čepič
*Jožef Stefan Institute*
*Jamova 39, 1000 Ljubljana, Slovenia*
*and*
*Faculty of Education, University of Ljubljana*
*1000 Ljubljana, Slovenia*
*e-mail: mojca.cepic@ijs.si*

The flexoelectric effect is a phenomenon where a space variation of the order parameter induces polarization. Chiral polar smectics are liquid crystals formed of chiral molecules and organized in layers. All phases in tilted chiral polar smectic liquid crystals have modulated structures and they are therefore good candidates for exhibiting the flexoelectric effect. The flexoelectric effect is less pronounced in the ferroelectric SmC$^*$ phase and in the antiferroelectric SmC$^*_A$. The flexoelectric effect is more pronounced in more complex phases: the three-layer SmC$^*_{\mathrm{FI1}}$ phase, the four-layer SmC$^*_{\mathrm{FI2}}$ phase and the six-layer SmC$^*_{\mathrm{6d}}$ phase.

In this chapter we consider several important aspects of the flexoelectric effect for chiral polar smectic liquid crystals and for the variety of phases. First, we discuss the reason for indirect interlayer interactions, which extend to more distant layers, and the lock-in to multi-layer structures. Second, although it was believed for a long time that polarization in tilted chiral smectics is always perpendicular to the tilt with the smectic layer normal, a component in the direction of the tilt may exist. And third, in multi-layer structures, the flexoelectrically induced polarization can be extremely large but is difficult to measure.

## Contents

## 5.1. Introduction

Flexoelectricity in liquid crystals has been studied for a long time. If molecules that form a liquid crystal have significant dipole moments, the phenomena related to flexoelectricity can be rather pronounced. In addition, liquid crystals can easily be manipulated by external fields and specially designed surfaces. Many recently synthesized liquid crystalline materials have extremely complex structures.[1–3] In liquid crystals only two types of deformation contribute to flexoelectricity:[4] the splay and the bend and both can occur with respect to the ordering of longer and shorter average molecular axes; however, the origins of the flexoelectrically induced polarization are present in stable structures as well. More complex stable structures can also have intrinsic splays[5,6] or other peculiar deformations. Although in the last ten years the zoo of extremely complex structures has filled many of the still-empty cages, we will consider only the more simple complex structures in this chapter. We will limit ourselves to chiral polar smectics, which have ferroelectric, ferrielectric and antiferroelectric phases.

The chapter is organized as follows: The second section discusses the prototype polar smectics: the ferroelectric liquid crystals. We discuss the structure of the ferroelectric phase, the theoretical explanation for it and we introduce the flexoelectric effect in chiral polar smectics. Next we introduce a new set of chiral polar smectics, the antiferroelectric liquid crystals, and we describe the structures of different phases found in these systems. We present the discrete theoretical modelling approach, which experimentally consistently describes the phases and their properties. Then we introduce the discrete form of the flexoelectric effect in these systems and show that without flexoelectricity no interactions of longer range would be significant and therefore no structures with longer periods than two layers would be stable. We discuss also a few phenomena that are related to the complexity of the structures, such as the existence of a longitudinal, i.e. parallel to the tilt, component of the polarization. Finally we conclude with a discussion about the observability of flexoelectric polarization.

## 5.2. Ferroelectric Liquid Crystals

Smectic liquid crystals have been known for a long time.[7] Elongated molecules organize in layers. Within a layer the molecules move freely and they are not positionally correlated. Each layer therefore behaves as a truly orientationally ordered two-dimensional liquid. A long-range order is present in the organization of the layers. In smectics, layers are straight and the normals of each layer are parallel. Two forms of smectics are of interest. In the smectic A phase (SmA) the long molecular axes are on average parallel to the layer normal. As this means that they are on average perpendicular to the smectic layer, the phase is often described as 'an orthogonal phase'. In the smectic C phase (SmC$^*$) the long molecular axes form an angle, usually small, with the layer normal on average. The latter were the basis of the symmetry consideration that led to the synthesis and discovery of ferroelectric liquid crystals.

Ferroelectric liquid crystals were first synthesized in 1974 after a discussion between theorists, chemists and experimentalists during a queue for lunch.[8,9] The theoretical idea proposed by Meyer that chiral molecules organized in the SmC$^*$ structure should have polar layers with a polarization perpendicular to the tilt and the layer normal, was almost immediately realized by the chemist Strzelecky. The era of polar liquid crystals began.

The idea was rather simple, as impressive discoveries usually are. If the molecule is chiral it differs from its mirror image. A molecule whose long axis, say along the $z$ axis, is tilted away from the layer normal rotates due to the thermodynamics. It rotates around its long axis; however, sometimes it also rotates around its short axes. There are two short axes and for symmetry reasons one is perpendicular to the tilt and the layer normal (along the $y$ axis) and the other axis is perpendicular to the tilt (along the $x$ axis) but is in the plane formed by the layer normal and the tilt (see Fig. 5.1a,b). The three coordinate axes $x$, $y$ and $z$ form a local 'molecular' coordinate system. For a chiral molecule all orientations are equal as long as the molecules are orthogonal to the layer. However, when molecules are tilted, not all the orientations of the molecule are equivalent any more. In Fig. 5.1b we can clearly see that the 'micro' molecular environment is different for different molecular orientations with respect to rotation around the long molecular axis. The molecule rotates from the orientation where the 'wings' of the tilted S shape are oriented to the inner part of the layer and toward neighbouring layers. These situations evidently differ energetically. As a consequence, the molecule spends more time in the one specific most favourable orientation on average. Due to symmetry reasons, asymmetric chiral molecules also have a dipole moment, which has a component perpendicular to the long axes. In Fig. 5.1c, we can see that the hindered rotation around the long molecular axis, with regard to flipping around the $y$ axis, results in a net polarization along the $y$ axis. The blue circle in Fig. 5.1 signifies that the molecule does not have an up-down symmetry. This symmetry is established at the structural level as the number of 'up' molecules is on average the same as the number of 'down' molecules.

The induced polarization is a result of averaging the three distributions presented in Fig. 5.1b and c, below. The component of the polarization along the $y$ axis remains; molecular rotations around the short axes cancel the components of polarization perpendicular to the $y$ axis on average.

The first phenomenological model to explain well, qualitatively and also semi-quantitatively, the behaviour of ferroelectric liquid crystals was suggested soon after their discovery.[10] It assumed that the nematic and smectic orders were constant and reduced the problem to considering the part of the free energy depending on the tilt and on the polarization. Below we will briefly describe this phenomenological description proposed by Pikin and Indenbom, but we will take a slightly different approach. We will focus on the polarization and especially on the flexoelectrically induced polarization, so we will consider the free energy in two steps: initially without polarization and then the changes that appear due to polarization.

### 5.2.1. *Phenomenological modelling of chiral tilted smectics*

Let us first consider a fictitious material, which has a tilted SmC$^*$ phase, whose molecules are chiral but not polar. Let us also keep in mind that such a material does not exist in reality. There exist materials, which at a certain, well-defined temperature, exhibit a sign reversal of their polarization. At exactly the temperature of the polarization reversal the polarization does not exist due to several reasons: the average of the coexistent chiral conformations is not polar or the hindrance of the rotation is such that the polarization coincides with one of the two axes ($x$ and $z$) or both. But this is not the case for a wider temperature range. The molecule itself always has some polarization because of its structural asymmetry.

The free energy is therefore described solely by the order parameter

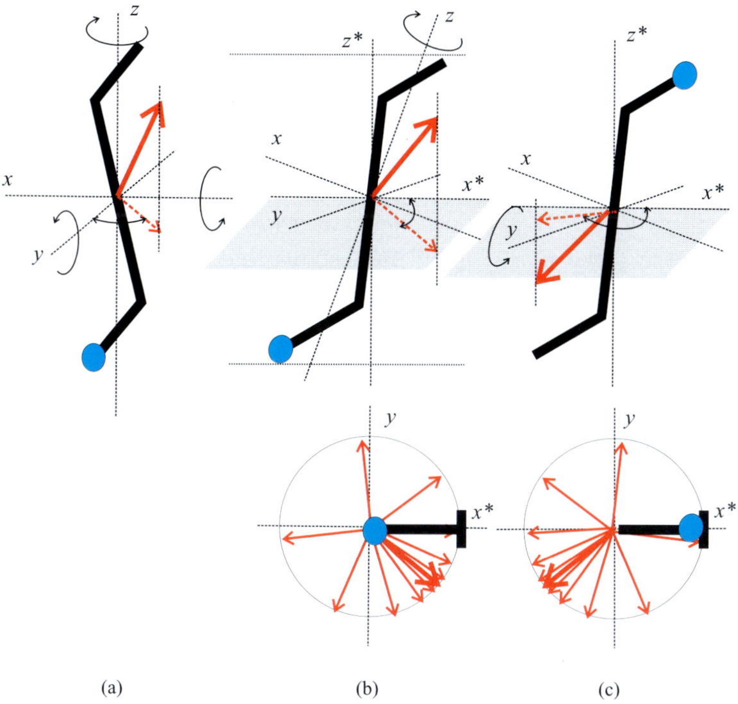


Fig. 5.1. Above: The red arrow marks the molecular dipole moment for a static molecule, the dashed arrow is its projection on the layer. Below: The T sign marks the average tilt. The thin arrows mark the instantaneous dipole moments of several molecules illustrating the distribution of dipoles. The thick arrow gives the average. The axes of the static coordinate system bound to the layer are marked by $x^*$, $y^*$ and $z^*$. For the non tilted molecule the axes with and without stars coincide. For the tilted molecule only the direction along the $y$ axis is the same as the static $y^*$ direction. The two coordinate systems ($x$, $y$, $z$ and $x^*$, $y^*$, $z^*$) are used for the explanation of rotation hindrances only. (a) Three axes of rotation for an 'average' molecule. (b) Above: A tilted molecule. Below: Rotation around the long axis ($z$) is hindered resulting in a polarization distribution with a net polarization. (c) Above: A molecule can flip around its short axis ($y$). Below: The polarization distribution for the flipped molecule.

describing tilted smectics. The two-dimensional tilt-order parameter is

$$\boldsymbol{\xi} = \{\xi_x, \xi_y\} = \{n_x\, n_z, -n_y\, n_z\}, \tag{5.1}$$

where $\boldsymbol{n}$ is a local director expressed in a coordinate system in which the direction of the $z$ axis corresponds to the smectic layer normal (Fig. 5.2). The combination of the components resumes the up-down symmetry of the director. The part of the free energy depending on the tilt is

$$G = \frac{1}{2}\, a_0\, \xi^2 + \frac{1}{4}\, b_0\, \xi^4 + \lambda \left( \xi_x\, \frac{\partial \xi_y}{\partial z} - \xi_y\, \frac{\partial \xi_x}{\partial z} \right) + \frac{1}{2}\, K \left( \frac{\partial \boldsymbol{\xi}}{\partial z} \right)^2 . \qquad (5.2)$$

The first two terms give the tilt-dependent free energy for an isolated layer. The first coefficient $a_0 = a(T - T_0)$ is the only temperature-dependent coefficient in Eq. (5.2). For simplicity we consider only continuous transitions to the tilted phase, found in systems that can be described by a positive coefficient $b_0$. The generalization to the discontinuous transition to the tilted phase[11] does not add any important new physics to the problem. There exist also systems where the first coefficient does not depend on temperature monotonously around the transition temperature;[12] this is a very rare situation where the transition takes place at the temperature where the Landau expansion for the temperature dependence of the leading coefficient $a_0$ has to include higher-order terms.

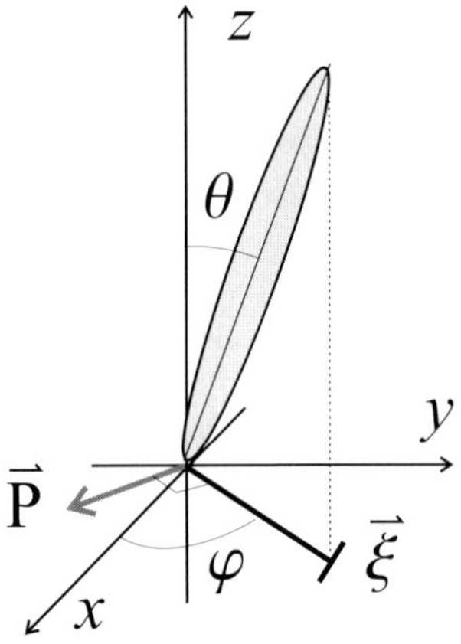


Fig. 5.2. Definitions of the tilt- and polarization-order parameters. The $z$ axis coincides with the smectic layer normal. The $x$ and $y$ directions are chosen arbitrarily. For the tilt-order parameter the nail notation is used.

The third and the fourth terms are there because the tilt in chiral systems is modulated. Neighbouring layers are not tilted in parallel and (usually) there is a small angle in the tilt direction. The angle between them that describes the difference between the direction of the tilts in neighbouring layers is called the *phase difference.* It is a consequence of the competition between elastic interactions favouring a parallel tilt orientation in neighbouring layers (the $K$ term) and the chiral interactions, which favour non-parallel tilts in neighbouring layers (the $\lambda$ term). The reason for the non-parallelism of the tilted molecules in neighbouring layers is the chiral molecular structure. Molecules forming liquid crystals are rather large and some parts of two different molecules are often much closer than their centres of masses. The van der Waals intermolecular interactions due to the polarizability of different links have to be considered: nearby parts are attracted or repelled more strongly than more distant parts. Here the internal molecular structure plays an important role. When two chiral molecules are set in two positions in space they orient in the most favourable orientation and the molecular eigenaxes are almost never parallel. When chiral molecules are tilted and ordered within a layer, the general orientation is not possible but the tendency to non-parallelism remains and a small phase difference between tilt directions in neighbouring layers is formed due to the competition between the elastic interactions ($K$) and the chiral interactions ($\lambda$).

When searching for a structure, we can start from an assumption of a constant tilt consistent with X-ray measurements. There is no reason for the tilt to vary from one layer to another as this is associated with changes in the layer thickness, which is energetically extremely disfavoured. Therefore the tilt is allowed to vary in direction only. The corresponding *ansatz* is

$$\boldsymbol{\xi} = \cos\theta \sin\theta\{\cos\phi(z), \sin\phi(z)\} \approx \theta\{\cos\phi(z), \sin\phi(z)\}. \qquad (5.3)$$

The angles $\theta$ and $\phi$ are defined in Fig. 5.2. The approximation $\cos\theta \approx 1$ and $\sin\theta \approx \theta$ is legitimate as the Landau expansion is valid only close to phase transitions where the order parameters are still small. Inserting Eq. (5.3) into Eq. (5.2) gives

$$G = \frac{1}{2}\left(a_0 + 2\,\lambda\,\frac{d\phi}{dz} + K\left(\frac{d\phi}{dz}\right)^2\right)\theta^2 + \frac{1}{4}\,b_0\,\theta^4. \qquad (5.4)$$

Minimization of the free energy with respect to the phase angle is straightforward

$$\frac{d^2\phi}{dz^2} = 0; \qquad \frac{d\phi}{dz} = q; \qquad \phi = qz + \phi_0. \qquad (5.5)$$

The phase angle therefore continuously increases or decreases. The structure given by Eq. (5.5) is a well-known helicoidally modulated structure. The tilt direction of chiral molecules organized in layers and tilted with the long axes away from the layer is almost always helicoidally modulated even without polarization. The wave number of the helicoidal modulation is

$$q = -\frac{\lambda}{K}. \tag{5.6}$$

Inserting this into Eq. (5.4) gives the simple form of the Landau free energy expansion for a single order parameter close to the continuous phase transition

$$G = \frac{1}{2}\left(a_0 - \frac{\lambda^2}{K}\right)\theta^2 + \frac{1}{4}\, b_0\, \theta^4. \tag{5.7}$$

Due to chiral interactions the phase with the non-zero order parameter is additionally stabilized and the transition temperature is (slightly) higher than in the absence of chiral interactions. The solution is well known

$$\theta = 0 \quad \text{for} \quad T > T_0 + \frac{\lambda^2}{a\,K},$$
$$\theta = \sqrt{-\frac{\left(a_0 - \frac{\lambda^2}{K}\right)}{b_0}} \quad \text{for} \quad T < T_0 + \frac{\lambda^2}{a\,K}. \tag{5.8}$$

So, the structure is helicoidally modulated and can be described as

$$\boldsymbol{\xi} = \theta\,\{\cos(qz + \phi_0), \sin(qz + \phi_0)\}, \tag{5.9}$$

where $q$ is given by Eq. (5.6) and $\theta$ is given by Eq. (5.8). Here we see that the structure is not uniform. We have intentionally omitted the polar properties of the molecules, which will be discussed later. No effects of polarization have yet been considered. Chiral molecules are always polar and therefore a material with a phase having modulated tilt is a good candidate for a strong flexoelectric effect. Let us now study the effects of the polar properties on the structure.

### 5.2.2. *Polar properties and flexoelectricity*

In the previous section we described terms that contribute to the free energy of the system, however, without considering interactions where polarization is involved. In this section we will take a closer look at interactions that appear due to the polarization or influence polarization, especially flexoelectricity.

Polarization is a secondary order parameter, which appears only when the molecules are tilted with respect to the smectic layer normal. It is defined as:

$$\boldsymbol{P} = \{P_x, P_y\}. \tag{5.10}$$

Polarization is actually a consequence of hindered rotation when the molecules are tilted. The microscopic reason is that when a molecule, which is not straight, becomes tilted, its rotation around the long axis becomes hindered as discussed before. We have to consider also rotations around the short axes or equivalency of the up and down molecular orientations. When all possible orientations are averaged, only the component of the polarization that is perpendicular to the tilt and the layer normal, remains (Fig. 5.1b–d). As the tilt is the reason for the polarization and the tilt is also associated with the layer thickness, the effect is reminiscent of piezoelectricity. Therefore it is called *piezoelectric polarization*. The name is a bit misleading as the compression of the layer does not necessarily mean the tilt is in one direction only. Molecules can tilt in any direction (as proposed for the de Vries SmA structure[13]), which does not lead to polarization within the layer.

How can we describe the piezoelectric effect phenomenologically? We usually assume that molecular rotation is only slightly hindered. This is also consistent with experimental data, as the magnitude of polarization is usually only up to 10% of the polarization that would be present if all of the molecular dipoles in the layer were ordered.[14] The polarization contribution to the free energy can therefore be approximated by the lowest order, which is linear. Namely, the opposite tilt induces the opposite polarization:

$$c_p\ (\boldsymbol{\xi} \times \boldsymbol{P})_z = c_p\ (\xi_x\ P_y - \xi_y\ P_x)\,. \tag{5.11}$$

The coefficient $c_p$ gives the properties of the material and is chiral by its nature. It has opposite signs for oppositely handed materials. Here $\boldsymbol{\xi}$ and $\boldsymbol{P}$ are considered as three-dimensional vectors having their $z$ component equal to zero.

Flexoelectricity is usually explained as being due to asymmetric molecular shapes. If a molecule is thicker on one side (drop shape) then in an unbent media the up and down orientations of the molecules are equally occupied. But if the material is bent, one of the two positions is more favourable. Molecules still rotate around the short axes; however, they spend more time in one position than in the other (Fig. 5.3a,b). If a molecule has a longitudinal component of polarization, the bent material is polarized as well. Similarly we can consider molecules, which are bent like 'banana' molecules.[15] Then, the two orientations (Fig. 5.3c) are not equivalent and rotation along the long molecular axis is hindered when the material is bent. The molecules spend more time in one of the orientations. If the polarization has a component in the bent plane perpendicular to the long molecular axis, the sample becomes polarized in this direction (Fig. 5.3d).

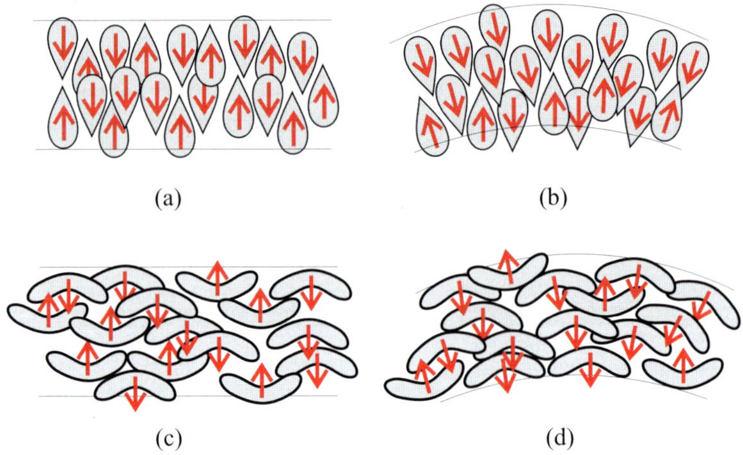

Fig. 5.3. (a) Asymmetric molecular shape does not influence the up-down distribution symmetry, while (b) in bent material the distribution is not symmetric. (c) Distribution of bent molecules is symmetric as long as the material is straight, but (d) distribution of the orientation is not symmetric when the material is bent.

In the SmC$^*$ phase the effects of the non-uniformity of the director on the polarization is not so straightforward. As mentioned already, the tilt direction precesses through the layer resulting in a helicoidally modulated tilt. The directions of the tilt above and below the considered layer are not equal. Because of the different directions, molecules from neighbouring layers affect the molecular rotation along the long axis differently. This may affect the magnitude or the direction of the polarization. The part of the polarization that is induced by the non-uniformity of the director in the structure reminds us of the non-uniformity of the order that appears due to externally applied forces and the consequent bending. The name 'flexoelectrically induced polarization' is therefore usually used for this phenomenon, although the material is not bent by the application of an external force.

The flexoelectric contribution to the free energy has to take into account both the non-uniformity of director and the polarization. As in the SmC$^*$ phase the non-uniformity appears due to the changes of the tilt direction, which can be expressed in the derivatives of the tilt. In the SmC$^*$ phase, where the modulation periods are long and the variation of the tilt direction from layer to layer is small, we can consider the contribution to the free energy using only the term linear in the tilt and the polarization order parameter:

$$\mu\left(P_x\,\frac{d\xi_x}{dz}+P_y\,\frac{d\xi_y}{dz}\right). \tag{5.12}$$

The coefficient $\mu$ gives the properties of the material and does not depend on chirality. The polarization part of the free energy density can therefore be written as

$$G_P=c_p\;(\xi_x\,P_y-\xi_y\,P_x)+\mu\left(P_x\,\frac{d\xi_x}{dz}+P_y\,\frac{d\xi_y}{dz}\right)+\frac{1}{2\epsilon}P^2. \tag{5.13}$$

The last term gives the electrostatic contribution to the free energy. The coefficient $\epsilon$ is always positive and the polarization is never the reason for the phase transition in chiral polar liquid crystals. The polarization $P$ is an improper order parameter and appears only when the proper order parameter – the tilt – is non-zero.

The bilinearity in the coupling between the tilt and the polarization enables simple elimination of the polarization. Minimization with respect to the polarization gives

$$\begin{aligned}P_x&=-\epsilon\,c_p\,\xi_y-\epsilon\,\mu\,\frac{d\xi_x}{dz},\\ P_y&=\epsilon\,c_p\,\xi_x+\epsilon\,\mu\,\frac{d\xi_y}{dz}.\end{aligned} \tag{5.14}$$

In Eq. (5.14) it can clearly be seen that two phenomena contribute to the polarization. The first component, the piezoelectric polarization, is defined by the tilt magnitude and is always strictly perpendicular to the local tilt direction. The second component, the flexoelectric contribution, depends on the derivative of the tilt. It can clearly be seen that the uniform derivative in Eq. (5.5) – the constant – also induces polarization in the same (or the opposite) direction as the piezoelectric contribution; however, when the derivative is not constant, the polarization may also have a component parallel to the tilt.

How can the polarization, or in more detail, both components of the polarization, influence the structure? Insert Eq. (5.14) into Eq. (5.13) and the contribution due to the polarization will be expressed using the tilt as

$$G_P=-\frac{1}{2}\,\epsilon\,c_p^2\,\xi^2-\epsilon\,\mu\,c_p\left(\xi_x\,\frac{\partial\xi_y}{\partial z}-\xi_y\,\frac{\partial\xi_x}{\partial z}\right)-\frac{1}{2}\,\epsilon\,\mu^2\left(\frac{\partial\boldsymbol{\xi}}{\partial z}\right)^2. \tag{5.15}$$

Comparing Eq. (5.2) and Eq. (5.15) we can see the same form of quadratic terms. Therefore the coefficients are only renormalized:

$$\begin{aligned}\tilde{a}_0 &= a_0 - \epsilon\, c_p^2 \\ \tilde{\lambda} &= \lambda - \epsilon\, c_p\, \mu \\ \tilde{K} &= K - \epsilon\, \mu^2 .\end{aligned} \tag{5.16}$$

The solutions differ only quantitatively. The transition temperature from the orthogonal to the tilted phase is higher, which is a usual consequence of additional interactions in systems. The chiral properties are changed as well as the elastic properties. Whether the polarization interactions shorten or extend the pitch modulation cannot be deduced from the equations. The ratio $q = -\frac{\tilde{\lambda}}{\tilde{K}}$ depends on the signs of $\lambda, c_p$ and $\mu$. The flexoelectric coefficient $\mu$ is non-chiral and has the same sign in chiral and non-chiral samples. On the other hand, changing the handedness of the system, both parameters $\lambda$ and $c_p$ change sign. But both of them can be of the same or the opposite signs, which is a material property.

We have seen that flexoelectricity (and piezoelectricity) influences the material properties expressed in the coefficients of Eq. (5.16) and consequently influences quantitatively the pitch, the phase transition temperature and the tilt. But there are no qualitative changes, i.e. no new stable phases with qualitatively different structures appear due to these two interactions. Below we shall see that in systems where the interactions are more complex the role of the flexoelectric phenomenon can be very significant and it is crucial for the stability of many structures.

## 5.3. Antiferroelectric Liquid Crystals

Antiferroelectric liquid crystals were discovered in 1989.[16,17] Their antiferroelectric properties were considered as a surprise as, at that time, nobody believed that significant changes of the tilt from layer to layer were possible. In addition to the surprising antiferroelectric properties of one of the phases, several additional phases within a rather narrow temperature region were found. Why such a rich variety of phases occurs within a narrow temperature region and what their structures are, has been a hot experimental and theoretical problem for a long time.[1] The main difference between the microscopic structures of the phases was the period of the basic structure. Various phases will be described in more detail later but here we only mention the periods. The SmC* phase is defined by the structure of a single layer and the difference between the tilt directions in neighbouring layers. In the $\mathrm{SmC}_A^*$ phase the basic period has two layers, the $\mathrm{SmC}_{\mathrm{FI1}}^*$ phase has a three-layer period and the $\mathrm{SmC}_{\mathrm{FI2}}^*$ phase has a four-layer period. Recently a structure with a six-layer period has been found in the $\mathrm{SmC}_{6\mathrm{d}}^*$ phase. Structures with longer periods can be stable only in systems where interactions between distant layers exist. Below we will show that flexoelectricity is crucial for interactions between more distant layers, i.e. the layers that are not nearest neighbours.

### 5.3.1. *Structures of phases*

Let us first briefly describe the phase structures, their macroscopic properties and the usual phase sequences. To date, six different tilted phases have been experimentally confirmed: the usual SmC$^*$ phase, the $\mathrm{SmC}_A^*$ phase, the $\mathrm{SmC}_{\mathrm{FI1}}^*$ phase, the $\mathrm{SmC}_{\mathrm{FI2}}^*$ phase, the $\mathrm{SmC}_{6\mathrm{d}}^*$ phase and the $\mathrm{SmC}_\alpha^*$ phase. The phases are listed with respect to an increasing commensurate periodicity except the last, the $\mathrm{SmC}_\alpha^*$ phase, which has a short period incommensurate with a number of layers in general. Let us also define the way in which the structures will be presented. For the sake of clarity we will avoid three-dimensional images but we will present the structures from a bird's-eye view. For example, Fig. 5.4 gives the structures of the SmC$^*$, $\mathrm{SmC}_A^*$ and the two $\mathrm{SmC}_\alpha^*$ phases with a pitch longer than four layers and a pitch shorter than four layers, respectively. The layer normal is parallel to the $z$ axis and the tilt and the orientation are given on the cone. For the projections of the tilt we will use the projection of the ellipses. The direction of the tilt can be found using the arrows in the circles as follows. In the bird's-eye view, the lower part of an 'average' molecule in a layer is always positioned in the centre of the circle (the tail of the arrow), the radius of the circle gives the magnitude of the tilt and the position of the other side (the head of the arrow) gives the tilt direction. The layers are numbered. The $j$th layer is numbered as 0, the $(j+1)$th layer is numbered 1, etc.

#### 5.3.1.1. *The ferroelectric SmC$^*$ phase*

The structure of the ferroelectric SmC$^*$ phase was described in detail in the previous section. We will repeat only the essentials. Molecules are tilted away from the layer normal, the tilt precesses around the cone from layer to layer but the direction differs between two layers only slightly. The tilts are almost synclinic. If the helicoidal modulation is neglected, we can say that

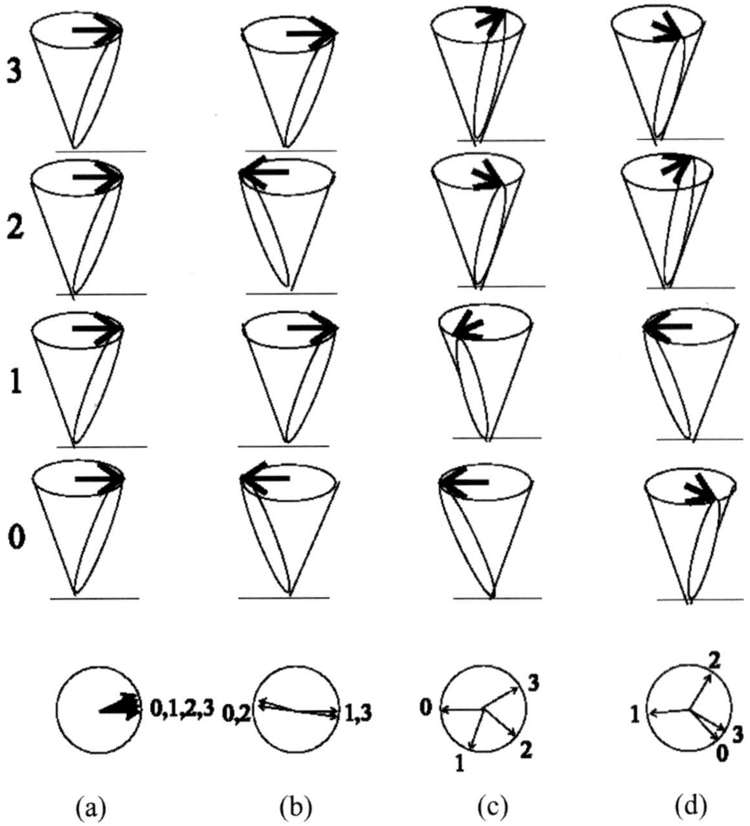


Fig. 5.4. Structures of (a) the SmC* phase, (b) the $SmC_A^*$ phase and (c,d) the $SmC_\alpha^*$ phase. An arrow is used instead of a nail symbol for the tilt directions for ease of presentation.

a repeating unit of the structure is a single layer. As the tilts are almost parallel, polarizations in neighbouring layers are also parallel. Hence the structure is locally (over a few layers) synclinic and ferroelectric; therefore the name. The structure can be described as a sequence of angles formed by tilt directions in neighbouring layers and only a single angle $\alpha$ is required where $\alpha \approx 0$ (Fig. 5.4a).

### 5.3.1.2. *The antiferroelectric $SmC_A^*$ phase*

The antiferroelectric phase is present in all materials that are called antiferroelectric liquid crystals, as the lowest temperature phase of the tilted

SmC* variants of phases. In two neighbouring layers the molecules are tilted in opposite directions (anticlinically) having therefore also antiparallel polarizations (Fig. 5.4b). Polarizations cancel out over two layers. The structure is rather resistant to the influence of an electric field as long as the field is not strong enough to induce the structure to change into the synclinic ferroelectric one. A double peak under a triangular voltage and a double hysteresis under an electric field can be observed. Therefore and also due to the local polarization order it is called antiferroelectric $SmC_A^*$ phase. This structure, although macroscopically very different from the SmC* structure can also be described by a single angle formed by the tilt directions in neighbouring layers, i.e. by a single phase difference $\alpha \approx \pi$.

#### 5.3.1.3. *The incommensurate $SmC_\alpha^*$ phase*

This structure, when it is present, appears always as the first stable tilted phase below the orthogonal SmA* phase. Tilts in all layers are equal and also all phase differences $\alpha$ are equal. From this point of view the phase is equal to the SmC* and the $SmC_A^*$ phase. The main difference in the macroscopic properties is the magnitude of the angle $\alpha$, which is neither close to zero nor close to $\pi$ (Fig. 5.4c and d). The structure is helicoidally modulated but has a very short pitch, which extends from a few tens of layers to shorter periods than four layers.[18–22] Due to the short pitch, the structure optically does not differ from the orthogonal phase and does not exhibit other properties like optical rotatory power, typical for other helicoidally modulated structures.

#### 5.3.1.4. *The antiferroelectric $SmC_{FI2}^*$ phase*

Another phase that appears above the antiferroelectric $SmC_A^*$ phase has a four-layer periodicity (Fig. 5.5a). The tilts in all layers are the same but the direction of the tilt in neighbouring layers forms a sequence of two different angles $\alpha$ and $\beta$. The sum of the two angles is close to $\pi$. The slight deviation $\Delta/2$ from $\pi$ results in a long helicoidal modulation. The polarization cancels out over four layers and the structure behaves antiferroelectrically.

#### 5.3.1.5. *The ferrielectric $SmC_{FI1}^*$ phase*

Its basic period is three layers and the whole structure can be described by the sequence of three phase differences between tilt directions in neighbouring layers: two $\alpha$s and a $\beta$. The sum of the three phase differences

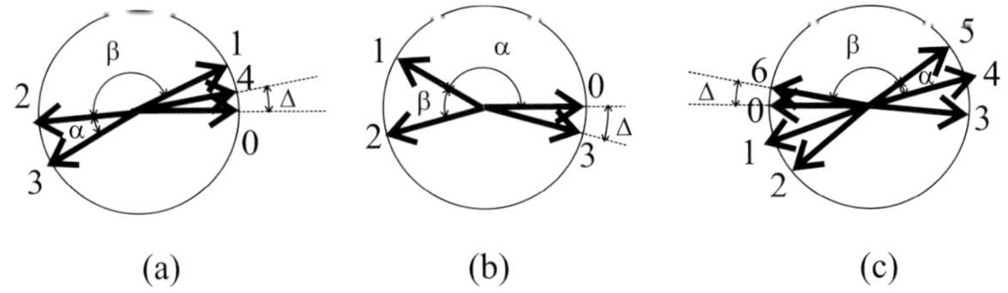


Fig. 5.5. Structures of (a) the $\mathrm{SmC}^*_{\mathrm{FI2}}$ phase with two interchanging phase differences in the sequence $\alpha, \beta$. The sum $2(\alpha + \beta)$ differs from $2\pi$ by an angle $\Delta$ resulting in a long helical modulation. (b) The $\mathrm{SmC}^*_{\mathrm{FI1}}$ phase and (c) the $\mathrm{SmC}^*_{\mathrm{6d}}$ phase with the three phase differences in the sequences $\alpha, \alpha, \beta$. The sum of the angles $(2\alpha + \beta)$ differs from $2\pi$ by $\Delta$ in the $\mathrm{SmC}^*_{\mathrm{FI1}}$ phase and from $\pi$ by $\Delta/2$ in the $\mathrm{SmC}^*_{\mathrm{6d}}$ phase.

$2\alpha + \beta \approx 2\pi$ (Fig. 5.5b). The slight incommensurability results in non-strictly parallel tilts (differing by $\Delta$) in two neighbouring structural units and a consequent long helicoidal modulation of the structure. The polarization in the basic period cancels out only partially and therefore the structure has ferrielectric properties. The phase is present in chiral samples only and it appears immediately above the antiferroelectric $\mathrm{SmC}^*_A$ phase.

### 5.3.1.6. *The six-layer $SmC^*_{6d}$ phase*

This structure was discovered only recently.[23] It appears below the $\mathrm{SmC}^*_\alpha$ phase in situations where the incommensurate helical pitch approaches six layers. Its basic period extends over six layers. It can also be described by the sequences of three phase differences: two $\alpha$s and a $\beta$. Unlike the $\mathrm{SmC}^*_{\mathrm{FI1}}$ phase the sum of the three phase differences $(2\alpha + \beta)$ differs from $\pi$ by $\Delta/2$. (Fig. 5.5c). Unfortunately, only resonant X-ray measurements exist for this structure. It would be interesting to have reports on optical and other properties as well.

How are all these phases positioned in the phase sequence lowering the temperature? The sequence described below is different only in some rare cases of re-entrant phases.[24] However, there is no material known at present where all the phases are present in the same sample. The most general phase sequence has been found in the prototype material 4-(1-methylheptyloxycarbonyl)phenyl 4′-octyloxybiphenyl-4-carboxylate (**MHPOBC**).[25] It has all phases mentioned but the recently discovered $\mathrm{SmC}^*_{\mathrm{6d}}$ phase. However, in the phase sequence of the material exhibiting the six-layer $\mathrm{SmC}^*_{\mathrm{6d}}$ phase[23] other phases are missing. In different materials various phases are missing, for example, in racemic **MHPOBC** only the SmC$^*$ and the SmC$^*_A$ phases remain. But the order of the sequence remains the same.

$$\mathrm{SmA} \leftrightarrow \mathrm{SmC}^*_\alpha \leftrightarrow \mathrm{SmC}^*_{6\mathrm{d}} \leftrightarrow \mathrm{SmC}^* \leftrightarrow \mathrm{SmC}^*_{\mathrm{FI2}} \leftrightarrow \mathrm{SmC}^*_{\mathrm{FI1}} \leftrightarrow \mathrm{SmC}^*_A$$

Which of the phases are important for flexoelectricity? As we shall see below, the flexoelectric effect is the main reason for the large variety of phases. The flexoelectric interaction is actually the reason for significant interactions with the more distant layers. In addition, phases with larger phase differences are a source of another phenomenon: the local polarization can also have a component parallel to the tilt direction of the polarization. However, to understand the richness of the phenomena, let us first focus on the phenomenological model, which describes all the phases above, their properties and the phase sequence.

### 5.3.2. *Discrete model*

The structures described above are much different from the SmC$^*$ structure, as the tilt direction can change from layer to layer significantly. It was hard to believe for a long time that such changes are possible because they were considered as energetically too unfavourable.

The speculation that such changes can occur was considered theoretically[26] before any structure was actually observed with the newly developed resonant X-ray scattering method.[18] If we wish to describe the free energy of the system, where the direction or magnitude of the tilt order parameter can change abruptly from layer to layer, the continuum description of the order parameter changes is no longer adequate. The reason for terms that are described by derivatives of the order parameter with respect to the spatial variables, are interactions between molecules in neighbouring layers. If changes are significant, derivatives are no longer adequate and interactions have to take into account the discreteness of the layers. The interactions have to be divided into intermolecular interactions within a layer and to interactions between molecules from different layers. The approach is usually called a discrete phenomenological approach as interactions between discrete layers are taken into account explicitly.[26]

Let us illustrate the relation between the continuous and the discrete description of the free energy, for example for the elastic contribution to the free energy. The expression

$$\frac{1}{2}K\left(\left(\frac{d\xi_x}{dz}\right)^2+\left(\frac{d\xi_y}{dz}\right)^2\right) \tag{5.17}$$

presents the elastic contribution to the free energy density. In the discrete form the derivative is replaced by an average of the differences between the interacting layers above and below the considered layer:

$$\frac{d\xi_x}{dz} = \frac{1}{2}\left(\frac{\xi_x\left((j+1)d\right) - \xi_x\left(jd\right)}{d} + \frac{\xi_x\left(jd\right) - \xi_x\left((j-1)d\right)}{d}\right). \tag{5.18}$$

Here $j$ counts the layers and $d$ is the layer thickness. Inserting Eq. (5.18) into Eq. (5.17) gives an additional term of the form $\xi^2$, which provides a different description of the elastic energy with a coefficient $K$ that is related to the elastic properties of the material:

$$-\frac{1}{4}\,\frac{K}{d}\,\left(\frac{1}{2}\Big(\boldsymbol{\xi}\left(jd\right)\cdot\boldsymbol{\xi}\left((j+1)d\right) + \boldsymbol{\xi}\left(jd\right)\cdot\boldsymbol{\xi}\left((j-1)d\right)\Big)\right), \tag{5.19}$$

with a formally negative sign. Usually in the discrete form the free energy density is transformed to the free energy per layer as an 'integral' over the thickness of one layer. Therefore $K$ is divided by $d$ only. The negative sign in Eq. (5.19) favours parallel tilts in neighbouring layers and is consistent with our understanding of the elastic energy costs. In the expression the position of the order parameter is marked by the coordinate of the layer centre of mass, for example $jd$. Such notation may become tedious and also gives a false impression that the order parameter is nicely defined in other positions, for example, between the layers, so it is better to introduce the layer order parameters. The $\boldsymbol{\xi}_j$ marks the tilt order parameter in the $j$th layer and similarly $\boldsymbol{P}_j$ marks the polarization in the $j$th layer.

The expression in Eq. (5.19) is not limited to small changes from layer to layer. Let us now write the free energy in a discrete form. We will intentionally leave a discrete description for flexoelectric interactions for more detailed discussion later. Here we give the form of the free energy and in a discrete system show how to search for structures that minimize the free energy and how to verify their stabilities. We discuss also questions that remain open if flexoelectricity is not considered. The free energy is now the sum of the energies specific for each ($j$th) layer

$$\begin{aligned}\sum_j G_j = \sum_j \quad & \frac{1}{2}\,a_0\,\xi_j^2 + \frac{1}{4}\,b_0\,\xi_j^4 + c_p\,\left(\boldsymbol{\xi}_j \times \boldsymbol{P}_j\right)_z + \frac{1}{2\epsilon_0}\,P_j^2 \\ & + \frac{1}{2}\,a_1\,\left(\boldsymbol{\xi}_j \cdot \boldsymbol{\xi}_{j+1}\right) + \frac{1}{2}\,f_1\,\left(\boldsymbol{\xi}_j \times \boldsymbol{\xi}_{j+1}\right)_z \\ & + \frac{1}{4}\,b_Q\,\left(\boldsymbol{\xi}_j \cdot \boldsymbol{\xi}_{j+1}\right)^2 + \frac{1}{2\epsilon_1}\,\left(\boldsymbol{P}_j \cdot \boldsymbol{P}_{j+1}\right).\end{aligned} \tag{5.20}$$

In the free energy Eq. (5.20) only interactions to nearest neighbouring layers (NN) are considered. From what we know today, it seems that early speculations were not correct, although they led to important results.[26–28] It was assumed that significant direct interactions due to polar layer properties exist for the next nearest layers. It was shown later that they could not be considered as important.[29] What are the meanings of the terms in Eq. (5.20)? The first four terms give intralayer interactions. They give interactions between molecules that are found within the same layer. We mark intralayer interaction with a subscript 0. The remaining terms give interactions between molecules in neighbouring layers. Other than the coefficient $b_Q$, all of them are marked by the subscript 1, which marks interactions to the first neighbouring layers. The term with the coefficient $a_1$ can be recognized from Eq. (5.19) and it gives the energy cost for the non-uniformity of the tilt magnitude or direction in the first approximation. The idea that the coefficient $a_1$ can also be positive, favouring antiparallel tilt alignments in neighbouring layers, was straightforwardly connected to the elasticity and the resulting 'negative' elastic constant was rather hard to accept. As the negative coefficient $a_1$ allows for an experimentally consistent explanation for phase structures, the concept of elasticity has to be reconsidered for complex structures in antiferroelectric liquid crystals. So, the coefficient $a_1$ can have either a negative or a positive sign. A negative sign favours synclinic tilts in neighbouring layers and a positive sign favours anticlinic tilts in neighbouring layers. The next term with coefficient $f_1$ gives chiral interactions between neighbouring layers. The origin of the chiral interactions is the chiral molecular shape and consequently the chiral symmetry of the van der Waals field that is formed around the chiral molecule. When another molecule from a neighbouring layer is within this field, the orientation with the lowest energy forms an angle with the tilt direction of the molecules that are the source of the van der Waals attractive forces. This expression has to be considered as a first approximation for this contribution. The term $b_Q$ gives quadrupolar coupling. Molecules can have quadrupolar moments and if so they tend to be tilted perpendicularly in neighbouring layers. This phenomenon contributes to the coefficient $b_Q$ positively. On the other hand, if molecules are flat, lath-like objects, diffusion through the layers is promoted when the short axes order within a layer, but at the same time, the van der Waals attraction prefers antiparallel tilts. This part contributes negatively to the coefficient $b_Q$. The prevailing interaction determines the sign of the coefficient. It seems that materials having $b_Q$ positive are extremely rare.[22,30] The final term $\epsilon_1$

gives polar interactions to neighbouring layers. The term is conceptually new as it accounts for the non-uniformity of the polarizations, which was not considered in the continuous approach.

How can we find structures that minimize the free energy? From the formal point of view there is only one structure that can be described by a constant tilt magnitude and a constant phase difference between neighbouring layers. The *ansatz* for $\boldsymbol{\xi}_j$ and the polarization $\boldsymbol{P}_j$, which is perpendicular to $\boldsymbol{\xi}_j$,

$$\begin{aligned} \boldsymbol{\xi}_j &= \theta\{\cos(j\alpha + \phi_0), \sin(j\alpha + \phi_0)\}, \\ \boldsymbol{P}_j &= P\{-\sin(j\alpha + \phi_0), \cos(j\alpha + \phi_0)\}, \end{aligned} \tag{5.21}$$

is inserted into Eq. (5.20) and gives

$$\begin{aligned} G = \sum_j \quad & \frac{1}{2}\, a_0\, \theta^2 + \frac{1}{4}\, b_0\, \theta^4 + c_p\, \theta\, P + \frac{1}{2\epsilon_0}\, P^2 \\ & + \frac{1}{2}\, a_1\, \theta^2 \cos\alpha + \frac{1}{2}\, f_1\, \theta^2 \sin\alpha + \frac{1}{4}\, b_Q\, \theta^4 (\cos\alpha)^2 \\ & + \frac{1}{2\epsilon_1} P^2 \cos\alpha. \end{aligned} \tag{5.22}$$

The free energy is equal for all layers, if surface effects are neglected. Therefore the generality is not lost if only one layer is considered. The minimization of the free energy with respect to the polarization leads to

$$P = -\frac{\epsilon_0\, c_p}{1 + \frac{\epsilon_0}{\epsilon_1} \cos\alpha}\, \theta. \tag{5.23}$$

Inserting the polarization back into the polarization part of the free energy and expanding the free energy with respect to the ratio $\frac{\epsilon_0}{\epsilon_1}$ gives

$$G_P = -\frac{1}{2}\, c_p^2\, \epsilon_0\, \theta^2 + \frac{1}{2}\, c_p^2\, \epsilon_0\, \frac{\epsilon_0}{\epsilon_1}\, \theta^2\, \cos\alpha. \tag{5.24}$$

The first term renormalizes the first coefficient $a_0$ and increases the transition temperature in the same way as in the continuous form of the free energy, Eq. (5.16), while the coefficient $a_1$ changes due to the piezoelectrically induced polarization. The sign of the contribution to $a_1$ is positive as polarizations favour antiferroelectric ordering in neighbouring layers. In the next step the free energy is minimized with respect to the tilt order parameter.

$$\theta^2 = -\frac{\tilde{a}_0 + \tilde{a}_1 \cos\alpha + f_1 \sin\alpha}{b_0 + b_Q \cos^2\alpha}. \tag{5.25}$$

The solution has a physical meaning only when the right-hand side of the equation is positive. Close to the transition temperature this must be verified. Inserting the solution in the free energy leads to

$$G = -\frac{1}{4}\frac{(\tilde{a}_0 + \tilde{a}_1 \cos\alpha + f_1 \sin\alpha)^2}{b_0 + b_Q \cos^2\alpha}, \tag{5.26}$$

which finally has to be minimized with respect to $\alpha$ numerically. From known $\alpha$ then $\theta$ and $P$ are also defined and from the Eq. (5.21) the structure is known in detail.

### 5.3.3. *Discrete form of flexoelectricity*

In the previous section flexoelectric interactions were not considered in the free energy. We have also seen that only three of the structures found in antiferroelectric liquid crystals can be explained with the form of the free energy presented in the previous section. Let us first consider the discrete form of the flexoelectric effect and its influences on the theoretical description of the structures. We shall see that the flexoelectric effect is a source of indirect interactions between more distant layers and consequently the reason for all structures that cannot be expressed by the single phase difference.

Flexoelectricity appears because interactions with molecules above and below the interacting layer are different. These interactions hinder rotation of the molecules in interacting layers in a different way, which affects the direction of the most favourable molecular orientation around the long molecular axis. As a consequence the direction and the magnitude of the polarization change. The flexoelectric interaction is described by the following contribution to the free energy

$$\frac{1}{2}\,\mu\,(\boldsymbol{P}_j \cdot \boldsymbol{\xi}_{j+1} - \boldsymbol{P}_j \cdot \boldsymbol{\xi}_{j-1})\,. \tag{5.27}$$

Elimination of the polarization from the free energy Eq. (5.20) with the additional term Eq. (5.27) gives surprising results. The detailed derivation of the elimination and the consequent simple form of the free energy are given in Takezoe *et al.*[1] and Čepič and Žekš.[31] Here we give only the results and comment on them.

The free energy of a single layer, limited to interlayer interactions only, has after elimination of the polarization a very simple form

$$G_{int,j} = \sum_k \quad \frac{1}{2}\,\tilde{a}_k\,(\boldsymbol{\xi}_j \cdot \boldsymbol{\xi}_{j+k}) + \frac{1}{2}\,\tilde{f}_k\,(\boldsymbol{\xi}_j \times \boldsymbol{\xi}_{j+k})_z$$

$$+\frac{1}{4}\,b_Q\,(\boldsymbol{\xi}_j \cdot \boldsymbol{\xi}_{j+1})^2 \,. \tag{5.28}$$

Interactions with more distant layers can be understood in the following way. Direct interactions extend to nearest neighbouring layers only. However, interactions with nearest layers influence the polarization flexoelectrically and these different polarizations now interact with neighbouring layers. Below we give the achiral interactions up to the third neighbouring layer and the chiral one up to the second neighbouring layer:

$$\begin{aligned}
\tilde{a}_1 &= a_1 + \left(\epsilon_0\, c_p^2 + \frac{1}{4}\epsilon_0\,\mu^2\right)\left(\frac{\epsilon_0}{\epsilon_1}\right),\\
\tilde{a}_2 &= \frac{1}{2}\epsilon_0\,\mu^2,\\
\tilde{a}_3 &= -\frac{1}{8}\epsilon_0\,\mu^2\left(\frac{\epsilon_0}{\epsilon_1}\right),\\
\tilde{f}_1 &= f_1 - 2\,\epsilon_0\,c_p\,\mu,\\
\tilde{f}_2 &= \epsilon_0\,c_p\,\mu\left(\frac{\epsilon_0}{\epsilon_1}\right).
\end{aligned} \tag{5.29}$$

Interactions are of indefinite range, but the magnitudes have to be compared. $\epsilon_0/\epsilon_1$ is the ratio between polar interactions in the layer ($\epsilon_0$) and between neighbouring layers ($\epsilon_1$). If layers are well distinguished, which is actually the case in antiferroelectric liquid crystals, this ratio is estimated to be about 0.1 to 0.2. From the expressions it can clearly be seen that interactions with more distant layers decrease as $(\epsilon_0/\epsilon_1)^k$, where $k$ denotes the layer which is the $k$th neighbour. If we keep only terms linear in the ratio $\epsilon_0/\epsilon_1$, the interactions are significant only to the third neighbours in achiral indirect interactions expressed in $\tilde{a}_k$ and to the second neighbours in chiral indirect interactions described by $\tilde{f}_k$.

Let us take a closer look at the nature of these interactions. The achiral bilinear interactions have rather simple 'desires'. If they are negative, they favour a parallel orientation in the interacting layers. If they are positive, they favour antiparallel orientations in the interacting layers. Chiral interactions are different. They have opposite signs in oppositely handed enantiomers. Interactions favour, in this basic linear approximation, perpendicular tilts in interacting layers. The favourable sense, i.e. the $\pm 90°$

sign, is defined by the sign of the coefficient. As direct interactions are significant only for nearest layers, the effective interactions have at least two different contributions: the contribution of direct, van der Waals or steric interactions, which were initially expressed solely in tilt order parameters, and indirect interactions, which are the consequences of the piezoelectrically or flexoelectrically induced polarization, which consequently interact. As there are more contributions, we can understand the rich variety of phases within a narrow temperature region. The temperature variations affect the smectic order, which increases with decreasing temperature. The smectic order affects all sources of intralayer and interlayer interactions. Different contributions to the effective achiral interaction between nearest layers change and may cancel out. This does not mean that the neighbouring layers do not interact, it means only that the interactions may cancel at a particular temperature and that the direction of the tilt in one layer does not influence the direction of the tilt in a neighbouring layer through the usually most important achiral interlayer bilinear coupling $a_1$. At this temperature interactions with more distant layers significantly influence the structure and the whole zoo of phases may appear.

The historical theoretical explanation of the structures found in antiferroelectric liquid crystals was actually different. Initially it was believed that anticlinic tilts in next nearest layers, which are the source of the competing interactions, are present because polar interactions with next nearest layers are strong enough. It was shown later that the flexoelectric effect is much more important[29] and that these interactions are most probably the main source of the competition.

An especially hard nut to crack was a theoretical understanding of the multi-layer periodicities. It was believed that the periodicity should be locked to the periods for which interactions, direct or indirect, are important. Below we will show that interactions with the nearest layers can only stabilize periods up to six layers away.

### 5.3.4. *Lock-in periodicities*

Let us now discuss the periodicities of the structures. The majority of the terms in the free energy are bilinear. They allow for large changes in the tilt direction or magnitude and they are scalars or the $z$ component of the cross product. For the correctness, a $z$ component equal to zero is added to the two-dimensional order parameters $\boldsymbol{\xi}_j$ and $\boldsymbol{P}_j$ as before in Eq. (5.11). To aid in understanding the structure formation, we now consider structures that are favoured by separated terms describing direct or indirect interactions.

Achiral bilinear interactions can be considered in the same way. When the sign of the coefficient is negative, it favours parallel orientations in interacting layers. When the sign of the coefficient is positive, the antiparallel orientation in interacting layers is favoured. These favourable orientations do not depend on the orientations of layers that are not included in the interactions. Chiral bilinear interactions favour a perpendicular orientation in interacting layers. Figure 5.6 gives the schematic structures that are favourable for a specific term. Grey ellipses mark interacting layers; transparent ellipses mark layers that do not interact with the layer 0 due to the coefficient given in the first column. Their orientation is not important for the free energy due to the corresponding coefficient. Bilinear terms with coefficients $\tilde{a}_k$ and $\tilde{f}_k$ are insufficient for the stabilization of commensurate structures. The lock-in occurs due to the additional biquadratic interaction $b_Q$, which simultaneously favours parallel- or antiparallel-oriented tilts in neighbouring layers for negative values and provides the energy barrier for out-of-plane orientations. For positive values of the coefficient $b_Q$ the lock-in does not occur.

Let us discuss the origin of these interactions and how they can change with temperature in order to understand the formation of different structures in different temperature ranges.

### 5.3.4.1. *Achiral interactions* $\tilde{a}_1$

The term $\tilde{a}_1$ gives achiral bilinear interactions to nearest neighbouring layers expressed in tilts:

$$\tilde{a}_1 = a_1 + \epsilon_0\, c_p^2 \left(\frac{\epsilon_0}{\epsilon_1}\right) + \frac{1}{4}\epsilon_0\, \mu^2 \left(\frac{\epsilon_0}{\epsilon_1}\right). \tag{5.30}$$

The three components have different origins. The first part $a_1$ gives direct steric and van der Waals interactions. Molecules that penetrate through layers promote the same direction of tilt in neighbouring layers, therefore they contribute to the $a_1$ negatively. Van der Waals attractive interactions favour antiparallel tilts in neighbouring layers, for which parts of the molecules are on average closer than for synclinic interactions. They contribute to $a_1$ positively. The next component gives direct interactions between polarizations that are induced by the piezoelectric effect. Polarization interactions usually favour the antiferroelectric ordering of the polarizations and consequent anticlinic ordering of the tilts. Therefore their

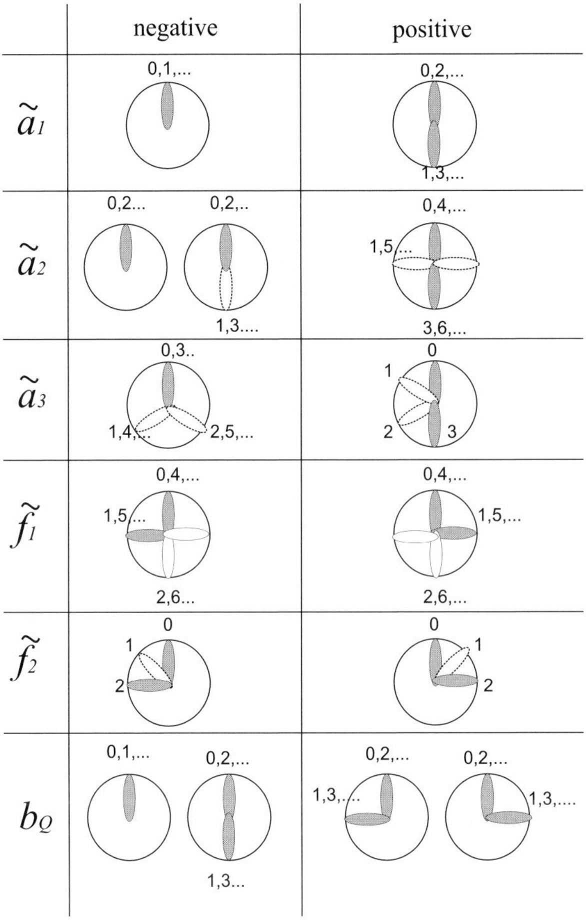


Fig. 5.6. Favourable structures for isolated interlayer interactions. Grey ellipses mark favourable orientations in layers that interact with layer 0. White ellipses mark layers that do not contribute to the interactions.

contribution to $\tilde{a}_1$ is positive. However, we can also find some rather unusual cases where the ferroelectric polarization in neighbouring layers is favoured.[27] The final part is the interaction of the flexoelectrically induced polarization. Piezoelectrically and flexoelectrically induced polarizations interact in the same way, therefore the discussion before is valid also for this part of the interaction. If the smectic order changes with decreasing temperature, the interlayer molecular diffusion decreases and positional interlayer correlations change.[27] Therefore specific contributions to $\tilde{a}_1$ vary with temperature changes and may cancel out and the coefficient $\tilde{a}_1$ may change sign.

If there were only interactions described by negative $\tilde{a}_1$, their structure would be synclinic ferroelectric. The basic period would consist of a single layer. For positive $\tilde{a}_1$ the structure would be antiferroelectric anticlinic (Fig. 5.6, first row).

#### 5.3.4.2. *Achiral interactions $\tilde{a}_2$*

The term $\tilde{a}_2$ gives the indirect achiral bilinear interaction to the next nearest neighbouring layers expressed in tilts:

$$\tilde{a}_2 = \frac{1}{2}\,\epsilon_0\,\mu^2. \tag{5.31}$$

The term is the result of the electrostatic free energy of the flexoelectrically induced part of the polarization. It can clearly be seen that this interaction originates in the flexoelectric effect and is always positive. The next nearest layers therefore favour an anticlinic orientation. The basic period of the structure, if this is the only interaction, would consist of four layers (Fig. 5.6, second row). In addition, if the flexoelectrically induced polarization is comparable to the piezoelectrically induced polarization, the indirect interaction with the next nearest layers can be surprisingly strong.

#### 5.3.4.3. *Achiral interactions $\tilde{a}_3$*

The term $\tilde{a}_3$ gives the indirect achiral bilinear interactions to the third nearest neighbouring layers expressed in tilts:

$$\tilde{a}_3 = -\frac{1}{8}\epsilon_0\,\mu^2\left(\frac{\epsilon_0}{\epsilon_1}\right). \tag{5.32}$$

The term is the result of the two interacting flexoelectrically induced polarizations in neighbouring layers. As the flexoelectric part of the polarization is determined by tilts in layers above and below the interacting layers, the neighbours are indirectly 'felt' over three layers. Interactions with the third nearest layers are not influenced by the sign of the flexoelectric coefficient $\mu$, as they depend on it quadratically. But the interlayer polar interactions can have either sign: positive favouring the usual antiferroelectric ordering and negative favouring the ferroelectric ordering.[27] So, we can in principle expect either: the favourable parallel ordering of the third neighbours and the consequent three-layer $\mathrm{SmC}^*_{\mathrm{FI1}}$ structure or the favourable antiparallel ordering of the third nearest neighbours and the consequent six-layer $\mathrm{SmC}^*_{6\mathrm{d}}$ structure (Fig. 5.6, third row).

#### 5.3.4.4. *Chiral interactions* $\tilde{f}_1$

The term $\tilde{f}_1$ gives the direct and indirect chiral bilinear interactions with neighbouring layers. Chiral interactions have different consequences than achiral ones. First, they depend on molecular handedness. Coefficients, which depend on these interactions linearly, change sign for the samples consisting of oppositely handed enantiomers. Second, the first approximation of the chiral interaction given by the cross product of the interacting layers favours a perpendicular arrangement of the interacting layers, where the sense of the angle is defined by the sign of the chiral interaction. So, the natural period in the absence of any other interaction is four times the number of layers which interact (Fig. 5.6, fourth row). As an illustration, $\tilde{f}_1$ favours four-layer wound structures while $\tilde{f}_2$ favours eight-layer wound structures. Third, chiral interactions are weaker than achiral ones for one order of the ratio $\frac{\epsilon_0}{\epsilon_1}$. So, the indirect chiral interactions with nearest layers are of the same order as achiral interactions with the next nearest layers, if we consider ratios only.

$$\tilde{f}_1 = f_1 - 2\,\epsilon_0\,c_p\,\mu. \tag{5.33}$$

In the expression Eq. (5.33) we can see that the flexoelectric effect when it is combined with the piezoelectric effect (the second part of the coefficient $\tilde{f}_1$) has similar effects as direct chiral interactions due to the van der Waals field having chiral symmetry around the chiral molecules given by $f_1$. We cannot distinguish between the two components as the piezoelectric coefficient $c_p$ and the coefficient describing direct chiral interactions $f_1$ probably depend equally (they are proportional) on the enantiomeric excess.

#### 5.3.4.5. *Chiral interactions* $\tilde{f}_2$

The final indirect bilinear interaction that is significant enough to be considered, is the interaction with the next nearest layers having a chiral character:

$$\tilde{f}_2 = \epsilon_0\,c_p\,\mu\left(\frac{\epsilon_0}{\epsilon_1}\right). \tag{5.34}$$

Its origin is in the combination of the chiral piezo-interaction and the achiral flexoelectric interaction. It favours a perpendicular orientation in interacting layers, which means a period of eight layers (Fig. 5.6, fifth row).

When interacting layers are considered, one does not take into account the orientations of intermediate layers. So, chiral interactions in principle favour structures that have periodicities of four times the interacting distances. Therefore they favour structures with a $\pm 90°$ difference in tilt directions for neighbouring layers ($\tilde{f}_1$) and the consequent four-layer periodicity or structures with a $\pm 90°$ difference in tilt direction for the next nearest neighbouring layers ($\tilde{f}_2$), which favours eight-layer periods. However, as they favour large angles between neighbouring and next neighbouring layers, they actually stabilize the $\mathrm{SmC}_\alpha^*$ phase with its short helicoidal modulations. It is not expected that chiral interaction to more distant layers will lead to commensurate structures with a period of eight layers.

5.3.4.6. *Quadrupolar biquadratic interactions* $b_Q$

This interaction has already been discussed. Here we consider favourable structures for the two possible signs of the coefficient (Fig. 5.6, sixth row). The negative sign is the reason for lock-in, while the positive sign of $b_Q$ is for interactions favouring perpendicular orientations in neighbouring layers, which combine with chiral interactions and additionally stabilize helicoidal modulations in the $\mathrm{SmC}_\alpha^*$ phases.

How does this correspond to experimental facts? Previously we discussed only solutions where the structure can be described by a single constant tilt $\theta$ and a single phase difference $\alpha$. This simple description allows for structures with three different phases: the $\mathrm{SmC}^*$, the $\mathrm{SmC}_A^*$ and the $\mathrm{SmC}_\alpha^*$ phases. Below we will consider structures where more phase differences are present as suggested by Fig. 5.5.

When searching for a solution, we can apply cycling boundary conditions[29] with the desired number of layers. Using this approach, we risk losing solutions that do not have commensurate periods or have periods longer than the cycling boundary conditions applied.

Another approach is to start with a large set of initial tilts and to run the minimization process numerically.[32,33] The risk is that the minimization procedure will lock onto metastable solutions due to the energetic barriers between structures with similar periods. When biquadratic interactions are important, the procedure may produce artificial structures.

The last procedure considers structures with particular sequences of phase differences.[1,28] We have to bear in mind that, for example, the sequence of two phase differences, say $\alpha$ and $\beta$, does not necessarily mean a commensurate structure. Only when the sum of the two angles is equal to the rational factor of $\pi$, can the periodicity with respect to the layers be recognized. Let us now use this approach for a theoretical consideration of three commensurate structures with periods of four, three and six layers described by a sequence of two or three phase differences.

### 5.3.4.7. *Period two: The SmC$^*_{FI2}$ phase*

The four-layer structure of the SmC$^*_{\rm FI2}$ phase is schematically drawn in Fig. 5.5a. It can be represented by the phase difference sequence $\alpha, \beta, \alpha, \beta, ...$ where the period in the sequence of the phase differences is two. The structure of the SmC$^*_{\rm FI2}$ phase can be described as:

$$\begin{aligned} \boldsymbol{\xi}_{2j} &= \theta\{\cos(j(\alpha+\beta)), \sin(j(\alpha+\beta))\}, \\ \boldsymbol{\xi}_{2j+1} &= \theta\{\cos(j(\alpha+\beta)+\alpha), \sin(j(\alpha+\beta)+\alpha)\}. \end{aligned} \tag{5.35}$$

Here the sum $(\alpha+\beta)$ is still arbitrary and the structure is not commensurate to any number of layers in principle. The average free energy Eq. (5.28) of the two layers using the *ansatz* Eq. (5.35) becomes

$$\begin{aligned} G_2 &= \frac{1}{2}\, a_0\, \theta^2 + \frac{1}{4}\, b_0\, \theta^4 + \frac{1}{4}\, \tilde{a}_1\, \theta^2 (\cos\alpha + \cos\beta) \\ &+ \frac{1}{2}\, \tilde{a}_2\, \theta^2 \cos(\alpha+\beta) + \frac{1}{4}\, \tilde{a}_3\, \theta^2 (\cos(2\alpha+\beta) + \cos(\alpha+2\beta)) \\ &+ \frac{1}{4}\, \tilde{f}_1\, \theta^2 (\sin\alpha + \sin\beta) + \frac{1}{2}\, \tilde{f}_2\, \theta^2 \sin(\alpha+\beta) \\ &+ \frac{1}{8}\, b_Q\, \theta^4 \left(\cos^2\alpha + \cos^2\beta\right). \end{aligned} \tag{5.36}$$

All tilts are constant, as all layers are found in the same environment with one neighbouring layer having the phase difference $\alpha$ and another having the phase difference $\beta$. In chiral systems the sum is $(\alpha+\beta) \approx \pi$ and the slight difference results in a long additional helicoidal modulation. To find the proper structure for a certain set of model coefficients, we have to minimize the following free energy first using the approximation that $\alpha + \beta = \pi$, which leads to the simplified form of the free energy:

$$G_2 = \frac{1}{2}\, a_0\, \theta^2 + \frac{1}{4}\, b_0\, \theta^4 - \frac{1}{2}\, \tilde{a}_2\, \theta^2 + \frac{1}{2}\tilde{f}_1\, \theta^2\, \sin\alpha + \frac{1}{4}\, b_Q\, \theta^4 \cos^2\alpha. \tag{5.37}$$

The solution is than obtained numerically using the three variable minimization procedure with first approximations for the sequence of the phase differences

$$\alpha = \frac{\tilde{f}_1}{b_Q\theta^2} \quad \text{and} \quad \beta = \pi - \frac{\tilde{f}_1}{b_Q\theta^2}. \tag{5.38}$$

For the tilt we use as the first approximation only the dependence on the interlayer interactions:

$$-\theta^2 = \frac{a_0}{b_0}. \tag{5.39}$$

The minimization usually results in a slight shift of the angles and the sum of the two angles $\alpha$ and $\beta$ differs by $\Delta/2$ from $\pi$ resulting in a long helicoidal modulation of the four-layer structure. When considering this solution, we have to be careful as the solution is often metastable. The free energy for the solution has to be compared to other possible solutions existing in the same temperature window. For example, if the solution for the structure of the $\mathrm{SmC}^*_\alpha$ phase is close to the six-layer period, we have to compare the solutions for $\mathrm{SmC}^*_{\mathrm{FI2}}$, $\mathrm{SmC}^*_\alpha$ and $\mathrm{SmC}^*_{\mathrm{6d}}$ to decide which has the lowest free energy and is therefore the stable one.

### 5.3.4.8. *Period three: The $SmC^*_{FI1}$ and the $SmC^*_{6d}$ phases*

The structures shown in Fig. 5.5b,c are formed by a sequence of three repeating phase differences $\alpha, \alpha, \beta, \alpha, \alpha, \beta, \ldots$. For the structure of the $\mathrm{SmC}^*_{\mathrm{FI1}}$ phase the sum $2\alpha + \beta \approx 2\pi$. For the structure of the $\mathrm{SmC}^*_{\mathrm{6d}}$ phase the sum $2\alpha + \beta \approx \pi$. However, both can be analysed with the same *ansatz* and the same simplified free energy. The symmetry requires different tilt magnitudes although we do not expect large differences.[34–36]

The structure of the three-layer $\mathrm{SmC}^*_{\mathrm{FI1}}$ phase and the structure of the recently discovered $\mathrm{SmC}^*_{\mathrm{6d}}$ phase can be written in the same way.

$$\begin{aligned} \boldsymbol{\xi}_{3j} &= \theta_0\{\cos(j(2\alpha+\beta)), \sin(j(2\alpha+\beta))\}, \\ \boldsymbol{\xi}_{3j+1} &= \theta_1\{\cos(j(2\alpha+\beta)+\alpha), \sin(j(2\alpha+\beta)+\alpha)\}, \\ \boldsymbol{\xi}_{3j+2} &= \theta_1\{\cos(j(2\alpha+\beta)+(\alpha+\beta)), \sin(j(2\alpha+\beta)+(\alpha+\beta))\}. \end{aligned} \tag{5.40}$$

The procedure to find the structure with a period three is the same as the procedure for finding the structure of the $\mathrm{SmC}^*_{\mathrm{FI2}}$ phase. The averaging of the free energy per layer has to be done over three layers. The phase difference $\beta$ has to be considered as $2\pi - 2\alpha$ for the three-layer $\mathrm{SmC}^*_{\mathrm{FI1}}$ phase structure and as $\pi - 2\alpha$ for the $\mathrm{SmC}^*_{6d}$ phase structure in the first approximation. For the first approximation of the tilt magnitude we can consider $\theta^2 \approx -\frac{a_0}{b_0}$; however, for the more detailed minimization of the free energy the four variable minimization must use a numerical solution for $\alpha$ and the consequent $\beta$ as starting points and the approximation for $\theta$ for both tilts $\theta_0$ and $\theta_1$. The solution can find metastable states and the free energies have to be compared with other possible solutions.

## 5.4. Flexoelectricity in Complex Structures

In the previous section it was shown that flexoelectricity is actually responsible for the interactions of longer range, i.e. for indirect interactions with more distant layers. In this section we will discuss the flexoelectric contribution to the polarization and a few examples where the polarization in a layer has a component along the tilt direction, i.e. a longitudinal component. It was believed for a long time that the polarization has only one possible component in chiral tilted smectics – the component perpendicular to the tilt. In structures where the tilt modulation is complex, this is not true. However, although the flexoelectric polarization in tilted chiral polar smectic structures can be very large, it influences the structures and is the main reason for competing interactions, so it is not possible for it to be detected by usual means such as dielectric response measurements or other straightforward effects of an electric field. We will also discuss the problem of observability of the flexoelectric component of the polarization. It seems that this component is extremely hard to detect experimentally.

### 5.4.1. *General direction of polarization*

The polarization in a specific tilted smectic layer formed of chiral liquid crystals always has two components, as already mentioned: the piezoelectric and the flexoelectric components. The piezoelectric component is to a first approximation:

$$\boldsymbol{P}_{j,\mathrm{piezo}} = \epsilon_0\, c_p\, \boldsymbol{n}_0 \times \boldsymbol{\xi}_j. \tag{5.41}$$

Here $\boldsymbol{n}_0$ is the layer normal. We consider only the first approximation as direct interactions between polarizations in neighbouring layers slightly change the magnitude of the polarization. However, they do not affect the direction of the polarization and therefore they do not introduce any new physics. The piezopolarization is strictly perpendicular to the tilt and the layer normal; the tilt order parameter and the polarization are mutually orthogonal, forming a left- or a right-handed system with a handedness that is opposite for left- and right-handed enantiomers (Fig. 5.7). On the other hand, the flexoelectric polarization depends on tilt orientations above and below the considered layer (again in the first approximation as above):

$$\boldsymbol{P}_{j,\mathrm{flexo}} = +\frac{1}{2}\,\epsilon_0\,\mu\,\left(\boldsymbol{\xi}_{j+1} - \boldsymbol{\xi}_{j-1}\right). \tag{5.42}$$

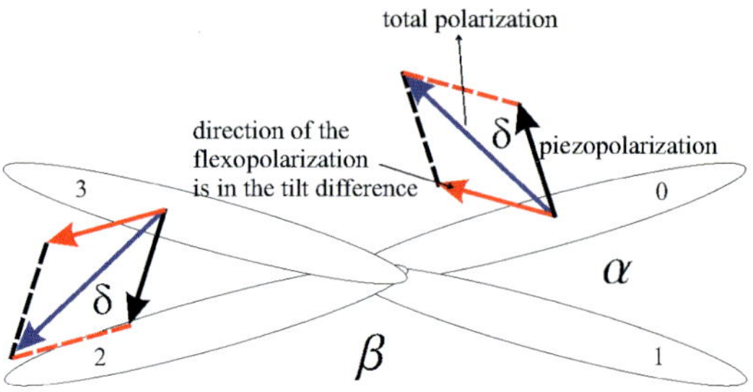


Fig. 5.7. Each layer in the four-layer structure of the $\mathrm{SmC}^*_{\mathrm{FI2}}$ phase has a piezoelectric and flexoelectric polarization. The flexoelectric polarization also has a longitudinal component. The angle $\delta$ marks the angle formed by the polarization direction and the direction perpendicular to the tilt.

The flexopolarization is parallel or antiparallel, depending on the sign of the flexoelectric coefficient $\mu$, to the difference in tilt directions in the layer above and below (Fig. 5.7). If the polarization also has a component parallel to the tilt direction in the layer, the projection of the polarization on the tilt order parameter gives the magnitude of the longitudinal component of the polarization:

$$P_{j,\|} = \frac{\boldsymbol{\xi}_j \cdot \boldsymbol{P}_j}{\mid \boldsymbol{\xi}_j \mid} = \frac{1}{2}\frac{\epsilon_0\mu}{\theta}\left(\boldsymbol{\xi}_j \cdot \boldsymbol{\xi}_{j+1} - \boldsymbol{\xi}_j \cdot \boldsymbol{\xi}_{j-1}\right). \tag{5.43}$$

It can clearly be seen that in structures having a constant phase difference ($\mathrm{SmC}^*$, $\mathrm{SmC}^*_A$, $\mathrm{SmC}^*_\alpha$) no longitudinal component of the flexoelectric polarization exists. In these systems the flexoelectric polarization could be large but it is always perpendicular to the tilt direction. For a longitudinal component, only structures of the $\mathrm{SmC}^*_{\mathrm{FI1}}$, $\mathrm{SmC}^*_{\mathrm{FI2}}$ and $\mathrm{SmC}^*_{\mathrm{6d}}$ phases are good candidates.

With respect to the difference, we can clearly distinguish three different situations. Structures like the $\mathrm{SmC}^*$, $\mathrm{SmC}^*_A$ and the $\mathrm{SmC}^*_\alpha$ phases can all be described by a single phase difference between neighbouring layers and also by a single phase difference between next nearest layers. In these systems, the flexoelectric polarization is always parallel or antiparallel to the piezoelectric polarization. It depends on the phase difference as

$$P_{j,\text{flexo}} = \epsilon_0 \, \mu \, \theta \, \sin \alpha. \tag{5.44}$$

It can clearly be seen that the flexoelectric contribution has the highest value for the phase difference 90°, which is characteristic for the $\text{SmC}_{\alpha}^{*}$ phase having four-layer periodicity. How can this be understood? The electrostatic energy that is associated with the polarization of layers favours large flexoelectric polarizations and the non-parallelism of neighbouring layers is favoured to achieve the antiparallelism or the closest to it in next neighbouring layers.

The same effect also occurs in more complex structures. Examples are structures with periods of four (two phase differences), three and six layers (three phase differences). Let us take a closer look at these structures as they reveal new phenomena related to flexoelectricity.

The four-layer $\text{SmC}_{\text{FI2}}^{*}$ phase structure is one such example. In Fig. 5.7 we can see the piezoelectric component of polarization and the flexoelectric component of polarization. The flexoelectric component of the polarization is in the direction of the tilt order parameter difference. Therefore the flexoelectric polarization also has a component that is parallel to the tilt order parameter. In systems where the flexoelectric polarization is significant, this component may be important. How large is it? To analyse the flexoelectric contribution to the polarization, the small deviation $\Delta$ of the sum $\alpha + \beta$ from $\pi$ can be neglected and we assume that $\beta = \pi - \alpha$. How large is the longitudinal component of polarization? Inserting Eq. (5.35) into Eq. (5.42) gives the following expression

$$\begin{aligned} P_{2j,\|} &= \frac{1}{2}\, \epsilon_0 \mu \, \theta \,(\cos\alpha - \cos\beta) \approx \epsilon_0 \mu \theta \sin\left(\frac{\alpha-\beta}{2}\right) = -\epsilon_0 \mu \, \theta \, \sin\alpha, \\ P_{2j+1,\|} &= \frac{1}{2}\, \epsilon_0 \mu \, \theta \,(\cos\beta - \cos\alpha) \approx \epsilon_0 \mu \, \theta \sin\left(\frac{\alpha-\beta}{2}\right) = \epsilon_0 \mu \theta \, \sin\alpha, \end{aligned} \tag{5.45}$$

where the equality $\beta = \pi - \alpha$ is taken into account.

Another structure, which possesses longitudinal polarization, is the three-layer structure of the $\text{SmC}_{\text{FI1}}^{*}$ and the $\text{SmC}_{\text{6d}}^{*}$ phases. One layer has a different environment (two neighbouring layers form the phase difference $\alpha$) than the other two (one phase difference $\alpha$ and another $\beta$) and therefore the tilt magnitude could be slightly different.[35,36] The flexoelectric polarization is proportional to the difference of tilt orders above and below the considered layer, which is related to splay, and also to the difference in the magnitude of the tilt, which is related to bend. Because the variation of the layer thickness is energetically extremely costly, we would not expect a large tilt variation in temperature regions where the tilt is already significant.

From symmetry reasons, two layers out of three have longitudinal components of polarizations, while the flexoelectric polarization in the third layer is strictly perpendicular to the tilt in the layer (Fig. 5.8).

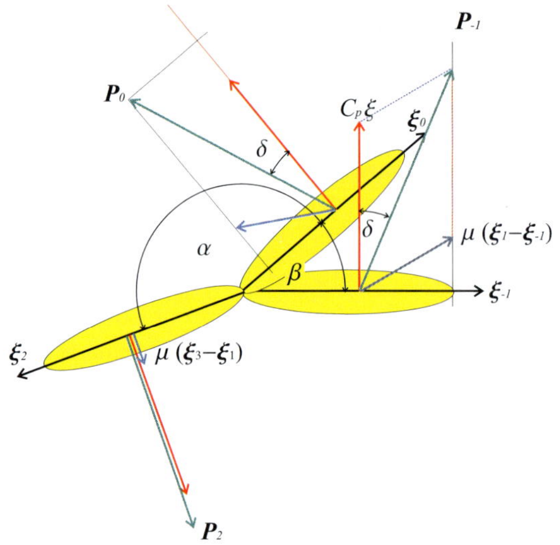


Fig. 5.8. Only two out of three layers in the three-layer structure of the $\mathrm{SmC}^*_{\mathrm{FI1}}$ phase have a longitudinal component of the flexoelectric polarization. The angle $\delta$ marks the angle formed by the polarization direction and the direction perpendicular to the tilt.

$$
\begin{aligned}
P_{3j,\|} &= \frac{1}{2}\,\epsilon_0\mu\,\theta\,(\cos\alpha - \cos\beta)\,, \\
P_{3j+1,\|} &= 0, \\
P_{3j+2,\|} &= \frac{1}{2}\,\epsilon_0\mu\,\theta\,(\cos\alpha - \cos\beta)\,.
\end{aligned}
\tag{5.46}
$$

As expected, in the layer where the phase differences above and below are equal no longitudinal polarization is present. The expressions can be further simplified. For the $\mathrm{SmC}^*_{\mathrm{FI1}}$ phase structure, $\beta = 2\pi - 2\alpha$ approximately and the longitudinal polarizations are approximately

$$
P_{3j,\|} = -P_{3j+2,\|} = \frac{1}{2}\,\epsilon_0\mu\,\theta\,\sin\left(\frac{3\alpha}{2}\right) \quad \text{and} \quad P_{3j+1,\|} = 0. \tag{5.47}
$$

For the $\mathrm{SmC}^*_{6\mathrm{d}}$ phase structure, $\beta = \pi - 2\alpha$ approximately and the longitudinal polarizations are approximately

$$P_{3j,\|} = -P_{3j+2,\|} = \frac{1}{2}\,\epsilon_0\mu\,\theta\,\cos\left(\frac{3\alpha}{2}\right) \quad \text{and} \quad P_{3j+1,\|} = 0. \tag{5.48}$$

In both cases the slight deviation $\Delta$ of the sum $2\alpha+\beta$ from $\pi$ or $2\pi$ and the slight difference of tilt magnitudes are not considered, therefore the magnitudes of the longitudinal polarizations are approximate. However, neither phenomenon contributes significantly to the magnitude of the longitudinal flexoelectric polarization.

### 5.4.2. *On the observability of flexoelectric polarization*

All consequences of the flexoelectric effect considered so far are indirect and are not directly observable. Flexoelectric interactions change the elastic constant (Eq. 5.16) in the SmC* phase, which influences the layer polarization (Eq. 5.42). The layer polarization influences intralayer interactions and interactions to nearest layers and is consequently a source of interactions to more distant layers. The flexoelectric interactions stabilize different structures with longer periodicities. But these are all indirect effects, which often cannot be separated from other effects. As the flexoelectric effect is of achiral origin, there is no simple way to isolate the influence of flexoelectricity on the structure or the macroscopic properties.

So, how can the flexoelectrically induced polarization be measured? It would be natural to study interactions with external electric fields. However, it is not easy to observe the influences of the field on the structure, to measure the static or dynamic polarization in the electric field and similar.

How can we account for the interaction of the flexoelectric polarization with the external electric field? The free energy contribution due to the polarization found in an external electric field is

$$-\left(\boldsymbol{P}_j \cdot \boldsymbol{E}\right). \tag{5.49}$$

The flexoelectric contribution to the layer polarization $\boldsymbol{P}_j$ is given in Eq. (5.42). For the interaction of a sample with an electric field due to the flexoelectrically induced polarization, we have to sum the contributions of all layers:

$$\sum_{j=1}^{N} -\left(\boldsymbol{P}_{j,\text{flexo}} \cdot \boldsymbol{E}\right) = \sum_{j=1}^{N} -\frac{1}{2}\,\epsilon_0\,\mu\,\left(\boldsymbol{\xi}_{j+1} - \boldsymbol{\xi}_{j-1}\right) \cdot \boldsymbol{E}. \tag{5.50}$$

As the field is the same in all layers, it can clearly be seen that the contributions remain only for surface layers, where no neighbouring layers of the same material exist. The bulk contribution cancels pairwise out and is exactly zero. This result is not new, as early papers on flexoelectricity mentioned that the flexoelectric effect is limited to a contribution at the surfaces.[37]

What are the other possibilities for measuring the flexoelectric effect? We could consider dielectric coupling in a high-frequency electric field.

$$\Delta\epsilon\left(\boldsymbol{\xi}_j \cdot \boldsymbol{E}\right). \tag{5.51}$$

The main reason for coupling is the anisotropy in the polarizability. As in tilted smectics the rotation about the molecular long axis is hindered and a tilted system is not uniaxial. The tilted unwound SmC$^*$ system has three different eigenaxes: the highest refraction index is for polarization along the average long molecular axes while the two smaller indices correspond to polarizations that are perpendicular to the tilt and the layer normal, i.e. along the piezoelectrically induced polarization, and the final refraction index is for polarization perpendicular to the tilt but in the plane of the tilt and the layer normal. If we assume that when the polarization is not perpendicular due to the flexoelectricity, then the direction of the eigenaxes also changes. Therefore the dielectric coupling of the complex structure could be slightly different than expected from simple geometry (Fig. 5.9). The problem is that this effect is extremely small, as it mainly influences the orientations of the eigenaxes for the two smaller refractive indexes.

On the other hand, as flexoelectricity has no chiral influence, we would expect that it could be observed by the imposed bend or twist. Unfortunately, for the surface polarizations the flexoelectric component and the simple polarization due to the broken symmetry at the surface (for example, dipoles tend to be oriented toward or away from the air in free-standing films) cannot be separated.

## 5.5. Conclusions

This chapter considers polarization in polar smectics in general and the flexoelectric polarization in polar smectics in particular. The existence of flexoelectric polarization has been known for a long time. The effects related to the phenomenon are more pronounced in systems formed of polar molecules including chiral polar smectics. Therefore it is not surprising that several important consequences of the flexoelectric phenomenon are present

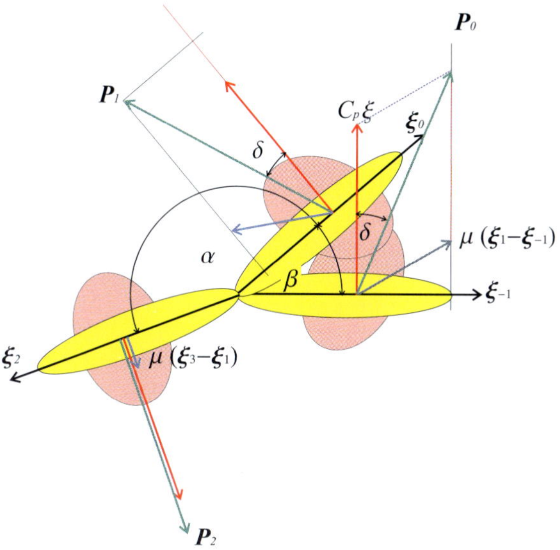


Fig. 5.9. The optical indicatrix in the layer with a longitudinal polarization is rotated. The optical indicatrix, averaged over three layers, should differ from that in systems where only perpendicular polarizations in the layer are present.

in these systems.

In the most simple chiral polar tilted smectics, ferroelectric liquid crystals, the flexoelectric phenomenon influences the structure of the SmC* phase only quantitatively. It affects the elastic and chiral couplings and consequently slightly changes the transition temperature to the tilted phase and the pitch of the helicoidal modulation.

In more complex chiral polar smectics, antiferroelectric liquid crystals, there are many consequences of the flexoelectric effect. It influences interlayer interactions and causes indirect interactions between more distant layers to appear and become important. The phenomenon is the reason for the appearance of commensurate structures that extend up to six layers. In addition, longitudinal polarization, i.e. the polarization that has a component parallel to the tilt, exists in more complex structures such as the $\mathrm{SmC}^*_{\mathrm{FI2}}$, the $\mathrm{SmC}^*_{\mathrm{FI1}}$ and the $\mathrm{SmC}^*_{\mathrm{6d}}$ phases. Unfortunately it seems that flexoelectric polarization cannot be detected separately from other phenomena by simple means. A way of measuring the flexoelectric contribution in tilted polar smectics still seems to be an open question.

## References


1. H. Takezoe, E. Gorecka and M. Čepič, Antiferroelectric liquid crystals: Interplay of simplicity and complexity, *Rev. Mod. Phys.* **82**(1), 897–937, (2010). doi: 10.1103/RevModPhys.82.897
2. H. Takezoe and Y. Takanishi, Bent-core liquid crystals: Their mysterious and attractive world, *Jpn. J. Appl. Phys.* **45**(2A), 597–625, (2006). doi: 10.1143/JJAP.45.597
3. C. Tschierske, Micro-segregation, molecular shape and molecular topology partners for the design of liquid crystalline materials with complex mesophase morphologies, *J. Mater. Chem.* **11**(11), 2647–2671, (2001). doi: 10.1039/B102914M
4. M.J. Stephen and J.P. Straley, Physics of liquid crystals, *Rev. Mod. Phys.* **46**(4), 617–704, (1974). doi: 10.1103/RevModPhys.46.617
5. N. Vaupotič and M. Čopič, Polarization modulation instability in liquid crystals with spontaneous chiral symmetry breaking, *Phys. Rev. E.* **72**(3), 031701/1–4, (2005). doi: 10.1103/PhysRevE.72.031701
6. N. Vaupotič, M. Čopič, E. Gorecka and D. Pociecha, Modulated structures in bent-core liquid crystals: Two faces of one phase, *Phys. Rev. Lett.* **98**(24), 247802/1–4, (2007). doi: 10.1103/PhysRevLett.98.247802
7. P. Oswald and P. Pieranski, *Smectic and Columnar Liquid Crystals: Concepts and Physical Properties Illustrated by Experiments. Liquid crystals book series*, Taylor & Francis, 2006.
8. R.B. Meyer, L. Liebert, L. Strzelecki and P. Keller, Ferroelectric liquid crystals, *J. Physique Lett.* **36**(3), L69–L72, (1975). doi: 10.1051/jphyslet:0197500360306900
9. R.B. Meyer, Ferroelectric liquid crystals; A review, *Mol. Cryst. Liq. Cryst.*, **40**(1), 33–48, (1977). doi: 10.1080/15421407708084469
10. S.A. Pikin and V.L. Indenbom, Thermodynamic states and symmetry of liquid crystals, *Sov. Fiz. Uspekhi* **21**(6), 487–501, (1978). doi: 10.1070/PU1978v021n06ABEH005557
11. C.C. Huang, S. Dumrongrattana, G. Nounesis, J.J. Stofko and P.A. Arimilli, Heat-capacity, tilt-angle, and polarization measurements near the smectic-A–chiral-smectic-C transition of one liquid-crystal compound, *Phys. Rev. A* **35**(3), 1460–1463, (1987). doi: 10.1103/PhysRevA.35.1460
12. V. Novotná, M. Glogarová, M. Kašpar, V. Hamplová, E. Gorecka, D. Pociecha and M. Čepič, Reentrant orthogonal smectic-A phase below a tilted smectic-C phase in a chiral compound, *Phys. Rev. E* **83**(2), 020701/1–4, (2011). doi: 10.1103/PhysRevE.83.020701
13. A. de Vries, The description of the smectic A and C phases and the smectic A–C phase transition of TCOOB with a diffuse-cone model, *J. Chem. Phys.* **71**(1), 25–31, (1979). doi: 10.1063/1.438123
14. I. Muševič, R. Blinc and B. Žekš, *The Physics of Ferroelectric and Antiferroelectric Liquid Crystals*, World Scientific, Singapore, London, 2000.

15. T. Niori, T. Sekine, J. Watanabe, T. Furukawa and H. Takezoe, Distinct ferroelectric smectic liquid crystals consisting of banana shaped achiral molecules, *J. Mater. Chem.* **6**(7), 1231–1233, (1996). doi: 10.1039/JM9960601231
16. A.D.L. Chandani, T. Hagiwara, Y. Suzuki, Y. Ouchi, H. Takezoe and A. Fukuda, Tristable switching in surface stabilized ferroelectric liquid crystals with a large spontaneous polarization, *Jpn. J. Appl. Phys.* **27**(5), L729–L732, (1988). doi: 10.1143/JJAP.27.L729
17. A.D.L. Chandani, E. Gorecka, Y. Ouchi, H. Takezoe and A. Fukuda, Antiferroelectric chiral smectic phases responsible for the tristable switching in MHPOBC, *Jpn. J. Appl. Phys.* **28**(7), L1265–L1268, (1989). doi: 10.1143/JJAP.28.L1265
18. P. Mach, R. Pindak, A.-M. Levelut, P. Barois, H.T. Nguyen, C.C. Huang, and L. Furenlid, Structural characterization of various chiral smectic-C phases by resonant X-ray scattering, *Phys. Rev. Lett.* **81**(5), 1015–1018, (1998). doi: 10.1103/PhysRevLett.81.1015
19. P. Mach, R. Pindak, A.M. Levelut, P. Barois, H.T. Nguyen, H. Baltes, M. Hird, K. Toyne, A. Seed, J.W. Goodby, C.C. Huang and L. Furenlid, Structures of chiral smectic-C mesophases revealed by polarization-analyzed resonant X-ray scattering, *Phys. Rev. E* **60**(6), 6793–6802, (1999). doi: 10.1103/PhysRevE.60.6793
20. D. Schlauf, Ch. Bahr and H.T. Nguyen, Structure of the chiral smectic-$C_\alpha^*$ phase, *Phys. Rev. E* **60**(6), 6816–6825, (1999). doi: 10.1103/PhysRevE.60.6816
21. P.M. Johnson, S. Pankratz, P. Mach, H.T. Nguyen and C.C. Huang, Optical reflectivity and ellipsometry studies of the Sm-$C_\alpha^*$ phase, *Phys. Rev. Lett.* **83**(20), 4073–4076, (1999). doi: 10.1103/PhysRevLett.83.4073
22. Z.Q. Liu, B.K. McCoy, S.T. Wang, R. Pindak, W. Caliebe, P. Barois, P. Fernandes, H.T. Nguyen, C.S. Hsu, S. Wang and C.C. Huang, Unique pitch evolution in the smectic-$C_\alpha$ phase, *Phys. Rev. Lett.* **99**(7), 077802/1–4, (2007). doi: 10.1103/PhysRevLett.99.077802
23. S. Wang, L.D. Pan, R. Pindak, Z.Q. Liu, H.T. Nguyen and C.C. Huang, Discovery of a novel smectic-$C^*$ liquid-crystal phase with six-layer periodicity, *Phys. Rev. Lett.* **104**(2), 027801/1–4, (2010). doi: 10.1103/PhysRevLett.104.027801
24. D. Pociecha, E. Gorecka, M. Čepič, N. Vaupotič, B. Žekš, D. Kardas and J. Mieczkowski, Reentrant ferroelectricity in liquid crystals, *Phys. Rev. Lett.* **86**(14), 3048–3051, (2001). doi: 10.1103/PhysRevLett.86.3048
25. E. Gorecka, D. Pociecha, M. Čepič, B. Žekš and R. Dabrowski, Enantiomeric excess dependence of the phase diagram of antiferroelectric liquid crystals, *Phys. Rev. E.* **65**(6), 061703/1–4, (2002). doi: 10.1103/PhysRevE.65.061703
26. M. Čepič and B. Žekš, Influence of competing interlayer interactions on the structure of the Sm $C_\alpha^*$ phase, *Mol. Cryst. Liq. Cryst.* **263**(1), 61–67, (1995). doi: 10.1080/10587259508033570

27. M. Čepič and B. Žekš, Electrostatic interlayer interactions in tilted polar smectic liquid crystals, *Mol. Cryst. Liq. Cryst.* **301**(1), 221–229, (1997). doi: 10.1080/10587259708041771
28. M. Čepič, E. Gorecka, D. Pociecha, B. Žekš and H.T. Nguyen, Theoretical and experimental study of the intermediate $SmC^*_{FI2}$ and the $SmC^*_{FI1}$ phases in antiferroelectric liquid crystals, *J. Chem. Phys.* **117**(4), 1817–1826, (2002). doi: 10.1063/1.1485732
29. A.V. Emelyanenko and M.A. Osipov, Theoretical model for the discrete flexoelectric effect and a description for the sequence of intermediate smectic phases with increasing periodicity, *Phys. Rev. E* **68**(5), 051703/1–16, (2003). doi: 10.1103/PhysRevE.68.051703
30. K. Susman, B. Žekš and M. Čepič, Theoretical analysis of continuous pitch evolution and reversed phase sequence in antiferroelectric liquid crystals, *Phys. Rev. E* **81**(3), 031701/1–4, (2010). doi: 10.1103/PhysRevE.81.031701
31. M. Čepič and B. Žekš, Flexoelectricity and piezoelectricity: The reason for the rich variety of phases in antiferroelectric smectic liquid crystals, *Phys. Rev. Lett.* **87**(8), 085501/1–4, (2001). doi: 10.1103/PhysRevLett.87.085501
32. P.V. Dolganov, V.M. Zhilin, V.K. Dolganov and E.I. Kats, Commensurate polar smectic structures with a two-component order parameter, *Phys. Rev. E* **82**(4), 040701/1–4, (2010). doi: 10.1103/PhysRevE.82.040701
33. P.V. Dolganov and V.M. Zhilin, Unwinding of the antiferroelectric helix in an electric field, *Phys. Rev. E* **81**(5), 051704/1–5, (2010). doi: 10.1103/PhysRevE.81.051704
34. P.V. Dolganov, V.M. Zhilin, V.E. Dmitrienko and E.I. Kats, Polar smectic subphases: phase diagrams, structures and X-ray scattering, *JETP Lett.* **76**(8), 498–501, (2002). doi: 10.1134/1.1533774
35. P. Fernandes, P. Barois, E. Grelet, F. Nallet, J.W. Goodby, M. Hird and J.S. Micha, Extension of the resonant scattering technique to liquid crystals without resonant element, *Eur. Phys. J. E* **20**(1), 81–87, (2006). doi: 10.1140/epje/i2006-10006-4
36. P.V. Dolganov, V.M. Zhilin, V.K. Dolganov and E.I. Kats, Ferrielectric smectic phase with a layer-by-layer change of the two-component order parameter, *JETP Lett.* **87**(5), 253–257, (2008). doi: 10.1134/S0021364008050068
37. R.B. Meyer, Piezoelectric effects in liquid crystals, *Phys. Rev. Lett.* **22**(18), 918–921, (1969). doi: 10.1103/PhysRevLett.22.918

# Chapter 6

# Flexoelectricity in Lyotropics and in Living Liquid Crystals

Alexander G. Petrov

*Institute of Solid State Physics, Bulgarian Academy of Sciences*
*72 Tzarigradsko chaussee, 1784 Sofia, Bulgaria*
*e-mail: director@issp.bas.bg*

The theory and experiments of lyotropic and biomembrane flexoelectricity are reviewed. Flexoelectricity is a reciprocal relation between electricity and mechanics in soft lyotropic systems, i.e., between curvature and polarization. Experimental evidence of model and biomembrane flexoelectricity (including the direct and the converse flexoelectric effects) is reported. The biological implications of flexoelectricity are underlined. Flexoelectricity enables membrane structures to function like soft micromachines and nanomachines, sensors and actuators, thus providing important input to nanoionics applications. Nanobio examples include membrane transport, membrane contact, mechanosensitivity, electromotility, hearing, nerve conduction, etc.

## Contents

## 6.1. Introductory Remarks

Modern liquid crystal physics investigates two types of object: thermotropic and lyotropic liquid crystals.[1] The general building principles of lyotropic liquid crystals fully apply to living liquid crystalline systems,[2] notably biomembranes. The celebrated 'fluid lipid-globular protein mosaic model'[3–5] implies that lipids are organized in a liquid crystalline bilayer (which is a basic structural unit of a lamellar lyotropic phase). The well-known guest-host concept of liquid crystal physics also applies: the lipid bilayer is the host in which integral proteins (guests) are embedded. Biomembranes with a high protein concentration,[6] where the lipids do not form a continuous bilayer but rather patch the spaces between proteins while the proteins themselves are arranged in a double-tiered pattern, are also well known (see the structure-function unitization model of biomembranes[7]). In both cases, though, the principles of liquid crystal physics (*viz.*, long-range translational disordering versus long-range orientational ordering of constituent chain or rod-like molecules) are valid.[1]

Recently, it has become increasingly clear that for the nanoscience of condensed matter (living matter included) mechanics plays a very important role. It is often so tightly coupled to electricity that they cannot be considered separately. Unlike microelectronics where as a rule only the electrons are moving while the nuclei stay at rest, in nanoelectronics both charges and nuclei are thought to move (in fact, in living matter the important charge carriers are ions, rather than electrons, thus nanoionics is the more relevant term here). Thus nanosystems perform much more like machines when compared to microsystems. The basic reason for this is the softness of the molecular media: liquid crystals, lyotropics, macromolecules, etc.

The first impression when looking at a high-magnification micrograph of a cell is that cell membranes are not flat but very often tend to be strongly curved on a nanometre-length scale. Unlike tensile or compressive strains, it is very easy to sustain curvature strains (i.e., torques) in a membrane because of the smallness of $K$, the curvature elastic modulus.[1] In many cases this curvature is dynamic.[1] These facts logically imply the involvement of flexoelectricity, which is a liquid crystal analogue of piezoelectricity. Piezoelectricity operates through a translational mechanical degree of freedom of a membrane (i.e., area stretching, thickness compression, etc.). Instead, flexoelectricity involves an orientational degree of freedom: the membrane curvature.

It was first thought that membranes might have piezoelectric properties 70 years ago in a discussion of the possible origin of inductance in the squid's giant axon.[8] The first realization that flexoelectricity plays a role in biological membrane ion transport is half that age.[9] Recently, experimental evidence for model and native membrane flexoelectricity has emerged, which will be reviewed below. The most up-to-date methods of membranology (patch clamp, atomic force microscopy (AFM), laser tweezers, confocal microscopy) are employed today. Therefore, the existence of flexoelectricity in most membrane processes, where curvature changes occur, is now widely accepted. Examples include the highly convoluted cristae of inner mitochondrial membranes in energized mitochondria, the edges of retinal rod outer segments and discs, the thylakoid membranes of chloroplasts, the brush borders of intestinal epithelial cells, the tips of spiculae in echinocytes, the microvilli between contacting cells, the stereocilia in the hearing organ, etc., etc.

Flexoelectricity is a basic mechano-electric effect that enables the nanometre-thick membranes of living matter to function like soft machines, thus converting the electrical stimuli of the living world into mechanical ones, and vice versa. It also allows, by using model nanomembranes, the construction of mechanosensors and actuators for nanoionics applications.

## 6.2. Curvature, Polarization and Flexoelectricity

### 6.2.1. *Flexoelectricity of lamellar phases*

Flexoelectricity is a basic mechano-electric phenomenon in liquid crystal physics.[10] The first hints for the existence of a curvature-polarization tensor in liquid crystals, although with different symmetry transformation properties, can be found in a manuscript by Freedericksz (1940) that made use of the tensorial analysis method of general relativity and was published only quite recently.[11] Meyer (1969) derived[10] an original expression for the bulk polarization of a three-dimensional deformed liquid crystal by employing a phenomenology as in solid piezoelectrics,[12] but with a different meaning for the deformations (orientational rather than translational ones). The polarization of a liquid crystal subjected to a splay or bend deformation of its director field $\mathbf{n}(\mathbf{r})$ is[10,13]

$$\mathbf{P}_V = e_1\mathbf{S} + e_3\mathbf{B}, \tag{6.1}$$

where $\mathbf{P}_V$ is the *bulk* electric polarization per *unit volume* (in C/m$^2$), $\mathbf{S} = \mathbf{n}\,\mathrm{div}\,\mathbf{n}$ is the splay vector and $\mathbf{B} = -\mathbf{n}\times\mathrm{curl}\,\mathbf{n}$ is the bend vector. In a local orthogonal coordinate system $(\mathbf{x},\mathbf{y},\mathbf{z})$ in which $\mathbf{n} = \mathbf{z}$ at the origin, $\mathbf{S} = (s_1 + s_2)\mathbf{n}$ and $\mathbf{B} = (b_1\mathbf{x} + b_2\mathbf{y})$. $s_1 = \partial n_x/\partial x$, $s_2 = \partial n_y/\partial y$ are the two components of splay and $b_1 = \partial n_x/\partial z$, $b_2 = \partial n_y/\partial z$ are the two component of bend (all expressed by the partial space derivatives of the liquid crystal director,[10] in m$^{-1}$). $e_1$ and $e_3$ are the splay and bend flexoelectric coefficients in C/m, respectively (for a recent review of thermotropic flexoelectricity see, e.g. Petrov[14]).

For a single, two-dimensional liquid crystal lipid bilayer, or a membrane, in a smectic A-like state, the director field is represented by the membrane normal $\mathbf{n}$. Flexoelectricity is then defined[1] as a curvature-induced *area* membrane polarization, or, conversely, as an electric field-induced membrane curvature. Lipids and proteins are oriented parallel to each other along the local membrane normal in the flat state. A curvature of the membrane surface leads, indeed, to a splay type deformation of the molecular local director, with a splay vector $\mathbf{S} = (s_1 + s_2)\mathbf{n}$, while a bend deformation along the membrane normal is not allowed because there is no third dimension. Then, obviously, the only polarization component points along the membrane normal.

Incidentally, for a curved surface, the splay components are equal to the local surface curvatures, i.e. $s_i = c_i = 1/R_i$, where $R_i$ are the local radii of curvature (since $\mathrm{div}\,\mathbf{n} = c_1 + c_2$ is invariant, we can take the sum of the two principal radii of curvature along any two directions orthogonal to the normal $\mathbf{n}$, e.g., the largest and the smallest). The phenomenological expression of Petrov (1975) for the only component of surface polarization along the membrane normal is[1,9]

$$P_s = f(c_1 + c_2). \tag{6.2}$$

$P_s$ is the electric polarization per *unit area* in C/m, $c_1$ and $c_2$ are the two principal membrane curvatures, the largest and the smallest, in m$^{-1}$ and $f$

is the area flexoelectric coefficient in C (coulombs), typically a few units of electron charge. Note that curvature is a symmetry-breaking operation and a curved membrane can be polarized according to Curie's principle. The flexocoefficient is positive if the polarization points out from the centre of curvature (Fig. 6.1).

Phenomenologically, there is a close analogy between flexoelectricity of thermotropics and lyotropics. Equations (6.1) and (6.2) are in correspondence since a bend deformation of the director is not allowed either in a two-dimensional bilayer or in a lamellar lyotropic phase. From dimensional arguments we have concluded that the three-dimensional to two-dimensional correspondence leads to $f = e_1 d$, where $d$ is the bilayer thickness. However, the molecular mechanisms of the two are very different, e.g., molecular shape asymmetry is not a precondition for dipolar membrane flexoelectricity (see Petrov[1] and below).

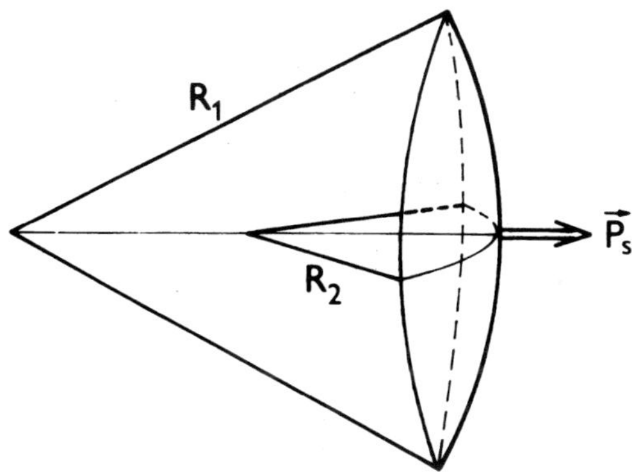


Fig. 6.1. Flexoelectric (curvature-induced) polarization $\mathbf{P_s}$ and the sign convention for the flexocoefficient $f$: for the case shown $f$ will be positive. $R_1$, $R_2$ are the principal radii of the bilayer curvature (from Petrov,[1] with kind permission from Taylor and Francis Group).

### 6.2.2. *Flexoelectricity of columnar phases*

Concluding this section on the phenomenology of flexoelectricity in lyotropic systems, we shall briefly discuss this effect in systems of cylindrical micelles: lyotropic nematics and hexagonal phases. The bending of these micelles is also a symmetry-breaking operation (Fig. 6.2). To describe bending-induced polarization we proposed a new phenomenological equation:[1]

$$\Delta P_L = g\left(\frac{1}{R}\right), \tag{6.3}$$

where $P_L$ is the polarization increment per unit length of the cylindrical micelle, with a radius of curvature $R$ and where $g$ is a linear flexocoefficient (with a dimension of Cm). The expected order of magnitude is $g \propto fd = 5 \times 10^{-28}$ Cm.

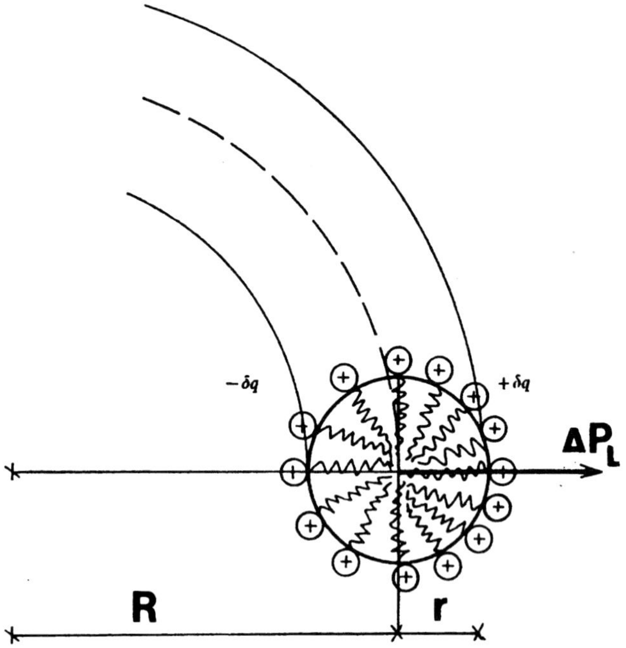


Fig. 6.2. Flexoelectric polarization of a cylindrical micelle. $R$ is the curvature radius, $r$ is the micelle radius, $P_L$ is the curvature-induced polarization per unit length (from Petrov,[1] with kind permission from Taylor and Francis Group).

### 6.2.3. *Direct and converse flexoeffect in lipid bilayers and membranes*

According to the Helmholtz equation an electric potential difference (voltage) appears across a polarized surface. From Eq. (6.2) the curvature-dependent part of this voltage is

$$\Delta U = P_s/\varepsilon_0 = (f^D/\varepsilon_0)(c_1 + c_2). \tag{6.4}$$

$f^D$ is the direct flexoelectric coefficient of a single bilayer and $\varepsilon_0$ is the electric constant. By imposing a predetermined total curvature and by measuring the curvature-induced voltage we can determine the flexoelectric coefficient of any given membrane or a lipid bilayer.

A lamellar lyotropic phase, essentially a stack of lipid bilayers, is a nano-heterogeneous medium and any attempt to comprehend its flexoelectricity in terms of bulk polarization and a bulk flexocoefficient would run into problems. The flexoelectric effect of the lamellar phase can most easily be described by summing up the flexovoltages of all consecutive curved bilayers that are connected in series by the interbilayer electrolyte.

As in the piezoelectricity of solids, flexoelectricity also manifests as a direct effect [$f^D$, Eq. (6.4)] and a converse effect [$f^C$, Eq. (6.5)], with an electric field-induced curvature:[1]

$$c_1 + c_2 = (f^C/K)E, \tag{6.5}$$

where $E$ is the transmembrane electric field and $K$ is the curvature elastic modulus. Equation (6.5) is valid for a tension-free membrane (or equivalently for zero lateral tension), which is the case in an osmotically balanced cell.

Exploring the molecular mechanisms of flexoelectricity is a central task of the liquid crystal approach in membrane biophysics.[1,13,15–21] The flexoelectric coefficient can be represented as an integral over the curvature derivative of the distribution of the normal component of polarization $P(z, c_+)$ across the membrane[16] ($c_+ = c_1 + c_2$ is the total curvature). Both direct and converse flexocoefficients can be expressed in this manner and these can be shown to be equal:[18]

$$f^D = f^C = \int \left.\frac{\partial P(z, c_+)}{\partial c_+}\right|_0 dz. \tag{6.6}$$

Model distributions including electric monopoles, dipoles and quadrupoles of lipids and proteins have been considered,[1,16] revealing their respective contributions to the total flexocoefficient $f$:

$$f = f^M + f^{Dip} + f^Q. \tag{6.7}$$

The relative amounts of these three contributions are dependent on the molecular structure of the lipids and proteins and the ionic conditions of the bathing electrolyte. If a membrane is asymmetric in the flat state then flexoelectricity adds to its total polarization a curvature-dependent increment.

Some final remarks concerning living membranes: an important feature of biomembranes is the strongly heterogeneous lipid composition of the liquid crystal membrane matrix. Studies of mixed lipid monolayers[22] demonstrate that the variation of the surface potential as a function of the area differs from that of pure monolayers: depending on the composition, $d\Delta V/dA$ could either be enhanced or weakened.

### 6.2.4. *Flexoelectricity and membrane lipids*

For lipids that are both charged and dipolar, it is convenient to express the sum of the two contributions to the flexocoefficient by the sum of the two components to the surface potential (Fig. 6.3).

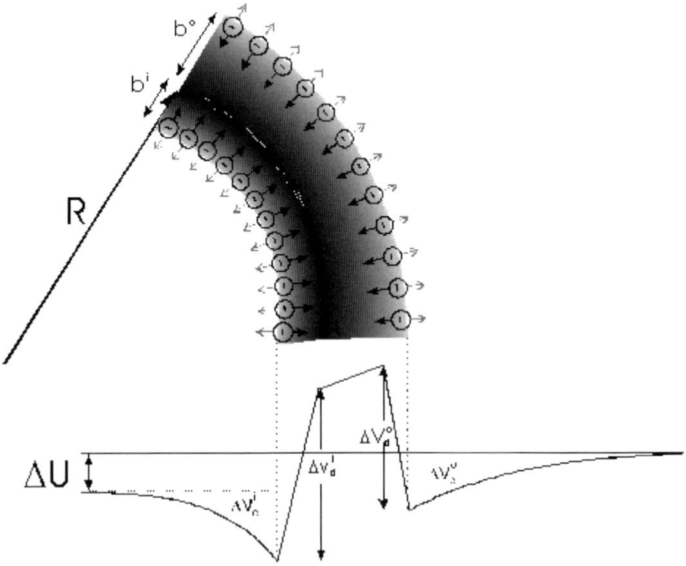


Fig. 6.3. Electric potential distribution across a curved asymmetric membrane bilayer. The charge $\Delta V_c^{o,i}$ and dipole $\Delta V_d^{o,i}$ components of the surface potential of each monolayer are indicated. The difference of these surface potentials is the flexoelectric voltage $\Delta U$, which is proportional to the membrane curvature $(2/R)$ (see text). $b^o$ and $b^i$ are the distances between the mechanically neutral surface of the bilayer and the corresponding aqueous interface. (Figure reprinted with kind permission of the publisher from Petrov and Sachs.[17] Copyright 2002 by the American Physical Society.)

For Debye lengths sufficiently shorter than half of the membrane thickness we derived a simple expression:[17]

$$f^M + f^{Dip} = -\varepsilon_0 A_0 \frac{d}{2}\left(\frac{d\Delta V^i}{dA^i} + \frac{d\Delta V^o}{dA^o}\right), \qquad (6.8)$$

where $A_0$ is the area per lipid molecule. In Eq. (6.8) the charge $(\Delta V_c^{o,i})$ and dipole $(\Delta V_d^{o,i})$ components of the double-layer surface potential of the outer $(o)$ and inner $(i)$ membrane surfaces are lumped into one (Fig. 6.3):

$$\Delta V^{o,i} = \Delta V_c^{o,i} + \Delta V_d^{o,i}.$$

The surface (Volta) potential $\Delta V$ is an experimentally measurable quantity in monolayers on a water/air interface or, which is more representative for half a membrane, on a water/oil interface.

In the more general asymmetric case the monopole contribution to Eq. (6.8) is

$$f^M = -\varepsilon_0\left[A^i \frac{dV_c^i}{dA^i}\left(b^i + \frac{\lambda_D^i}{2}\right) + A^o \frac{\Delta V_c^o}{dA^o}\left(b^o + \frac{\lambda_D^o}{2}\right)\right], \qquad (6.9)$$

$$\lambda_D^{o,i} = \sqrt{\varepsilon_w \varepsilon_0 kT/2e^2 n^{o,i}}. \qquad (6.10)$$

$\lambda_D$ is the Debye screening length of the outer (inner) monovalent electrolyte, $b^o$ and $b^i$ are the distances between the neutral mechanical surface of the *bilayer* and the corresponding membrane interfaces. Using Eq. (6.9) we can estimate $f_{Dip}$ or $f^M + f^{Dip}$ from experimentally measured surface potential versus area curves $\Delta V(A)$ (see the examples in Petrov[1]). Typical results are around $1 \times 10^{-20}$ C for uncharged lipids and around $2 \times 10^{-20}$ C for charged lipids.[1]

### 6.2.5. *Flexoelectricity and membrane proteins*

Integral proteins could provide a large contribution to the curvature-induced membrane polarization. Both dipolar and quadrupolar contributions could show up, more pronounced than those of lipids. The reason for such an expectation is the very large dipole moment measured for some proteins. Theory also demonstrates that such big molecules with no spherical symmetry may have a very large anisotropy of the quadrupole moment.

The dipole contribution is preconditioned by free lateral diffusion of proteins. This is a basic assumption in the fluid-mosaic model of biomembranes.[3–5] A large number of unidirectionally oriented conical and dipolar proteins should accumulate in the curved membrane regions, giving rise to an additional flexopolarization.[21] Following Petrov[1] the resulting flexocoefficient for this case is $f^{PDip} = 1.6 \times 10^{-19}$ C.

Furthermore, monolayer measurements[22] demonstrate the variation of the dipole moment of peripheral proteins under stretching or compression. According to us,[1] this implies the possibility of 'bimorph' flexoelectricity of peripheral proteins (an analogue of the piezoelectricity of a bimorph plate), especially if these are symmetrically adsorbed over the two membrane interfaces, as suggested in the Danielli–Davson model.[23]

The quadrupolar contribution is mostly expected in membranes with a high protein concentration,[6] where ordered arrays of integral proteins exist. Examples of this type include the purple membranes of *Halobacterium halobium*,[24] the inner mitochondrial membrane,[25] etc. The presence in biomembranes of extended domains of tightly packed globular proteins in a double-tiered pattern is a basic idea in the structure-function unitization model of biomembranes.[7] The estimated flexocoefficient[1,16] of an array of identical double-tiered quadrupolar proteins is substantial: $f^{PQ} \geq 4.5 \times 10^{-20}$ C as a lower limit.

The curving of purple membrane fragments in an external electric field or under surface charge asymmetry[26] could be considered as experimental evidence for the existence of the converse quadrupolar flexoeffect in bacteriorhodopsin protein arrays.

## 6.3. Experimental Studies of Flexoelectricity in Biomembranes

### 6.3.1. *Theoretical remarks*

Several authors accidentally found in the early 1970s the first experimental hints concerning the generation of AC currents by bilayer lipid membranes subjected to oscillating gradients of hydrostatic pressure.[27,28] However, no proper explanation of these observations was offered at that time, apart from a pressure-driven current due to a selective pushing of ions of a given sign through the membrane.[27] Subsequently, we calculated the suggested pressure-driven currents and noted that they are much weaker and do not match either the amplitude or the phase of the experimental data. Furthermore, we considered the oscillations of the membrane curvature in these experiments and explained the effect as a displacement current due to the oscillating flexoelectric polarization of the curved membrane.[9,13,15,29]

The AC oscillating pressure with a sound frequency was generated by a compression loudspeaker or a piezo-sounder.[29,30] Subsequently, a pulsed pressure was used generated by a pressure clamp including a pressure servo (see Section 6.3.3 below).

Let us assume that due to the pressure difference the membrane shape becomes a spherical segment of radius $R$:

$$c_1 = c_2 = \frac{1}{R} \tag{6.11}$$

and that the membrane curvature oscillates in time:

$$c_1 + c_2 = c(t) = 2c_m \sin \omega t, \tag{6.12}$$

where $\omega$ is the angular frequency of oscillations and $c_m = 1/R_m$ is the maximal curvature attained. Then, following Eq. (6.2) we can obtain a time-dependent flexopolarization. This polarization leads to a transmembrane AC voltage difference, Eq. (6.4), that is a first harmonic with respect to curvature oscillations and can be measured by two electrodes connected to a very high impedance electrometer [open circuit, i.e. zero current clamp of a patch clamp amplifier (see Fig. 6.4, where a detailed interpretation of the elicited voltage is given)]. Its amplitude is

$$U_f = \frac{f}{\varepsilon_0} 2c_m. \tag{6.13}$$

A displacement current due to the oscillating flexopolarization can be measured by two electrodes that are effectively shorted out via a low-impedance ammeter (shorted circuit, zero voltage clamp of a patch clamp amplifier, see Fig. 6.4). This displacement current through the meter can be calculated by adopting an equivalent circuit (Fig. 6.4a), containing an AC voltage generator $U_f$ (describing the oscillating flexoelectric voltage, Eq. (6.13)) and a capacitor $C_0$ (where $C_0$ is the membrane capacitance), while membrane resistance $R_m$ is assumed to be very large: $I_f = d(CU)/dt$. Its first harmonic amplitude is

$$I_f = f \frac{C_0}{\varepsilon_0} 2c_m \omega. \tag{6.14}$$

In this way, by measuring $U_f$ and $c_m$, or $I_f$, $C_0$ and $c_m$, we can determine experimentally the flexocoefficient $f$. Evaluation of the membrane curvature can be performed electrically from the second harmonic of the membrane capacitance current with a non-zero voltage clamp (the 'condenser microphone' effect) and supposing spherical curvature.[30] The actual curvature $c_1 + c_2$ of a black lipid membrane (BLM) can be measured interferometrically.[31–34]

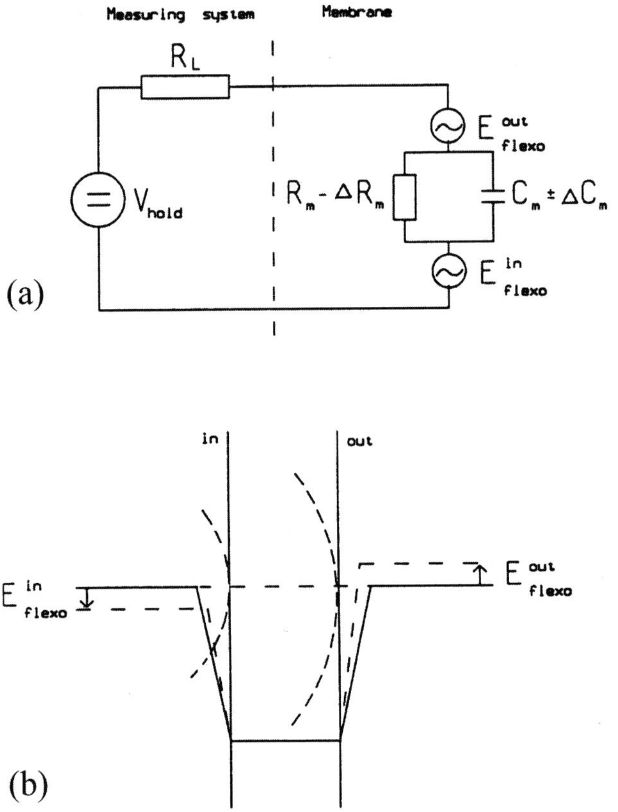


Fig. 6.4. The upper panel (a) shows the equivalent circuit of an oscillating membrane connected to a system for recording the flexoelectric response (normally a patch clamp amplifier). The lower panel (b) shows a graph of the potential distribution across a planar and a curved membrane, related to the variation of the surface dipole moments (*cf.* Fig. 6.1). $V_{hold}$ is the holding potential of a voltage clamp (respectively the voltage measured by the amplifier with a current clamp), $R_L$ is the input resistance of the recording system ($R_L$ is small in the voltage clamp regime, but infinitely large in the zero current clamp regime), $C_m$ is the membrane capacitance (and $\Delta C_m$ is its variation during oscillatory changes of the membrane area) and $R_m$ is the membrane resistance (and $\Delta R_m$ its variation during channel opening). $E^{out}$ and $E^{in}$ are two electromotive force generators that modulate the surface potentials of the two surfaces of the curved membrane (see (b) for a positive flexoelectric coefficient, negative surface charge and zero intramembrane field, i.e., zero current clamp). Since the two flexoelectric generators operate in counter phase they can be combined in one generator to produce a potential difference of about 1 mV for a flexoelectric coefficient of $1 \times 10^{-20}$ C and for a membrane radius of curvature of 1 $\mu$m (from Petrov,[1] with kind permission from Taylor and Francis Group).

The flexoelectric coefficients evaluated in this manner are listed in Table 6.1.

Table 6.1. Flexoelectric coefficients of bilayer lipid membranes made from different lipids under various ionic conditions.[1] 1–3: electrical estimation of curvature; 4–7: interferometric measurement of curvature, sign determination of $f$ not attempted. All experiments except 7 concern the direct flexoeffect. All data refer to the high-frequency range above 200–300 Hz, i.e., to the blocked flexocoefficient (from Petrov,[1] with kind permission from Taylor and Francis Group).

| Lipid; Electrolyte | Partial charge per lipid head $\beta$ (%) | Flexocoefficient ($\times 10^{-19}$ C) | Reference |
|---|---|---|---|
| 1. Phosphatidyl ethanolamine from *E. coli*; 50 mM KCl | −3.7% | −25.5 (±72.5%) | Petrov and Sokolov[30] |
| 2. Phosphatidyl choline from egg yolk; 0.1 M NaCl | −0.4% | +26.5 (±50%) | Derzhanski *et al.*,[35] recalculated[1] |
| 3. Phosphatidyl choline from egg yolk; 0.1 M NaCl + 1 mM $UO_2^{2+}$ | ca. 100% | −120 (±10%) | Derzhanski *et al.*,[35] recalculated[1] |
| 4. Phosphatidyl choline from egg yolk; 0.1 M KCl | −18.8% | 13 (±0.2%) | Todorov *et al.*,[31] recalculated[1] |
| 5. Glycerol monooleate (synthetic); 0.1 M KCl | 0 | 0.43 (±0.5%) | Todorov *et al.*,[31] recalculated[1] |
| 6. Phosphatidyl serine from bovine brain; 0.1 M KCl | ca. −100% | 20 (±25%) | Todorov *et al.*,[31] recalculated[1] |
| 7. Phosphatidyl serine from bovine brain; 0.1 M KCl + 1 mM $UO_2^{2+}$ | ca. 100% | 151 (±19%) | converse flexoeffect; Todorov *et al.*[34] data, fitted in Petrov[1] |

### 6.3.2. *The patch clamp technique and the direct flexoeffect of native membranes: Sine excitation*

The patch clamp method involves sealing small patches of native membranes at the tips of glass micropipettes by gentle suction of the pre-cleaned membrane. It is well described in the literature.[36] In the tip dip technique[37,38] model phospholipid membranes form on patch clamp pipettes. The investigation of the flexoelectric properties of model and native membranes[39–43] is a novel application of the patch clamp technique. It uses the ability to manipulate the patched membrane curvature on a micrometre scale by varying the pipette pressure. The AC oscillating pressure with a sound frequency is generated by a compression loudspeaker or a piezo-sounder, and is applied to the inside of the patch pipette by a flexible pipeline using the pressure outlet of the pipette holder.[1] There is a theoretical analysis of the oscillating pressure manipulation of patch-clamped membranes.[42] The radius of a patched membrane at the end of a cylindrical tip was calculated and it was concluded that a tension-free, flaccid patch is essential for the manifestation of substantial oscillating curvature. The effect of patch tension on the flexoresponse has been demonstrated.[39]

The experimental set-up is described elsewhere.[16,39,40] The first experimental demonstration of the flexoelectric response of a native membrane using a locust muscle was carried out in 1989 using a single lock-in amplifier.[39] The accumulation and averaging of current traces by a computer was also employed. Unlike model membranes native patches produce much noisier flexocurrents. In these experiments strong interference from channel activity of membrane patches was observed and therefore a channel-free patch may be an advantage for such measurements. Apart from the application of oscillating pressure, the membrane treatment and patching procedure closely followed the standard protocols (see above paragraph). Patches in either cell-attached, inside-out or outside-out configurations were studied. The tip diameters could not exceed 1 $\mu$m, though, because with larger tips giga-seal formation was precluded.

The data reported in Petrov[41] *et al.* were obtained using a dual lock-in amplifier (measuring the amplitude and phase of the flexocurrent separately) with an inside-out patch oscillating in the high-frequency range (150–500 Hz). Oscillation frequencies were selected so that the standing waves of the sound pressure were maximal at the pipette holder end of the plastic tube. Keeping the oscillating pressure amplitude constant, a linear increment of the first harmonic amplitude of 30 fA/Hz (rms) was observed. An extrapolation of the amplitude-frequency relation passed through zero. Thus, this was an observation of the flexocurrent due to displacement [Eq. (6.9)]. The evaluation of the membrane curvature with patches needs optical monitoring of the patch geometry because the 'condenser microphone' effect is much weaker than with BLM. In the absence of such optical imaging we could only estimate the curvature radius as equal to the tip radius, i.e., 0.7 $\mu$m (with an error of 50%). Thus, we calculated[1] for a locust muscle membrane a flexocoefficient of $2.5 \times 10^{-18}$ ($\pm 50\%$) C. This is a quite substantial value.

In confirmation of earlier results[39] we also observed in locust patches containing electrically gated $K^+$ channels a dramatic amplification of the flexoelectric current during channel opening.[41] The channels were opened by applying membrane voltages larger than 20 mV. The holding potential of the voltage clamp was increased in steps up to 50 mV and we observed, apart from the rapid jumps of the flexocurrent on a fast time scale (40 mV), a steady amplification of the first harmonic current amplitude of more than 50 times on slow time scale of minutes. The initial very low flexocurrent of 5 fA (rms) can be fully recovered with the holding potential brought back to zero, and could be reamplified with larger negative holding potentials that also open the channels. This striking effect can be understood as a transition from a detailed to a global neutrality regime of the monopole flexopolarization during channel opening:[1] curvature-induced transport of charges along open channels and across the whole membrane thickness leaves each one of the half-spaces charged with respect to the other and creates very large dipoles, situated at the same time across the low dielectric constant membrane core; therefore the enhancement of the flexocoefficient can reach two orders of magnitude depending on the electric coupling parameter.[1]

### 6.3.3. *The patch clamp technique and the direct flexoeffect of native membranes: Pulse excitation*

In the earlier studies reviewed above, oscillating pressures of various frequencies, from 20 to 600 Hz, were used and the amplitude and phase of the steady AC flexoelectric response currents were measured using lock-in amplifiers. In contrast, transient measurements of pressure pulse elicited voltages and currents of membrane patches were undertaken in a recent investigation.[18] Pressures are regarded as positive and suctions negative. An accumulation and averaging routine of pulsed response traces was employed; the number of individually applied pulses was up to 100. A control for the patch curvature was achieved by recording transients of capacitance changes.[44]

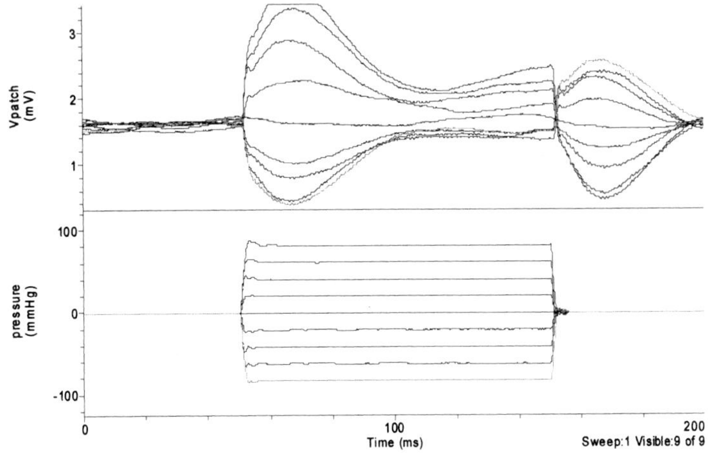


Fig. 6.5. Averaged voltage response of an astrocyte inside-out patch subjected to a pulse sequence of 100 ms duration with averaging of corresponding traces. The pulse amplitude ranges between $-1.07 \times 10^5$ Pa and $+1.07 \times 10^5$ Pa (1 mm Hg = 133.3 Pa). The patch current is clamped to zero by the fast current clamp of the Axopatch 200B. Activated adult astrocytes were used. An Axopatch 200B (Axon Instruments, CA) was used for patch clamping, while experimental protocols and data acquisition were controlled by Axon Instruments pClamp8 software via a Digidata 1322A acquisition system. Voltages were sampled in the fast current clamp of the Axopatch 200B. All potentials are defined with respect to the pipette interior. Pressure and suction were applied to the pipette by a HSPC-1 pressure clamp (ALA Scientific Instruments, NY) controlled by the pClamp software. Up to 100 consecutive responses (voltages or currents) to various pulse protocols were collected and averaged in real time. Off-line data analysis was performed with Clampfit and Origin 6.1 software. (Reprinted from Petrov.[18] Copyright 2006, with permission from Elsevier.)

Figure 6.5 displays the averaged voltage response of an astrocyte inside-out patch subjected to a pulse sequence of 100 ms duration, which alternates between $-1.07 \times 10^5$ Pa and $+1.07 \times 10^5$ Pa, with averaging of corresponding traces. The patch current is clamped to zero. A negative pressure produces a negative voltage inside the pipette (the head stage electrode), while a positive pressure produces a positive voltage. On switching on the pulse, a rapid jump of the trans-patch voltage (rise time less than 1 ms) was followed by a gradual rise resulting in a peak of 0.5 to 1.2 mV some 15 ms after the pulse onset, depending on the pulse amplitude. Furthermore, the voltage gradually decayed to roughly zero for negative pulses in about 50 ms, while for positive pulses it tended to increase again after passing through a minimum. This behaviour may reflect the dissipation of the voltage by a pressure-induced trans-patch current. The elicited voltage amplitude versus pulse amplitude reveals an initial linear region and saturation at extreme pressures (probably due to a saturation of the patch curvature). On switching off the pulse, a rapid jump of the trans-patch voltage (rise time less than 1 ms) of opposite sign to that on switching on was followed by a mirror image of the switching on transient with respect to the $x$ axis.

The patch response to negative pressure pulses of a constant amplitude of $1.07 \times 10^5$ Pa, but starting from an initial value varying between $+5.33 \times 10^4$ Pa and $-5.33 \times 10^4$ Pa, i.e. from a *pre-curved* patch state to a state of the same or *opposite* sign of curvature, was also demonstrated. The voltage response reveals a striking dependence of the fast initial jump on the starting and final sign of curvature, being minimal when both are negative, and maximal when switching from negative to positive. That is, the fast initial jump is related to a fast relaxation of the patch curvature, i.e. to flexoelectricity.

Finally, for a rat astrocyte membrane $f = 6.2\text{–}8.9 \times 10^{-21}$ C was estimated,[18] which is a rather low value, but probably reasonable in view of the lack of specific electromechanical activity of such cells. According to the interrelation of the signs of the pressure and voltage difference (a negative pressure produces negative voltage inside the pipette while a positive pressure produces a positive voltage), the sign of the rat astrocyte membrane flexocoefficient was found to be negative.

### 6.3.4. *The AFM technique and the converse flexoeffect of native membranes: Sine excitation*

The electrically stimulated membrane motions in cells using a whole cell voltage clamp were first investigated using AFM.[45] A patch pipette holding a HEK (Human Embryonic Kidney) cell was attached to a tubular piezo ceramic used for $x$,$y$,$z$ scanning. The voltage-clamped HEK293 cell membranes (under a 10 mV peak-peak AC carrier stimulus) with an AFM cantilever pressed against the membrane, moved the tip several nanometres normal to the plane of the membrane. The holding potential $V_h$ and the AC carrier voltage were applied to the cell by the patch clamp amplifier (EPC 7). The cantilever movement was translated into a voltage, $V_{det}$, by the laser and quadrant detector. The output was proportional to the height difference $\Delta h$ of the surface. The frequency dependence of the voltage-induced membrane movements of six HEK cells normalized to the amplitude of the lowest frequency was studied. The mean sensitivity at the lowest frequency was $0.15 \pm 0.05$ nm/mV$_{\rm pp}$ (at a cantilever stiffness of 0.01 N/m). The decrease of the signal amplitude at higher frequencies was affected by the detection system (with a resonance frequency of 2 kHz). These movements tracked the voltage at frequencies $> 1$ kHz with a phase lead of 60–120°, as expected for a displacement current. The tip displacement was outward with depolarization, implying a positive sign for the flexocoefficient. From the estimations made by Mosbacher *et al.*[45] we could infer a value of $1 \times 10^{-19}$ C for the flexocoefficient of the HEK293 membrane. This value is lower than that of the locust muscle, which is not surprising in view of the marked mechano-electric behaviour of the muscle membrane.

### 6.3.5. *The AFM technique and the converse flexoeffect of native membranes: Pulse excitation*

Further experimental results using pulsed excitation have recently been obtained.[46] Experiments were performed with whole-cell voltage-clamped HEK293 cells. The cell membrane was indented using the AFM cantilever of a modified Quesant Nomad AFM with a force of typically 0.46 nN. For each experiment, the authors averaged 50 repetitions of the cantilever displacement associated with 50 ms hyperpolarizing and depolarizing pulses of 100 mV (from a holding potential of $-60$ mV). Positive displacements represent the AFM moving into the cell. Rectangular pulses of linearly increasing amplitude produced membrane displacements in linear proportion to the voltage pulse amplitude ($\sim$ 1 nm/100 mv). Hyperpolarizing the membrane increased the local curvature around the tip as it is moving inward, i.e., the flexocoefficient of the HEK293 membrane is positive (as in Mosbacher *et al.*[45]) at normal ionic strength. Interestingly, a sign reversal was found at lower ionic strengths of the bath, below 10 mM.

The authors[46] offered an explanation of this novel finding using the Lippmann equation for membrane tension in the presence of an electric field. In a symmetrically charged membrane the Lippmann effect predicts tension changes that are quadratic with respect to the transmembrane voltage; the authors[46] showed for the first time that with asymmetrically charged membranes the Lippmann tension, which is a sum of two interfacial tensions, has a leading term that is linear with respect to the voltage as with flexoelectricity.

However, the torque-producing converse flexoeffect is roughly governed by the difference of the two interfacial tensions rather than by their sum. In Fig. 6.3 for the converse flexoeffect $\Delta U$ is an externally applied voltage difference. We see that $\Delta U$ modifies the surface potential of the two monolayers in opposite directions, thus producing a difference in their interfacial tensions. Such a difference will ultimately induce a surface torque that will curve the membrane. A Poisson–Boltzmann treatment[46] showed that this difference will also be linear with voltage, thus providing a model of the monopole contribution to the flexoeffect that depends on surface charges alone. The actual symmetry of the AFM problem is such that both the non-zero tension and the torque variations of the distended membrane will move the AFM tip.

### 6.3.6. *AFM and confocal microscopy techniques*

In the following we will review a purely flexoelectric treatment of this problem.[17] Let us assume that the indentation of an osmotically balanced cell by an AFM tip produces a shape like a logarithmic funnel[47,48] (where at each point $c_1 + c_2 = 0$) where the tip is covered by a spherical cap of radius $R$ (Fig. 6.6A). Producing a logarithmic funnel shape does not cost any curvature energy.[48] The only part of the membrane that produces an elastic restoring torque, equivalent to a normal force, is the spherical segment at the tip of the funnel. In the presence of a transmembrane field $E = \Delta U/d$, the segment electrical enthalpy contains (to first order) the flexoelectric term $-2fE/R$, which will modify the radius with respect to the zero field case ($\Delta R = \Delta h$):

$$\frac{\Delta R}{R} = \frac{R}{d}\frac{f}{K}\Delta U. \tag{6.15}$$

To evaluate this effect we extended the theory of flexoelectricity and curvature elasticity to asymmetric membranes.[17] The ratio $f/K$ governing the converse flexoelectric effect in Eq. (6.15) markedly depends on a decrease of the ionic strength because of the concomitant increase of the Debye length [*cf.* Eq. (6.9)]. When the ionic strength on one or both sides of the membrane tends to zero the charge contribution [Eq. (6.9)] diverges like the corresponding Debye length [Eq. (6.10)]. Notably, the charge (monopole) contribution to the flexocoefficient (being small at normal ionic strength) can become so large as to overcompensate the dipole contribution because it has an opposite sign. The expected nanometre displacement of an indented membrane surface, Eq. (6.15), where $f$ is calculated according to Eq. (6.9), with an explicit dependence on the ionic concentration via the Debye length, is shown in Fig. 6.6b. We can see that by decreasing the ionic strength of one bath (the right abscissa of Fig. 6.6b), the original positive sign due to the dipole flexocoefficient (determining the slope of the corresponding line on the butterfly surface) is reversed due to the monopole contribution, in agreement with the experimental data.[46]

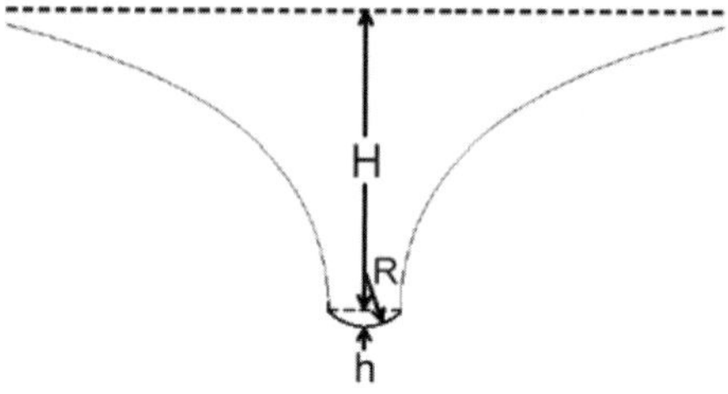


(a)

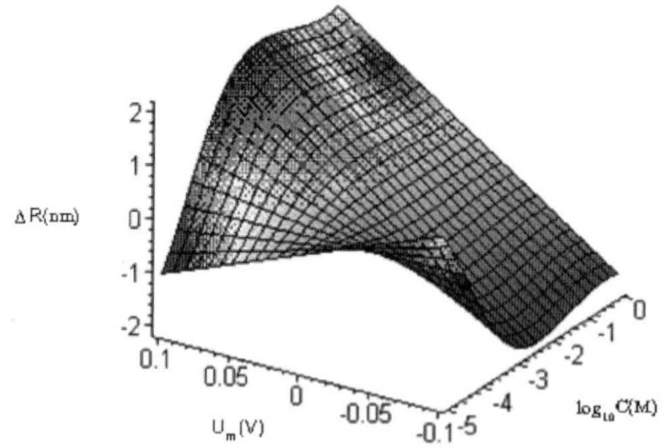


(b)

Fig. 6.6. (a) Cross section of an indented membrane with a spherical cap on a rod approximating an AFM tip. (b) Displacement of a membrane indented with an AFM tip having a curvature radius $R = 100$ nm, as a function of membrane voltage ($-100$ mV to 100 mV) and ionic strength. (Figures reprinted with kind permission of the publisher from Petrov and Sachs.[17] Copyright 2002 by the American Physical Society.)

In order to avoid the effect of membrane tension when pressing an AFM tip against the membrane, we can use the method of confocal microscopy, although with a lower sensitivity. Consider a spherical cell of radius $R$ and area $A_0$ that is whole-cell patch clamped. Apply a negative holding potential $\Delta U$ to the patch pipette, i.e. to the cell interior. In this case for a positive flexocoefficient the flexoelectric torque $t = -fE$ will tend to curve the membrane in the opposite direction to its spherical curvature, i.e., to reduce the overall curvature by increasing the cell radius from $R$ to $R + \Delta R$ and the cell area from $A_0$ to $A$. This will be counterbalanced by a stretching elasticity $K_S$ of the membrane. For the new equilibrium sphere with area $A$ and increment of the radius $\Delta R$:[18]

$$\frac{\Delta R}{R} = \frac{f}{2\sqrt{KK_S}}\frac{\Delta U}{d}. \tag{6.16}$$

Taking $f = 1 \times 10^{-18}$ C, $K = 1 \times 10^{-20}$ J, $K_S = 0.1$ N/m, $\Delta U = 0.1$ V and $d = 5$ nm, we get from Eq. (6.16) that $\Delta R/R = 31.6\%$, i.e., for $R = 10$ $\mu$m, $\Delta R$ will be 3.16 $\mu$m. Such a large change is easily measurable even with phase contrast microscopy. The tiny effect for a 100 times smaller flexocoefficient $f = 1 \times 10^{-20}$ C (a 32 nm increment of radius), should then be measurable by confocal microscopy without any contact with the membrane.

The question arises then why such large displacements were not found by AFM and what is the actual origin of the difference between the two calculations: the local Eq. (6.15) and the global Eq. (6.16). Perhaps this is due just to pressing the AFM tip against the membrane, which limits the flexoresponse to a much smaller membrane area around the tip, as compared to the total membrane area responding in the second case [in Eq. (6.15), $R$ is a much smaller radius of a local membrane invagination around the tip, and it enters the result as a scaling factor]. The authors[45] observed similar movements when they pressed the *flat* part of the AFM cantilever against the cell.

### 6.3.7. *The optical tweezers technique*

Using a novel experimental technique that combines optical trapping with patch clamp and fluorescence photometry,[49] it was shown that native biological membranes are capable of electrically induced piconewton-level force generation.[50] With this system, optical tweezers were used to form a plasma membrane tether from a patch-clamped cell while fluorescence photometry was used to measure the electrically induced forces on the membranes. Force generation is dependent on the membrane tension and the transmembrane electrical potential. Membranes are capable of high-frequency electromechanical transduction at more than 3 kHz. The force generation is attributed to flexoelectric changes of the tether radius (as above), leading to concomitant changes of the tether length. Notably, this electromechanical transduction is present even in cell membranes lacking the specialized motor protein, prestin.

## 6.4. Flexoelectricity and Cell Biology

### 6.4.1. *Electromechanics and mechano-electricity in cell membranes*

Flexoelectricity[51] is a fundamental property of liquid crystals, relating their mechanical and electrical degrees of freedom. In a membrane system with just two degrees of freedom, there are flexoelectric effects as discussed above, both direct and converse. The well-known fact that the inner mitochondrial membrane is quite smooth in the non-energized state, while in energized mitochondria it is extensively curved,[52–54] is quite indicative to us of a possible role for the flexoeffect in energy transformation. Thus, flexoelectricity could be the missing link in the hypothesis of conformational coupling during oxidative phosphorylation in mitochondria.[55] The charging of the membrane capacitor could then originate from curvature-induced flexo-polarization.

As we have seen above, in membranes (living and model) flexoelectricity provides a linear relation between membrane curvature and membrane polarization or transmembrane voltage and membrane bending stress. It is thus closely related to mechanosensitivity and mechanotransduction, basic features of living systems.[42]

A model using the direct flexoelectric effect for the transformation of mechanical into electrical energy in the hearing process in stereocilia has been proposed.[42]

Recently, the converse flexoeffect (transforming electrical into mechanical energy) was employed to describe the electromotility of an outer hair cell (OHC) membrane[56] (Fig. 6.7 and Fig. 6.8). Electromotility plays a central role in the process of mechano-amplification, which is vital for hearing high-frequency sounds.[57,58]

The elemental motile unit of an OHC strikingly resembles the flexoelectric domain structure[10] whose calculated period $\pi K_{11}(e_1 E)^{-1}$ also depends linearly on the inverse electric field (Fig. 6.9). Such a repetitive arc and pillar nano-architecture, containing *sharp* points at the confluence of any two adjacent arcs, is inherently polar and enhances the flexoelectric mechanism (e.g., a sinusoidal-shaped membrane will not enhance flexoelectricity: while one of the half-waves is reduced the opposite one will be extended and vice versa). This arc motif is repeated a few thousand times along the

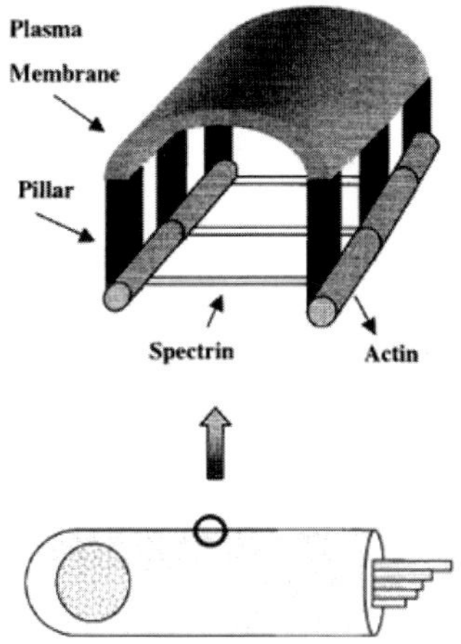


Fig. 6.7. Nanomechanical model for the converse flexoeffect in an outer hair cell and a schematic of the cell. These cells are cylindrically shaped with lengths ranging from 20 to 90 $\mu$m arranged along the cochlea and with a radius of 4–5 $\mu$m. A hair bundle, composed of stereocilia, is located at the apex of the cell. The lateral wall is the source of electromotility and it appears smooth under a light microscope. When examined with electron microscopy, the lateral wall appears corrugated. The folds in the membrane appear to terminate at pillar proteins that extend to the cytoskeleton. The cytoskeleton is composed of actin filaments cross-linked by spectrin molecules. (Reprinted from Raphael *et al.*[56] Copyright 2000, with permission from Elsevier.)

OHC membrane; this is how a *nanometre* displacement at the end of a single arc is amplified by three orders of magnitude resulting in a *micrometre* displacement at the cell end. Membrane arcs are also found in several other electromotile cells, e.g. *Oscillatoria*, *Flexibacter BH3*.[59] Thus, a membrane arc terminating on protein pillars is a basic nanoscale unit of a unique linear flexoelectric motor, which at the microscopic scale looks like a piezoelectric one.[56] However, such an apparent piezomotor has an enormous effective piezocoefficient, which has no analogue in organic materials. For example, while a cow femur has a piezocoefficient $c_{12} = 0.08 \times 10^{-12}$ C/N, quartz has $c_{12} = 2.3 \times 10^{-12}$ C/N and the best piezoelectric ceramic PZT has $c_{12} = 400 \times 10^{-12}$ C/N, the outer hair cell's effective piezocoefficient is $c_{12} = 20000000 \times 10^{-12}$ C/N $= 20$ $\mu$m/V[60,61] (the linear elongation of a 50-$\mu$m-long OHC is measured as 2 $\mu$m/100 mV).

We have explained[18] this huge piezoelectricity at the micrometre scale by considering the structure as a flexoelectric one at the nanoscale.[62] For $f = 1 \times 10^{-21}$ C (a *very* small value), we still obtained $c_{12}/c_{22} = 0.08$ pC/$\mu$m, which is *well* within the experimental limits of 0.03–0.1 pC/$\mu$m.

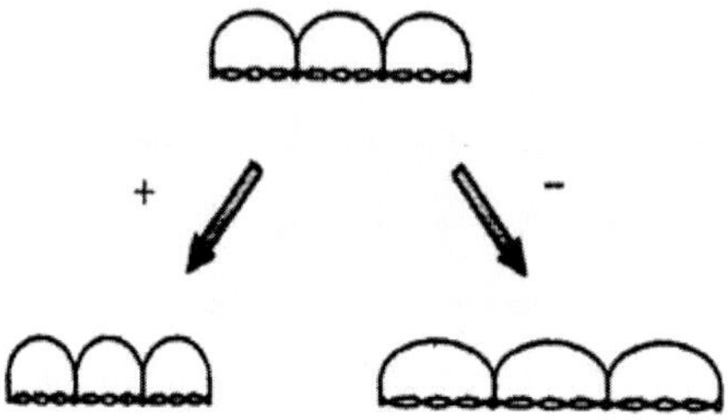


Fig. 6.8. Curvature changes in the elemental motile unit cause extension of the spectrin molecules attached to the pillar proteins. Three units are shown in the figure. A membrane depolarization (+) leads to a decrease in the radius of curvature and a shortening of the cell while hyperpolarization (−) leads to an increase of the radius of curvature and cell lengthening. (Reprinted from Raphael *et al.*,[56] Copyright 2000, with permission from Elsevier.)

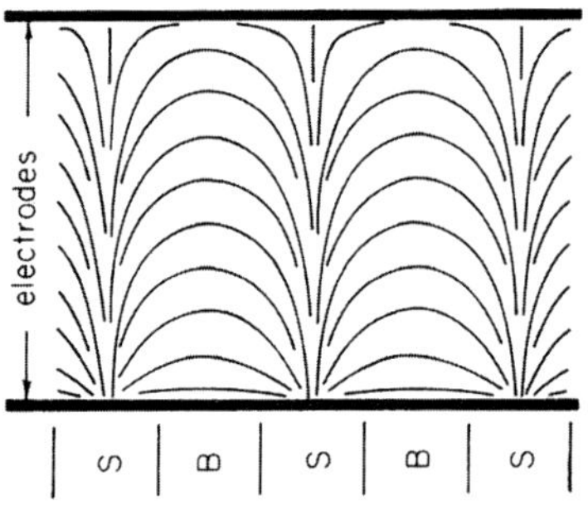


Fig. 6.9. Micromechanical model for the converse flexoeffect in a nematic liquid crystal. Cross section of a field-induced flexoelectric domain pattern in a nematic between plane-parallel electrodes, containing alternating regions of splay (S) and bend (B) of the director field. The calculated period of this pattern is inversely proportional to the electric field (*cf.* Fig. 6.8). (Figure reprinted with kind permission of the author and the publisher from Meyer.[10] Copyright 1969 by the American Physical Society.)

Furthermore, by taking the length elastic modulus as $c_{22} = 1 \times 10^2$ m/N[60,61] we calculated for the apparent piezocoefficient a value of $c_{12} = 8 \times 10^{-6}$ C/N = 8 $\mu$m/V. This is also close to the experimental value of 20 $\mu$m/V, showing that even a very weak membrane flexoelectricity at the nanolevel combined with a thousandfold repetition of the elemental motile unit, can explain the huge apparent piezoelectricity of OHC at the microlevel.[62]

Many other important membrane functions require the creation of membrane curvature (e.g., endocytosis, virus budding, cell and organelle morphogenesis, cell movement by pseudopodia, and cell contact). The out-of-plane fluctuations of membranes (e.g., the flicker phenomenon in erythrocytes[63]) provide a typical example of a fluctuating membrane curvature. This could cause an attractive flexoelectric force at short distances, which is important for cell-cell contact.[64] In each of these phenomena curvature-induced polarization is surely present and an acknowledgement of its presence can be of crucial importance for a deeper understanding of what is occurring.[1] The existence of flexoelectricity in passive and active membrane ion transport was postulated some time ago[9,13] and has recently been corroborated.[1,65] Therefore, the prospect of flexoelectricity being seen as an intricate relation between membrane mechanics and membrane electrodynamics is rather high.

### 6.4.2. *Opto-mechano-electricity*

An optical degree of freedom in membrane systems was recently explored in a flexoelectric context. In an opto-mechano-electric membrane system the number of possible combinations of the three degrees of freedom (serving either as inputs or outputs) is greatly increased. Moreover, because of the existence of these degrees of freedom in some natural membranes, new hints regarding the structure-function relation in photosynthetic membranes, retinal rods and discs, and other photoactive membranes can be obtained. In a membrane system with three degrees of freedom the possible energy interconversion effects, involving two or three degrees of freedom simultaneously, have already been classified.[1] The total number of such effects amounts to at least 12, because if there are three degrees of freedom additional subclasses are possible with respect to input/output combinations. At present, just a few of these effects have been experimentally demonstrated in membrane systems.

Black lipid membranes modified with semiconducting nano-sized CdS particles ('nanomembranes') serve as models for photosensitive natural cell membranes, where CdS crystallites mimic membrane proteins.[66] Nanomembranes are known to display photoelectric effects. Like some biomembranes, they are capable of transforming optical energy into electrical energy. The photoelectric effect, i.e., the photo-induced charge separation across a nanomembrane gives rise to a photocurrent or photovoltage.[67,68] This effect has also been described as the 'chemical capacitance effect'.[69] The converse effect is the electroluminescence of a membrane, i.e., light generation by a transmembrane voltage. This effect has been observed in natural photosynthetic membranes.[70]

Periodic variations of membrane curvature in these CdS nanomembranes lead to the discovery of a new effect, 'photoflexoelectricity'.[71] A mechanism for photoflexoelectricity in nanomembranes was been proposed and compared to experiment.[1]

This first observation of photoflexoelectricity in a biomimetic system allows us to speculate upon its possible existence in natural (e.g. photosynthetic) membranes as well. The existence of a converse effect (i.e., a photopotential-induced membrane curvature on illumination) also seems highly probable. The connectivity between membrane shape (curvature), light intensity and the membrane potential could be related to the mechanism responsible for phototaxis and other photoadaptive phenomena at the cell membrane level.

### 6.4.3. *Chemi-opto-mechano-electricity: Membrane machines*

A further, chemical degree of freedom has not been particularly discussed so far, but it is clear that it is inherent in all the others, not least because all coupling coefficients (elastic, flexoelectric, dielectric, photoelectric, etc.) are functions of membrane chemistry.[1] It can be used for the construction of flexoelectric membrane sensors.[1,30,35,72,73] The same considerations are valid for the thermal degree of freedom: the coupling coefficients are temperature dependent, if nothing else. By establishing the general principles of membrane mechanisms, membranologists will be able to prepare examples of 'protocells' (or membrane machines[74]), based on giant vesicular membrane structures with several degrees of freedom, as above. The first evidence for cyclic behaviour under steady illumination of a giant lipid vesicle, with stress-sensitive alamethicin channels and with a ferricyanide solution inside, has been found.[75] Suppressed fluctuations of the vesicle over long time periods, with intermittent short periods of extensive fluctuation activity, have been observed (Fig. 6.10).

Mechanical tension may be due to a converse-flexoelectric area dilatation induced by the photopotential, as analysed above [*cf.* Eq. (6.16)]. The photopotential is due to the pH gradient because of the ferricyanide photochemical reactions. The increase of membrane tension above a certain threshold can open the ion channels thus dissipating the pH gradient returning the membrane tension back to zero (at this instant extensive form fluctuations, *cf.* Fig. 6.10, are thermally excited), so that the channels close and the cycle repeats. Under this scenario it is clear that the whole photo-

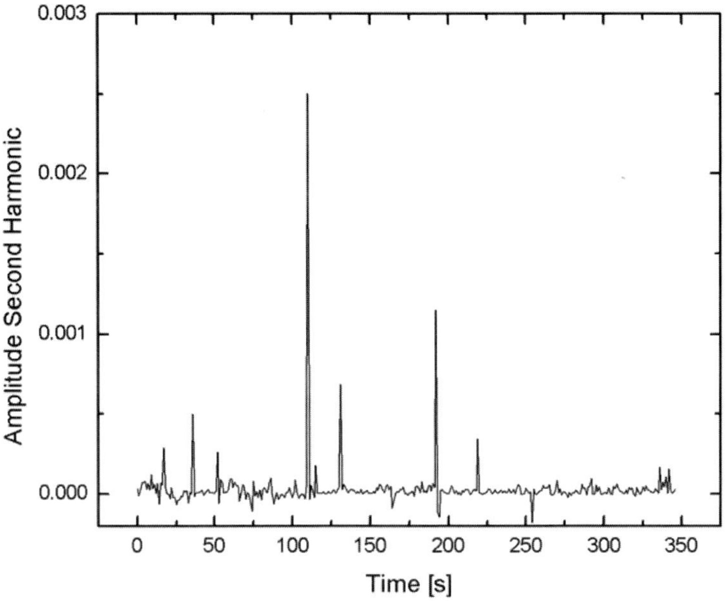


Fig. 6.10. Time dependence of curvature fluctuations of a giant lipid vesicle with stress-sensitive alamethicin channels in its membrane. Inside the vesicle there is a ferricyanide solution undergoing a photochemical reaction under illumination, which produces a pH gradient and a photopotential across the membrane. The graph shows the second Legendre polynomial amplitude of the angular autocorrelation function of the vesicle radius as a function of time. Brief episodes (peaks) of extensive thermal fluctuations in a tension-free membrane are separated by long periods of a tensed, non-fluctuating vesicle membrane. (V. Vitkova, A.G. Petrov, unpublished.)

chemi-flexoelectro-mechanical cycle will be repeated again and again, as long as there is sufficient energy in the form of light.

### 6.4.4. *Flexoelectricity and nerve conduction*

As mentioned, the first idea that membranes might have piezoelectric properties emerged in 1941 in a discussion of the possible origin of inductance in the squid's giant axon.[8] In 1980 it was discovered that the propagation of the action potential in a squid axon is accompanied by a propagation of a mechanical disturbance along the nerve membrane.[76] Apart from the rapid pressure change inside the axon[77] flexoelectricity of the nerve membrane was also considered as a reason for its surface displacements.[78] Applying the advanced method of optical coherence tomography the authors[78] were able to measure membrane displacements in the nanometre range with a subnanometre accuracy. The time evolution of the optical signal was roughly synchronous with the electrical action potential.

The flexoelectric displacement to a first approximation is given by Eq. (6.16) without the factor 2 in the denominator because of the cylindrical geometry of the axon. The estimation below, Eq. (6.16), demonstrates that a broad range of displacement amplitudes and signs is available, even at very modest flexoelectric values. The pressure effect can be described by a modified Laplace derivation where the membrane tension $\sigma$ in the free energy variation $\sigma\Delta A - p\Delta V = 0$ is represented by the lateral elastic stress:

$$\sigma = K_S\left(\frac{\Delta R}{R_0}\right),$$

where $R_0$ is the tension-free value of the axon radius. Thus, together with Eq. (6.16) we obtain for the amplitude of the membrane displacement:

$$\left(\frac{\Delta R_{ap} - \Delta R_r}{R_0}\right) = \frac{R_0}{K_S}(p_{ap} - p_r) + \frac{f}{\sqrt{KK_S}}\frac{\Delta V_{ap} - \Delta V_r}{d}, \qquad (6.17)$$

where $p_{ap}$ and $p_r$ are the pressures at the action potential and at rest, while $\Delta V_{ap}$ and $V_r$ are the action potential and resting potential, respectively. The sign of the potentials is determined by the charge sign of the axon interior. The unstressed value of the axon radius $R_0$ is for a zero membrane potential. The pattern of positive versus negative displacements along the axon length may be related to a pattern of positive versus negative values of the flexocoefficient. There is no need to invoke an isothermal phase transition of the membrane lipids in order to explain surface displacements.[79] The peculiarities of the propagation of single pulses of flexoelectric waves along nerve fibres need further attention. An application of the theory describing the periodic operation of the stereocilia as flexoelectric motors of the inner ear[80] [based on a system of coupled electromechanical equations, Equations (6.7) and (6.8) in Breneman *et al.*[80]] could also be attempted with respect to pulse propagation in nerve membranes.

## 6.5. Conclusion

Flexoelectricity is a reciprocal relation between the electrical and mechanical properties of liquid crystalline biostructures. We have underlined in this chapter the role of a specific mechanical degree of freedom, the membrane curvature, in this respect. Due to this mechanical degree of freedom, membranes are able to function like soft machines. We have stressed the importance of the flexoelectric phenomenon in the interrelation of the membrane curvature with the electrical, optical, thermal and chemical degrees of freedom of membranes. The flexoelectric concepts reviewed here could also be of primary importance in the construction of artificial membrane machines ('protocells'), as well as electromechanical sensors and actuators for nanoionics.

## References

1. A.G. Petrov, *The Lyotropic States of Matter: Molecular Physics and Living Matter Physics*, Gordon & Breach Publishers, New York, 1999.
2. G.H. Brown and J.J. Wolken, *Liquid Crystals and Biological Structures*, Academic Press, NY-SF-L, 1979.
3. P. Bothorel and C. Lussan, On a biological membrane model based on lipid-protein interactions, *C. R. Hebd. Séances Acad. Sci.* **266C**(26), 2492–2494, (1968).
4. S.J. Singer and G.L. Nicolson, The fluid mosaic model of the structure of cell membranes, *Science* **175**(4023), 720–731, (1972). doi: 10.1126/science.175.4023.720
5. S.J. Singer, The structure and function of membranes: A personal memoir, *J. Membrane Biol.* **129**(1), 3–12, (1992). doi: 10.1007/BF00232051
6. P. Sitte, Biomembranes: High protein concentration, *Naturwiss.* **66**(6), 315–316, (1979). doi: 10.1007/BF00441279
7. D.E. Green, S. Ji and R.F. Brücker, Structure-function unitization model of biological membranes, *Bioenergetics* **4**(1-2), 253–284, (1973). doi: 10.1007/BF01516061
8. K.S. Cole, Rectification and inductance in the squid giant axon, *J. Gen. Physiol.* **25**(1), 29–51, (1941).
9. A.G. Petrov, Flexoelectric model for active transport. In ed. J. Vassileva, *Physical and Chemical Bases of Biological Information Transfer*, Plenum Press, New York-London, 1975, p. 111.
10. R.B. Meyer, Piezoelectric effects in liquid crystals, *Phys. Rev. Lett.* **22**(18), 918–921, (1969). doi: 10.1103/PhysRevLett.22.918
11. V.K. Fredericksz and L.S. Polak (1940), Towards the Theory of Anisotropic Liquids. In eds. A.S. Sonin and V.Ja. Frenkel, *V.K. Freedericks (1885-1944)*. Moskwa, Nauka, Fizmatlit, 1995. pp. 142–170.
12. W.P. Mason, *Piezoelectric Crystals and their Application to Ultrasonics*, D. van Nostrand Co.Inc., T-NY-L, 1950.
13. A.G. Petrov, S.A. Seleznev and A. Derzhanski, Principles and methods of liquid crystal physics applied to the structure and functions of biological membranes, *Acta Phys. Pol. A* **55**(3), 385–405, (1979).

14. A.G. Petrov, Measurements and interpretation of flexoelectricity. In eds. D.A. Dunmur, A. Fukuda and G.R. Luckhurst, *Physical Properties of Liquid Crystals: Nematics Vol. 1*, EMIS Datareviews Series, Inst. Electrical Engineers, UK, 2001. pp. 251–264.
15. A.G. Petrov and A. Derzhanski, Membranes and lyotropic phases on some problems in the theory of elastic and flexoelectric effects in bilayer lipid membranes and biomembranes, *J. Phys. Colloques* **37**(C3), C3-155–C3-160, (1976). doi: 10.1051/jphyscol:1976330
16. A.G. Petrov, Flexoelectricity of model and living membranes, *Biochim. Biophys. Acta* **1561**(1), 1–25, (2001). doi: 10.1016/S0304-4157(01)00007-7
17. A.G. Petrov and F. Sachs, Flexoelectricity and elasticity of asymmetric biomembranes, *Phys. Rev. E* **65**(1), 021905/1–5, (2002). doi: 10.1103/PhysRevE.65.021905
18. A.G. Petrov, Electricity and mechanics of biomembrane systems: Flexoelectricity in living membranes, *Anal. Chim. Acta* **568**(1), 70–83, (2006). doi: 10.1016/j.aca.2006.01.108
19. A.G. Petrov, Flexoelectricity and mechanotransduction. In ed. O.P. Hamil, *Current Topics in Membranes, vol. 58: Mechanosensitive channels*, Elsevier/Academic Press, 2007. pp. 121–150.
20. F. Sachs, W.E. Brownell and A.G. Petrov, Membrane electromechanics in biology, with a focus on hearing, *MRS Bulletin* **34**(9), 665–670, (2009). doi: 10.1557/mrs2009.178
21. A.G. Petrov and I. Bivas, Elastic and flexoelectric aspects of out-of-plane fluctuations in biological and model membranes, *Progress in Surface Sci.* **16**(4), 389–511, (1984). doi: 10.1016/0079-6816(84)90016-9
22. R. Mutafchieva, *Equilibrium and Dynamic Properties of Monomolecular Layers as Models of the Alveolar Surface.* PhD Thesis, Bulgarian Academy of Sciences, Sofia, 1985.
23. J.F. Danielli and H. Davson, A contribution to the theory of permeability of thin films. *J. Cell. Comp. Phys.* **5** (4), 495–508, (1935). doi: 10.1002/jcp.1030050409
24. R. Henderson, The purple membrane from *Halobacterium halobium*, *Ann. Rev. Biophys. Bioeng.* **6**, 87–109, (1977). doi: 10.1146/annurev.bb.06.060177.000511
25. Eds. J.B. Robinson and D.S. Parsons, *Biological Membranes*, Clarendon Press, Oxford, 1975.
26. A. Zhivkov, *Optical and Electrooptical Studies of Purple Membranes*, Ph.D. Thesis, Institute of Physical Chemistry, Bulgarian Academy of Sciences, Sofia, 1995.
27. V.I. Passechnik and V.S. Sokolov, Permeability change of modified bimolecular phospholipid membranes accompanying periodical expansion, *Biofizika (Moscow)* **18**, 655–660, (1973).
28. A.L. Ochs and R.M. Burton, Electrical response to vibration of a lipid bilayer membrane, *Biophys. J.* **14**(6), 473–489, (1974). doi: 10.1016/S0006-3495(74)85928-X

29. A. Derzhanski, A.G. Petrov and Y.V. Pavloff, Curvature induced conductive and displacement currents through lipid bilayers, *J. Physique Lett.* **42**(6), L-119–L-122, (1981). doi: 10.1051/jphyslet:01981004206011900
30. A.G. Petrov and V.S. Sokolov, Curvature-electric effect in black lipid membranes, *Eur. Biophys. J.* **13**(3), 139–155, (1986). doi: 10.1007/BF00542559
31. A.T. Todorov, A.G. Petrov, M.O. Brandt and J.H. Fendler, Electrical and real-time stroboscopic interferometric measurements of bilayer lipid membrane flexoelectricity, *Langmuir* **7**(12), 3127–3137, (1991). doi: 10.1021/la00060a036
32. A. Todorov, *Experimental Investigations of Direct and Converse Flexoelectric Effect in Bilayer Lipid Membranes*, PhD Thesis, Syracuse University, 1993.
33. A. Todorov, A.G. Petrov and J.H. Fendler, Flexoelectricity of charged and dipolar bilayer lipid membranes studied by stroboscopic interferometry, *Langmuir* **10**(7), 2344–2350, (1994). doi: 10.1021/la00019a053
34. A. Todorov, A.G. Petrov and J.H. Fendler, First observation of the converse flexoelectric effect in bilayer lipid membranes, *J. Phys.Chem.* **98**(12), 3076–3079, (1994). doi: 10.1021/j100063a004
35. A. Derzhanski, A.G. Petrov, A. Todorov and K. Hristova, Flexoelectricity of lipid bilayers, *Liq. Cryst.* **7**(3), 439–449, (1990). doi: 10.1080/02678299008033820
36. B. Sakmann and E. Neher, Chapter 21, Geometric parameters of pipettes and membrane patches. In eds. B. Sakmann and E. Neher, *Single Channel Recording*, Plenum Press, New York-London, 1983. pp. 637–650.
37. R. Coronado and R. Latorre, Phospholipid bilayers made from monolayers on patch-clamp pipettes, *Biophys. J.* **43**(2), 231–236, (1983). doi: 10.1016/S0006-3495(83)84343-4
38. T. Schürholz and H. Schindler, Formation of lipid–protein bilayers by micropipette guided contact of two monolayers, *FEBS Letters* **152**(2), 187–190, (1983). doi: 10.1016/0014-5793(83)80376-7
39. A.G. Petrov, R.L. Ramsey and P.N.R. Usherwood, Curvature-electric effects in artificial and natural membranes studied using patch-clamp techniques, *Eur. Biophys. J.* **17**(1), 13–17, (1989). doi: 10.1007/BF00257141
40. A.G. Petrov, P.N.R. Usherwood and B.A. Miller, Mechanoelectricity of guest-host membrane systems: Lipid bilayers containing ion channels, *Mol. Cryst. Liq. Cryst.* **215**(1), 109–119, (1992). doi: 10.1080/10587259208038515
41. A.G. Petrov, B.A. Miller, K. Hristova and P.N.R. Usherwood, Flexoelectric effects in model and native membranes containing ion channels, *Eur. Biophys. J.* **22**(4), 289–300, (1993). doi: 10.1007/BF00180263
42. A.G. Petrov and P.N.R. Usherwood, Mechanosensitivity of cell membranes, *Eur. Biophys. J.* **23**(1), 1–19, (1994). doi: 10.1007/BF00192201
43. A.G. Petrov, Charge transfer processes in model and biological membranes: Defect and mechano-electric aspects; statics and dynamics, *Mol. Cryst. Liq. Cryst. A* **292**(1), 227–234, (1997). doi: 10.1080/10587259708031933
44. T.M. Suchyna and F. Sachs, An introduction to cell motility for the physical scientist, *Physical Biology* **1**(1), T1–T10, (2004). doi: 10.1088/1478-3967/1/1/T01

45. J. Mosbacher, M. Langer, J.K.H. Horber and F. Sachs, Voltage-dependent membrane displacements measured by atomic force microscopy, *J. Gen. Physiol.* **111**(1), 65–74, (1998). doi: 10.1085/jgp.111.1.65
46. P.-Ch. Zhang, A.M. Keleshian and F. Sachs, Voltage-induced membrane movement, *Nature* **413**(6854), 428–431, (2001). doi: 10.1038/35096578
47. W. Helfrich, Amphiphilic mesophases made of defects. In eds. R. Balian, M. Kleman and J.P. Poirier, *Physics of Defects*, North-Holland Publ. Co., Amsterdam, 1981. pp. 716–755.
48. A.G. Petrov and M.M. Kozlov, Curvature elasticity and passage formation in lipid bilayers. Lattice of passages, *Compt. Rend. Acad. Bulg. Sci.* **37**, 1191, (1984).
49. F. Qian, S. Ermilov, D. Murdock, W.E. Brownell and B. Anvari, Combining optical tweezers and patch clamp for studies of cell membrane electromechanics, *Rev. Sci. Instrum.* **75**(9), 2937–2942, (2004). doi: 10.1063/1.1781382
50. B. Anvari, F. Qian, F.A. Pereira and W.E. Brownell, Prestin-lacking membranes are capable of high frequency electromechanical transduction. In eds. A.L. Nuttall, T. Ren, P. Gillespie, K. Grosh and E. de Boer, *Auditory Mechanisms: Processes and Models*, World Scientific, London, 2005. pp. 270–276.
51. P.G. de Gennes and J. Prost, *The Physics of Liquid Crystals (2nd ed.)*, Clarendon Press, Oxford, 1993.
52. C.R. Hackenbrock, Ultrastructural bases for metabolically linked mechanical activity in mitochondria: II. Electron transport-linked ultrastructural transformations in mitochondria, *J. Cell Biol.* **37**(2), 345–369, (1968). doi: 10.1083/jcb.37.2.345
53. R.A. Harris, J.T. Penniston, J. Asai and D.E. Green, The conformational basis of energy conservation in membrane systems. II. Correlation between conformational change and functional states, *Proc. Natl. Acad. Sci. USA* **59**(3), 830–837, (1968).
54. D.E. Green and J.A. Young, Energy transduction in membrane systems, *Am. Sci.* **59**(1), 92–100, (1971).
55. A.L. Lehninger, *Biochemistry. The Molecular Basis of Cell Structure and Function*, Worth Publs., Inc., NY, 1972.
56. R.M. Raphael, A.S. Popel and W.E. Brownell, A membrane bending model of outer hair cell electromotility, *Biophys. J.* **78**(6), 2844–2862, (2000). doi: 10.1016/S0006-3495(00)76827-5
57. V.I. Passechnik, Mekhanizmi ulitki organa slukha. In ed. P.G. Kostyuk, *Accounts Sci. Techn. Human and Animal Physiol. Ser. 39*, VINITI, Moscow, 1988. pp. 6–121.
58. W.E. Brownell, A.A. Spector, R.M. Raphael and A.S. Popel, Micro- and nanomechanics of the cochlear outer hair cell, *Ann. Rev. Biomed. Eng.* **3**, 169–194, (2001). doi: 10.1146/annurev.bioeng.3.1.169
59. W.E. Brownell, Membrane based motor mechanisms. 1st World Flexoelectric Congress, SUNY-Buffalo, 2001.
60. K. Iwasa, A two-state piezoelectric model for outer hair cell motility, *Biophys. J.* **81**(5), 2495–2506, (2001). doi: 10.1016/S0006-3495(01)75895-X

61. X.-X. Dong, M. Ospeck and K. Iwasa, Piezoelectric reciprocal relationship of the membrane motor in the cochlear outer hair cell, *Biophys. J.* **82**(3), 1254–1259, (2002). doi: 10.1016/S0006-3495(02)75481-7
62. A.G. Petrov, Membrane flexoelectricity at nanoscale level, 2nd World Flexoelectric Congress, Rice-Houston, 2003.
63. H. Strey, M. Peterson and E. Sackmann, Measurement of erythrocyte membrane elasticity by flicker eigenmode decomposition, *Biophys. J.* **69**(2), 478–488, (1995). doi: 10.1016/S0006-3495(95)79921-0
64. I. Bivas and A.G. Petrov, Flexoelectric and steric interactions between two bilayer lipid membranes resulting from their curvature fluctuations, *J. Theor. Biol.* **88**(3), 459–483, (1981). doi: 10.1016/0022-5193(81)90277-0
65. A.G. Petrov, Flexoelectricity and ion channels: A confirmation of the flexoelectric model for ion transport, *Cell. & Molec. Biol. Lett.* **2**, suppl. 1, 231–253, (1997).
66. J.H. Fendler, *Membrane-Mimetic Approach to Advanced Materials*, Springer-Verlag, Berlin, 1994.
67. S. Baral and J.H. Fendler, Cadmium sulfide-mediated photoelectric effects in bilayer lipid membranes, *J. Am. Chem. Soc.* **111**(5), 1604–1614, (1989). doi: 10.1021/ja00187a012
68. X.K. Zhao, S. Baral, R. Rolandi and J.H. Fendler, Semiconductor particles in bilayer lipid membranes. Formation, characterization, and photoelectrochemistry, *J. Am. Chem. Soc.* **110**(4), 1012–1024, (1988). doi: 10.1021/ja00212a005
69. D. Mauzerall, H. Gerischer and J.J. Katz (Eds.), *Light-Induced Charge Separation in Biology and Chemistry*, Dahlem Konferenzen, Berlin, 1979. p. 241.
70. V. Brumfeld, I.R. Miller and R. Korenstein, Electric field-induced lateral mobility of photosystem I in the photosynthetic membrane, *Biophys. J.* **56**(3), 607–614, (1989). doi: 10.1016/S0006-3495(89)82707-9
71. M. Spassova, A.G. Petrov and J.H. Fendler, Photoflexoelectric effects in bilayer lipid membranes, *J. Phys. Chem.* **99**(23), 9485–9490, (1995). doi: 10.1021/j100023a028
72. K. Sun, Toward molecular mechanoelectric sensors: Flexoelectric sensitivity of lipid bilayers to structure, location, and orientation of bound amphiphilic ions, *J. Phys. Chem.* **101**(33), 6327–6330, (1997). doi: 10.1021/jp971546j
73. A. Zheliaskova, S. Naidenova, Y. Marinov, I.R. Mellor, P.N.R. Usherwood and A.G. Petrov, Detection of heavy metal ions ($Cd^{2+}$ and $Hg^{2+}$) by their influence on flexoelectricity of patch clamped membranes, *Compt. Rend. Acad. Bulg. Sci.* **54** (12), 53–56, (2001).
74. R. Lipowsky, From membranes to membrane machines. In eds. D. Reguera, J.M.G. Vilar and J.M. Rubi, *Statistical Mechanics of Biocomplexity. Lecture Notes in Physics, vol. 527*, Springer, Berlin, 1999. pp. 1–23.
75. A.G. Petrov, Flexoelectric effects in biological membranes, Invited lecture, Rank Prize Symposium, Grasmere, UK, 2004.
76. K. Iwasa and I. Tasaki, Mechanical changes in squid giant axons associated with production of action potentials, *Biochem. Biophys. Res. Commun.* **95**(3), 1328–1331, (1980). doi: 10.1016/0006-291X(80)91619-8

77. I. Tasaki and K. Iwasa, Rapid pressure changes and surface displacements in the squid giant axon associated with production of action potentials, *Jpn. J. Physiol.* **32**(1), 69–81, (1982). doi: 10.2170/jjphysiol.32.69
78. T. Akkin, D. Landowne and A. Sivaprakasam, Optical coherence tomography phase measurement of transient changes in squid giant axons during activity, *J. Membrane Biol.* **231**(1), 35–46, (2009). doi: 10.1007/s00232-009-9202-4
79. T. Heimburg, Die Physik von Nerven, *Physik Journal* **8**(3), 33–40, (2009).
80. K.D. Breneman, W.E. Brownell and R.D. Rabbitt, Hair cell bundles: Flexoelectric motors of the inner ear, *PLoS ONE* **4**(4), 1, (2009). doi: 10.1371/journal.pone.0005201

# Chapter 7

# Applications of Flexoelectricity

Per Rudquist and Sven T. Lagerwall

*Microtechnology and Nanoscience, Chalmers University of Technology*
*41296 Göteborg, Sweden*
*e-mail: per.rudquist@chalmers.se, sven.lagerwall@mac.com*

There have been a number of attempts to utilize the flexoelectric effect for electro-optic devices. Most of these have been aimed at new display modes, sometimes inspired by the twisted nematic (TN) cell and often with the aim of attaining bistability. The proposed modes so far do not consider smectic materials (where the flexoelectric phenomena are more complex) but deal with nematics or cholesterics (chiral nematics with a sufficiently short pitch to be negatively uniaxial). While success so far has been modest, two of the proposed devices stand out: the flexoelectro-optic device using cholesterics and the zenithal bistable device (ZBD) using non-chiral nematics. The first is a fast analogue device with several remarkable and unique properties. It is characterized by symmetric switching and a flexoelectric polarization is induced at every switching step. As its physics is relatively transparent and analytically accessible it is discussed here at some length, in particular because it gives valuable clues to both the potential and the problems of this device, which has a good chance to become a viable technology in the years to come. The second device is based on the direct flexoelectric effect, which means that a bulk polarization of a particular sign is essentially inverted in the switching operation. It has just reached the market, after a long time of experiments and modelling, in the form of a passively driven display with memory. Its physics is very complex and it can only be modelled to some extent in two dimensions and there is hardly any analytical theory that satisfactorily describes its global working mechanism. However, the basic ideas are simple and beautifully illustrate the physics of flexoelectricity and, as we will describe, the realization of a viable device represents no small engineering achievement.

## Contents



## 7.1. Introduction

The first electro-optic device based on flexoelectricity was proposed by Dozov *et al.* in 1992.[1] It used a hybrid-aligned nematic (HAN) cell having a distortion-induced flexoelectric polarization **P** in the hybrid plane and parallel to the surfaces (Fig. 7.1). An electric field **E** applied in the plane of the cell and normal to **P** induces a twist of the hybrid director structure resulting from the torque $\mathbf{P} \times \mathbf{E}$. The amount of twist and its sign is determined by the amplitude and direction of the electric field, resulting in a polar, analogue electro-optic effect. The zero-field director plane along one of the crossed polarizers gives a dark state while the field-induced twisted structure optically acts like a twisted nematic cell. It is similar to an in-plane switching (IPS) device,[2] but in this case there is a polar effect. Ideally a HAN flexoelectric device will use a material with the dielectric anisotropy $\Delta\varepsilon = \varepsilon_{\parallel} - \varepsilon_{\perp} = 0$ and it is an elegant and so far the only proposed device geometry using a pure flexoelectric effect, i.e. where the switching torque is solely based on the coupling between an existing distortion-induced flexoelectric polarization and the electric field. However, the range of induced twist is limited as electrohydrodynamic instabilities occur for larger fields.[3]

In this chapter we will focus on two other electro-optic applications of

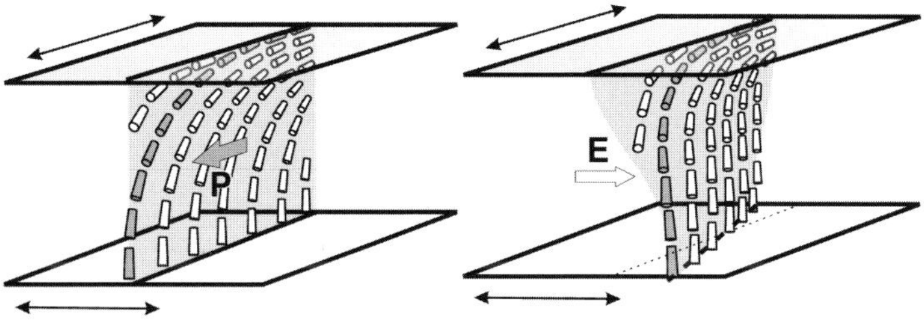


Fig. 7.1. The field-induced twist in a hybrid-aligned device, based on the torque between **E** and the flexoelectric polarization **P** permanently induced in the splay-bend due to the hybrid alignment. The arrows indicate the crossed polarizers.

flexoelectricity. One is the Zenithal Bistable Device (ZBD) in which the flexoelectric effect has been successfully exploited in a nematic memory display. In addition to the torque from dielectric anisotropy it uses the *direct flexoelectric effect*, i.e. the separation of charges due to a strain in the director field. The resulting polarization density couples directly to an applied electric field by which the director configuration can be switched between two stable states. In this device a splay-bend orientational distortion is generated near one of the confining surfaces, which has a convenient boundary condition combined with a grating profile.

The second application uses the *converse flexoelectric effect*, i.e. a field-induced splay-bend distortion, to generate a fast, symmetric and thresholdless linear electro-optic effect in a cholesteric liquid crystal.

These two examples illustrate that it is important whether the structure already possesses a polarization **P** or if **P** is field-induced. If we ignore the possibility of a spatially non-uniform order parameter (which is likely only very close to a surface) the contribution to the free energy density in the presence of a field is, in the first case, $f = -\mathbf{P} \cdot \mathbf{E}$, and the torque on the director is $\boldsymbol{\Gamma} = \mathbf{P} \times \mathbf{E}$, where

$$\mathbf{P} = e_1\mathbf{n}(\nabla \cdot \mathbf{n}) - e_3\mathbf{n} \times (\nabla \times \mathbf{n}), \tag{7.1}$$

$e_1$ and $e_3$ are the splay and bend flexoelectric coefficients, and $\mathbf{n}$ is the director.

The simple expression for $f$ means that the response time will be inversely proportional to the field, $\tau \propto 1/(PE)$ when there is flexoelectric coupling . This is the same dynamics as in ferroelectric liquid crystal (FLC) switching. In the cholesteric case the expression for $f$ is somewhat more complicated and, as we will see below, it turns out that $\tau$ is independent of $E$.

According to Eq. (7.1) $P$ is zero for the two cases of uniform director fields and pure twist. Hence both cases can serve as a 'zero state' as far as flexoelectric excitations are concerned. It is important to note that a twist is not associated with a polarization (i.e. $e_2$ is identically zero, *cf.* Fig. 7.2). An unstrained nematic has a centre of symmetry (centre of inversion). On the other hand, none of the elementary deformations – splay, twist or bend – have a centre of symmetry. According to Curie's principle they could then be associated with the separation of charges analogous to the piezo-effect in solids. This is true for splay and bend but not for twist because of an additional symmetry in that case: if we twist the adjacent directors in a nematic on either side of a reference point, there is always a two-fold symmetry axis along the director of the reference point. In fact, any axis perpendicular to the twist axis is such an axis. Due to this symmetry no vectorial property can exist perpendicular to the director. In other words, a twist does not lead to the separation of charges. This is the reason why twist states appear naturally in liquid crystals and are extremely common. It also means that an electric field cannot induce a twist just by itself in the bulk of a nematic. If anything it *reduces* the twist. A twist can only be induced in a situation where a field turns the director out of a direction that has previously been fixed by boundary conditions (which, for instance, happens in the pixels of an IPS display).

The linear electro-optic effect in a cholesteric, i.e. a hard-twisted chiral nematic (the helical pitch must be less than the wavelength of visible light) was a very original proposal for using the flexoelectric effect in a new display, shutter or modulator device. The patent application by R.B. Meyer and J.S. Patel dates from 1987 and was granted in 1990.[4] The physics was developed in a series of papers[5–8] by these authors and later elaborated by others.[9–21] It is now commonly called the *flexoelectro-optic effect*.

While it is true that this effect has not yet led to a commercial application, its potential is still quite striking. Because of its unusually rich and illuminating physics and the fact that it historically precedes the memory device, we will start with its description, in Section 7.2, and then give an account of the development towards nematic memory devices in Section 7.3. That work was recently given its first commercial applications in different forms of the so-called zenithal bistable device display, discussed at length in Section 7.4. This is a display in which the flexoelectric effect is used to control the switching between two stable states, even if most of the switching is performed by a normal dielectric torque ($\propto \Delta\varepsilon$). Their bistability is not as symmetric as in the FLC displays but, on the other hand, they are by far not as asymmetric as in cholesteric bistable displays.

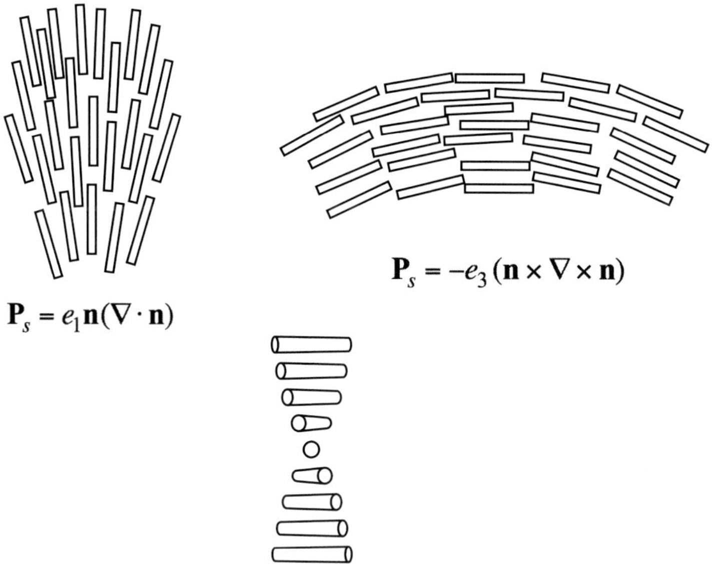


Fig. 7.2. Oseen's three elementary elastic deformations do not have a centre of symmetry. Therefore one might think they should be accompanied by an induced polarization analogous to the piezo-effect in solids. This is partly true. A splay violates the invariance under $\mathbf{n} \to -\mathbf{n}$, hence the $\mathbf{P}$ vector is in the direction of $\mathbf{n}$. The invariance holds for bend and twist but in a bend deformation $\mathbf{P}$ is perpendicular to $\mathbf{n}$. This is not possible in a twist because this deformation has a two-fold rotation axis $\perp\mathbf{n}$ that inverts any $\mathbf{P}\perp\mathbf{n}$. Thus a nematic has only two flexoelectric coefficients.

## 7.2. The Flexoelectro-optic Effect

In two of his earliest papers on liquid crystals,[22,23] Meyer investigated how electric and magnetic fields act on the anisotropy of the electric and magnetic susceptibility in a cholesteric liquid crystal. This effect is quadratic in the field.

If an electric field is applied perpendicular to the helical axis, the helix starts to unwind and, for a sufficiently high value of the field, will be completely unwound. The unwinding of a cholesteric helix is a dramatic and spectacular event and this probably masked the less spectacular event that takes place long before in the same geometry, namely that the field tilts the helix axis in a plane perpendicular to the field. This is the same as saying that the optic axis (the axis of the oblate cholesteric indicatrix) tilts in that plane. The effect is linear in the electric field and therefore arises from a different coupling mechanism, which will be the subject of a fairly comprehensive discussion below. The linearity means that the rotation direction changes with the polarity of the field. The induced rotation is proportional to the applied field until it reaches values where the quadratic coupling starts to unwind the helix. The geometry of the structural change is illustrated in Fig. 7.3. The electric field $E$ makes the director turn through an angle $\phi(E)$. The optic axis is perpendicular to the director in short-pitch cholesterics hence the optic axis turns by the same amount. When the helical axis, which is the optic axis, is parallel to the electroded glass plates we have an optical retarder that can be switched azimuthally in the plane of the cell.

What happens physically is that the electric field $\mathbf{E}$ induces a polarization $\mathbf{P}$ proportional to $\mathbf{E}$ (a dielectric effect). This polarization is generated in the form of a periodic splay-bend deformation in the director field, which is a polarized state. The deformation appears in any plane perpendicular to the new optic axis. In their first paper Patel and Meyer state that their analysis was inspired by Bouligand's beautiful discovery, while working with biological tissue,[24] that if you cut a cholesteric structure through a plane obliquely to the helix axis, a periodic splay-bend pattern (an 'arc pattern' in the tissue) will occur in that plane. This is different from what is shown in Fig. 7.3 because what you see in the Bouligand case is the *projection* of the director field not the director field itself. Thus there are no polar effects involved. Here this pattern appears in the director field and the splay-bend deformations are coupled to a local polarization of the liquid crystal via the flexoelectric coupling.[5] Now, the flexoelectric effect means that polarization is induced by curvature. In the inverse (or converse) flexoelectric effect curvature is induced by an electric field. The field-induced deflection of the optic axis is thus an inverse flexoelectric effect. Like the quadratic effect this linear effect is dielectric in nature while not related to the dielectric anisotropy. It is a field-induced director distortion that intrinsically carries a dipolar moment. Let us compare the two effects. The normal (quadratic) effect is quite familiar. Here a non-zero value of $\Delta\varepsilon$ gives a torque $\mathbf{P} \times \mathbf{E} \propto E^2$ as long as $\mathbf{P}$ is not parallel to $\mathbf{E}$. In the second case, imagine the field-free cholesteric state as full of splay and bend fluctuations of all length scales that correspond to dipole fluctuations, which sum to zero on average. When we now apply an external field, a torque $\propto E$ acts such that these fluctuations became *biased* to a non-zero $\mathbf{P}$. When we increase the field a new torque appears that only vanishes when the director distortion reaches a new equilibrium configuration where the elastic energy

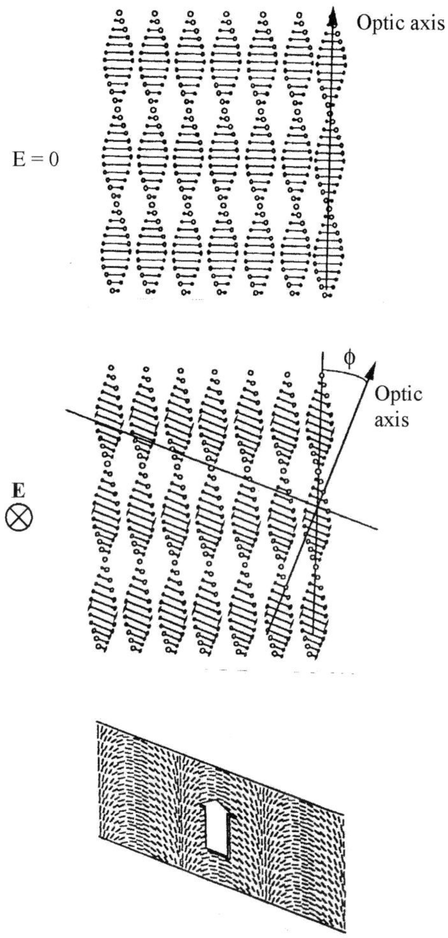


Fig. 7.3. The deviation of the optic axis in a cholesteric (hard-twisted chiral nematic, $p < \lambda$, where $p$ is the cholesteric pitch and $\lambda$ is the wavelength of light) when an electric field **E** is applied perpendicular to the helical axis. The cholesteric geometry allows a flexoelectric polarization to be induced in the direction of **E**. The plane containing the director, which is perpendicular to the page in the middle figure and is shown in the lower figure, illustrates the splay-bend distortion and the corresponding polarization that arises. (After Rudquist,[15] inspired by Meyer and Patel.[4,5])

in the distortion matches the electric energy. This takes a characteristic time $\tau$, which we will derive below.

The deviation $\phi(E)$ is linear in $E$ and tilts of more than 30° can be achieved in the linear region with response times in the range 20 $\mu$s to

100 $\mu$s. It is faster than dielectric switching and works both ways: if we reverse the sign of $E$ the optic axis deviates in the opposite direction, $-\phi$. This is in-plane switching, which should be ideal for driving with conventional thin film transistors (TFTs), see Section 7.2.3 below. The transmission of such a device between crossed polarizers is $T = T_0 \sin^2(2[\phi_0 + \phi(E)]) \sin^2(\pi d \Delta n/\lambda)$, where $T_0$ is the intensity after the first polarizer, $\phi_0$ is the angle between the zero-field optic axis and the polarizer, $\phi(E)$ is the field-induced tilt of the optic axis, $d$ is the cell thickness, $\Delta n$ is the birefringence of the short-pitch structure and $\lambda$ is the wavelength of the light. Finally, $\phi(E)$ turns out to be practically independent of temperature, which is a fairly unique property for a liquid crystal effect.

The flexoelectro-optic effect should also be very interesting in a phase modulator, e.g. a spatial light modulator. The continuous switching of a retarder with constant birefringence gives a continuous phase-only modulation with circularly polarized light.

### 7.2.1. *Basic theory*

In order to understand better the application potential of this electro-optic effect and its fundamental differences from nematic electro-optics based on dielectric anisotropy, and in particular to appreciate the many remaining challenges in its practical application in a device, we will give a derivation of the dynamical properties. The formalism is slightly different from that used by some authors[5–8] and more similar to other descriptions.[15,25] In the presence of an electric field $\mathbf{E}$ the free energy density is written, if we neglect the influence from the surfaces,

$$f = \frac{1}{2}K_1(\nabla \cdot \mathbf{n})^2 + \frac{1}{2}K_2(\mathbf{n} \cdot \nabla \times \mathbf{n})^2 + \frac{1}{2}K_3(\mathbf{n} \times \nabla \times \mathbf{n})^2 - \mathbf{E} \cdot \mathbf{P}, \quad (7.2)$$

where the $K_i$ are the Oseen–Frank constants and $\mathbf{P}$ is given by Eq. (7.1). We have here also neglected the dielectric term since it is quadratic in the field and furthermore as we require $\Delta\varepsilon$ to be very small. If we let the helix axis be along the $z$ direction, then $\mathbf{n} = (\cos kz, \sin kz, 0)$ where $k = 2\pi/p$ is the cholesteric wave number and $p$ is the pitch. If we now apply the field in the $y$ direction, the optic axis tilts by the angle $\phi$ and the director components change to $\mathbf{n} = (\cos kz \cos\phi, \sin kz, -\cos kz \sin\phi)$, which we insert into Eq. (7.2). With the use of Eq. (7.1) and after putting the average values of $\sin^2 kz$ and $\cos^2 kz$ equal to 1/2 we find for the total free energy

$$f = \frac{1}{4}(K_1 + K_3)k^2 \sin^2\phi + \frac{1}{2}K_2k^2(1 - \cos\phi)^2 - \frac{1}{2}(e_1 - e_3)Ek\sin\phi. \quad (7.3)$$

Taking $\partial f/\partial\phi = 0$ we finally find the value of tilt that minimizes the free energy. The result is, using the abbreviations $\frac{1}{2}(K_1 + K_3) = K$ and $\frac{1}{2}(e_1 - e_3) = e$,[a]

$$\tan\phi = \frac{eE}{K_2k} + (1 - K/K_2)\sin\phi. \quad (7.4)$$

For moderate deflection angles $\tan\phi \approx \sin\phi$ and we find that in this limit the dependence on the twist vanishes,

$$\tan\phi = \frac{eE}{Kk}. \quad (7.5)$$

The twist constant $K_2$ does not appear in this simple expression, valid for reasonably small tilt, and to first order the deformation occurs at constant helical pitch ($k$ = const). In a practical device like a display, however, the tilt has to be at least 22.5° to provide 100% modulation in geometry '1' of Fig. 7.4, or about 30–35° in geometry '2' to give an acceptable transmission in the preferred, symmetric drive.

The advantage of geometry '2' is the easy DC-balanced addressing accomplished by simple field reversal between subsequent (sub)frames. In '1', the zero-field state gives 50% transmission, while dark and bright states are produced with positive and negative $E$, respectively, as in a surface-stabilized FLC device. For such large deflections of the optic axis the twist deformation becomes important, see Section 7.2.4. There is no unwinding in the field and therefore there are no hysteretic effects (at least in a material that has not been polymer stabilized). In the small angle limit Eq. (7.5) further reduces to

$$\phi = \frac{eE}{Kk} \quad \text{or} \quad \phi = aE. \quad (7.6)$$

In the same limit the free energy expression, Eq. (7.3), reduces to

$$f = \frac{1}{2}Kk^2\phi^2 - eEk\phi. \quad (7.7)$$

[a] The expression $\frac{1}{2}(e_1 - e_3) = e$ is confusing and misleading, not least because it looks like an anisotropy. It is due to an unfortunate sign convention in the expression for the bend vector. This was discussed in Rudquist and Lagerwall[14] where the opposite sign convention is proposed. With this convention the average flexoelectric coefficient $e$ is given by the more natural expression $\frac{1}{2}(e_s + e_b) = e$.

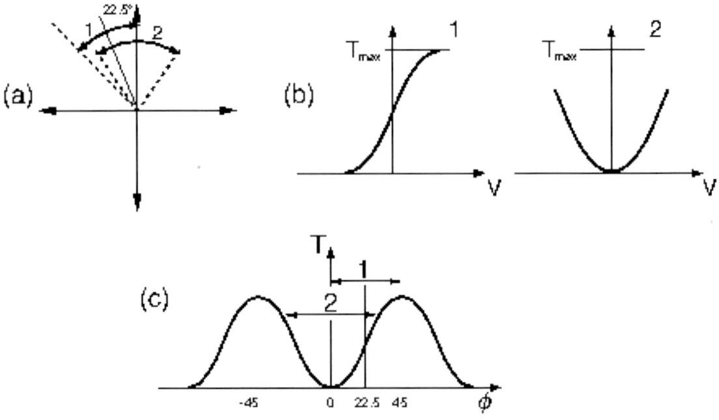


Fig. 7.4. (a) The two basic geometries '1' and '2' for the flexoelectro-optic effect in the in-plane switching mode. In '1' the zero-field optic axis is at $\phi_0 = 22.5°$ and full intensity modulation is achieved for $-22.5° \leq \phi(E) \leq +22.5°$. In '2' the zero-field optic axis is along the polarizer ($\phi_0 = 0°$), which gives the dark state, while $\phi(\pm E)$ gives symmetric bright states. The transmission voltage curves of '1' and '2' (b) are given by $T = T_0 \sin^2(2\phi)\sin^2(\pi d\delta n/\lambda)$ with $\phi = \phi_0 + \phi(E)$ (c).

This means that the flexoelectric torque $\Gamma$ will contain two contributions:

$$\Gamma = -\frac{\partial f}{\partial \phi} = ekE - Kk^2\phi, \tag{7.8}$$

and is thus different in nature from the quadratic dielectric torque $\Gamma_{\text{diel}} \propto E^2$. At the onset of the applied field ($\phi = 0$) $\Gamma \propto E$, but decreases when the optic axis swings out; it is zero when the optic axis has reached its equilibrium value according to Eq. (7.6). The response $\phi(E)$ is thus in both directions a pure relaxation to the equilibrium value given by the actual value of $E$. This means that only one response time $\tau$ characterizes the dynamics, not the usual pair $\tau_{\text{ON}}$ and $\tau_{\text{OFF}}$ that appears when quadratic effects are involved, as in conventional nematic liquid crystal displays (LCDs).

After any perturbation the tilt variable $\phi$ will relax towards the new equilibrium with a speed that is proportional to the rate of change of the free energy with respect to this variable, with a kinetic coefficient having the dimension of an inverse viscosity. This will give us the dynamics:

$$\frac{\partial \phi}{\partial t} = -\gamma^{-1}\frac{\partial f}{\partial \phi}. \tag{7.9}$$

From Eq. (7.7)

$$\gamma \frac{\partial \phi}{\partial t} = -Kk^2\phi + eEk. \tag{7.10}$$

This can be recognized as a Debye equation

$$\gamma \frac{\partial \phi}{\partial t} + A\phi = BE, \tag{7.11}$$

often written, by introducing a characteristic time $\tau = \gamma/A$, in the form

$$\tau \frac{\partial \phi}{\partial t} + \phi = (B/A)E = \frac{e}{Kk}E. \tag{7.12}$$

The saturated value of the induced tilt ($\partial\phi/\partial t = 0$) is found to be $\phi_s = (B/A) = eE/Kk$ in conformity with Eq. (7.6). The induced tilt as a function of time is given by the solution of Eq. (7.12):

$$\phi(t) = \phi_s(1 - e^{-t/\tau}) \tag{7.13}$$

with

$$\tau = \frac{\gamma}{Kk^2}. \tag{7.14}$$

When the field is removed the induced tilt relaxes back to zero with the same characteristic time:

$$\phi(t) = \phi_s e^{-t/\tau}. \tag{7.15}$$

The important point is that Eq. (7.12), which describes this motion, just shows that the rate of change of $\phi$ is proportional to the angular difference between the initial and the final state of $\phi$. In fact, to explicitly show this, the equation can also be written as

$$-\frac{\partial \phi}{\partial t} = (\phi - \phi_0)/\tau. \tag{7.16}$$

Thus, the response time of a flexoelectro-optic device does not depend on the history of switching or on the final state, in contrast to conventional nematic devices. This means that there is only one time constant involved, regardless of whether the switching occurs between grey levels or between extreme states.

The flexoelectric coupling is not chiral, so what is the role of chirality in this case? The answer is that the helically twisted state is the only one that is flexoelectrically 'neutral' (there is no local polarization related to twist) and therefore the only state from where a splay-bend deformation can increase continuously from zero in a symmetric fashion independent of the direction of **E**, while allowing for a homogeneously space-filling splay-bend. How it increases is illustrated in Fig. 7.5.

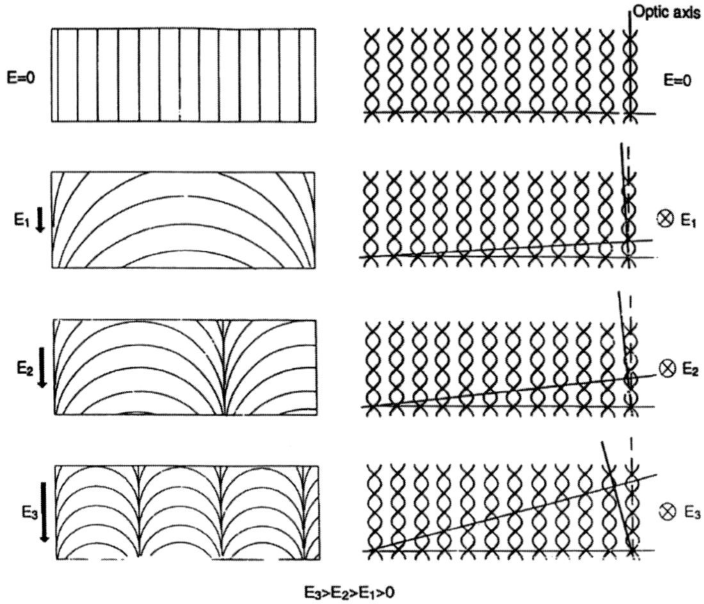


Fig. 7.5. Increasing curvature in response to an increasing electric field: the thresholdless field-induced periodic splay-bend deformation. The pattern in the director field is shown to the left (in any oblique cut perpendicular to the optic axis) and the optic axis deflection with increasing field is shown to the right. (From Rudquist *et al.*,[13] reproduced with kind permission of Taylor & Francis, http://www.tandfonline.com.)

### 7.2.2. *How can we optimize the performance?*

In order to appreciate the limitations in performance we will summarize: according to Eq. (7.6) for small deflections the optic axis rotates by an angle $\phi$ proportional to $E$:

$$\phi = \frac{eE}{Kk},$$

where $e = (e_1 - e_3)/2$, $K$ is the average of the splay and bend elastic constants ($K_1$, $K_3$) and $k$ is the helical wave number. As we can see, the deflection is opposed by the splay-bend elasticity ($K$) and is also the smaller the more hard-twisted ($k$) the cholesteric. The twist elastic constant $K_2$ is not relevant for small $\phi$ for which the deflection takes place at constant twist.

In the same approximation, according to Eq. (7.14), the switching time is given by

$$\tau = \frac{\gamma}{Kk^2},$$

where $\gamma$ is the rotational viscosity. As we can see, compared with Eq. (7.6) there is a trade-off between the maximum deflection for a particular field and the switching speed. Thus the value of $k$ has to be optimized. From Eq. (7.6) it appears possible to increase the deflection $\phi$ by making the cholesteric pitch longer (smaller $k$) but this is limited by the basic condition $k > 2\pi/\lambda$. If the pitch is lengthened there are diffraction effects and finally there is a fingerprint texture, which is of course useless for electro-optics. Moreover, as evident from Eq. (7.14), reducing $k$ means that the switching speed rapidly decreases (quadratically).

An interesting feature of Eq. (7.6) is that for a cholesteric with a temperature-independent pitch, $\phi$ is temperature independent, because both $e$ and $K$ are proportional to $S^2$, where $S$ is the nematic order parameter containing the temperature dependence.[15] In Eq. (7.14), $\gamma$ and $K$ only partly compensate each other in this respect. As it is possible to make a cholesteric with constant $k$ independent of temperature, we can see that the flexoelectro-optic effect will have an induced tilt $\phi$ that is temperature independent and a dynamics that is weakly temperature dependent and, in addition, independent of the applied field. The latter is true as long as $\Delta\varepsilon \approx 0$.

### 7.2.3. *The magnitude of the flexoelectric polarization*

The fast analogue flexoelectro-optic effect, which is at least ten times faster than the nematic modes used today, is very attractive for high speed active matrix displays. In this respect it is interesting to calculate the flexoelectrically induced polarization when the optic axis is deflected by the angle $\phi$. The polarization will be used to estimate the charge that flows, i.e. the current generated in the switching process. This is of importance if we want to use thin-film transistors to drive the flexoelectro-optic effect in some device.

The induced polarization $\mathbf{P}$ is given by Eq. (7.1). Inserting the same vectorial expression for $\mathbf{n}$ as for Eq. (7.2) we find for the two contributions

$$\mathbf{n}(\nabla\cdot\mathbf{n}) = (k\sin kz\cos kz\sin\phi\cos\phi, k\sin^2 kz\sin\phi, -k\sin kz\cos kz\sin^2\phi) \tag{7.17}$$

and

$$\mathbf{n}\times\nabla\times\mathbf{n} = (-k\sin kz\cos kz\sin\phi\cos\phi, k\cos^2 kz\sin\phi, -k\sin kz\cos kz\sin^2\phi). \tag{7.18}$$

Averaging over the whole system as before we find a simple expression for $\mathbf{P}$:

$$\mathbf{P} = \frac{1}{2}(e_1 - e_3)k\sin\phi\,\mathbf{y}. \tag{7.19}$$

As expected, the flexoelectric polarization is entirely in the $y$ direction. With $\frac{1}{2}(e_1 - e_3) = e$ its magnitude is given by

$$P = ek\sin\phi. \tag{7.20}$$

If we insert the values $\phi = 30°$, $k = 2\pi/p = 2\pi/0.5\ \mu\text{m}^{-1}$ and $e = 5$ pC/m we find a value of $P \approx 3$ nC/cm$^2$. This very low value should make TFT driving possible in spite of the relatively high voltage that has to be applied (20–50 V). In comparison, FLC materials typically have ten times higher polarization (but ten times lower driving voltage), while antiferroelectric materials have considerably higher polarization values.[26]

### 7.2.4. *Does the twist really stay constant during switching?*

A glance at Fig. 7.6 immediately shows that the twist cannot stay constant during the switching process. The helix is stretched when the optic axis swings out, i.e. the pitch $p$ gets longer, the wave vector $k$ shorter and the twist constant $K_2$ becomes relevant. To be precise, even a first-order elasticity should be taken into account, but it will here be ignored. We have solved the full Eq. (7.4) numerically for different values of $K_2$ around 10 pN and for different values of $K_2/K$ and the results are shown in Fig. 7.7. It can be seen that the angular deflection is smaller for high fields relative to what would have been expected from a simple extrapolation of the linear part. The deviation from linearity is largest in the one-elastic-constant approximation. On the other hand it can also be seen that the linearity is excellent even at very high tilts, provided that the value of $K_2$ is relatively small compared to $K$.

While the linearity is thus not a problem, the fact that the periodic pitch changes during switching is a problem. At high tilt this will lead to the appearance of ripples in the homogeneous structure. This is very similar to the instability appearing in the electroclinic switching in the smectic A*

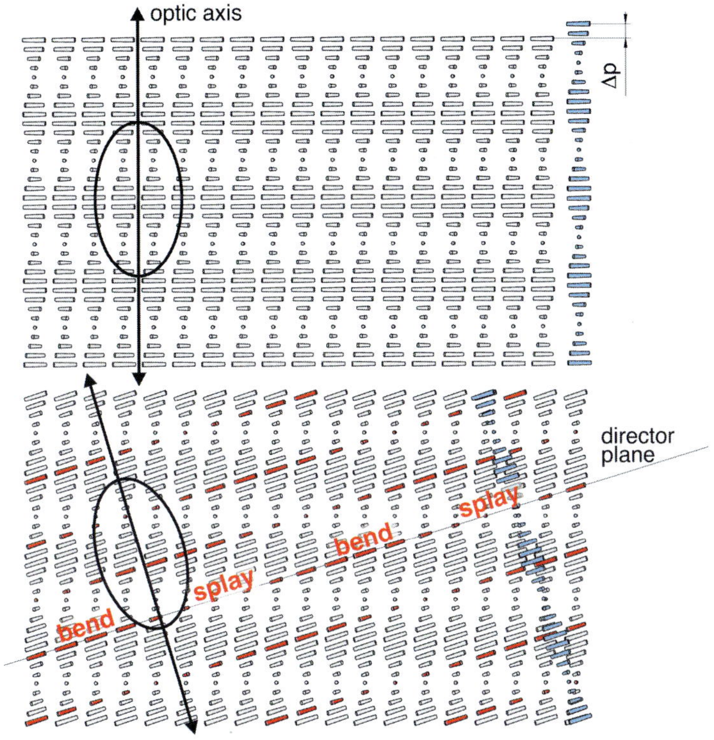


Fig. 7.6. For large induced tilts of the optic axis the helix is 'stretched' ($\Delta p$) and $K_2$ becomes important, *cf.* Eq. (7.4). The stretched helix is indicated in blue in the lower figure and compared to the non-distorted helix in the upper figure.

phase,[27,28] where the period in the layer spacing decreases but the effect is the same.

### 7.2.5. *A complicated surface condition*

The major practical problem in exploiting the flexoelectro-optic effect lies in the requirement for the helix axis to lie parallel to the cell surfaces. As shown in Figs 7.8 and 7.9, neither the homeotropic nor the planar boundary condition will really support this requirement. A large number of procedures have been tried,[15] including a combination of various surface treatments with cooling or shear under AC fields. Evidently a stable uniform lying helix (ULH) structure would require periodic boundary conditions in the form of stripes of alternating homeotropic and planar anchoring as discussed by Patel and Meyer.[4] The use of surface gratings as periodic

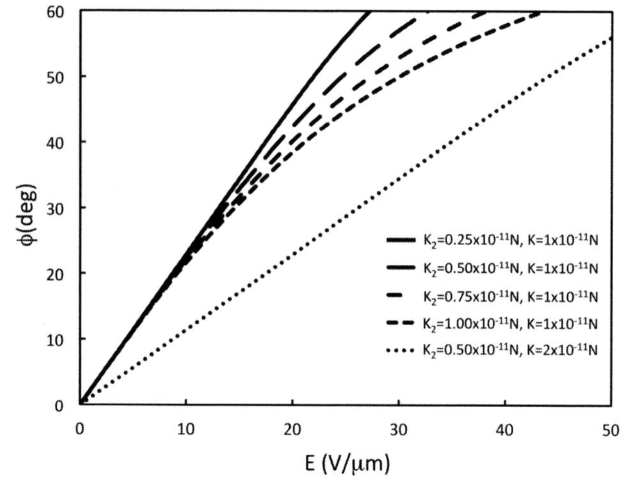


Fig. 7.7. Induced tilt as a function of the applied field for different combinations of $K$ and $K_2$. The calculations used a flexoelectric coefficient of $e = 5$ pC/m and a pitch of $p = 0.5$ $\mu$m. $K_2/K = 0.25$, 0.5, 0.75, 1 and 0.25 from left to right, but in the last case $K$ is doubled. For $K_2/K \leq 0.25$ there is almost perfect linearity.

boundary conditions for the alignment of a cholesteric helix in the ULH geometry was first studied by Rudquist[15] but the alignment quality obtained, without the use of electric fields or mechanical shear, was not satisfactory. If electric fields are to be applied in the alignment process, a finite, positive dielectric anisotropy is required, at least in the temperature and frequency range where the alignment procedure is carried out. Periodic planar or homeotropic boundary conditions could possibly also be created through photo-patterning on self-assembled monolayers (SAMs) by interference illumination of two UV laser beams. Even with periodic boundary conditions the lying helix structure will probably need to be polymer-stabilized.[29] However, attempts so far using periodic boundary conditions have been unsuccessful in terms of alignment quality. Neither have experiments with gratings of suitable periodicity been successful. Seemingly successful cell prototypes have been fabricated but the long-term stability could not be demonstrated. There is first of all a strong tendency for the obtained texture to transform into the stable Grandjean texture with time. The use of $\Delta\varepsilon > 0$ to stabilize the lying helix might hamper the electro-optic performance as the dielectric unwinding of the helix will result in a birefringence change influencing the transmitted light intensity long before the helix is

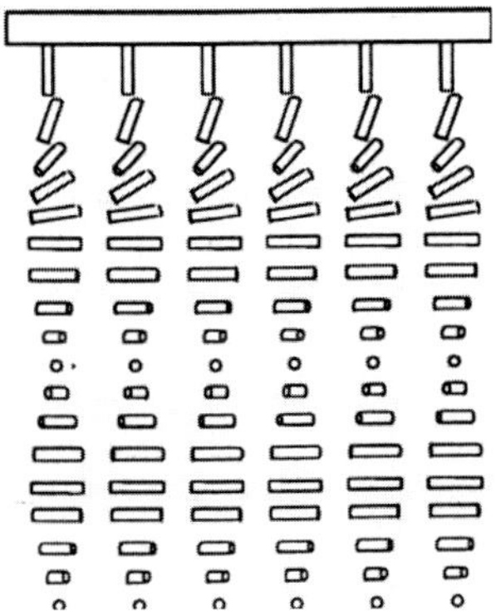

Fig. 7.8. Even homeotropic boundary conditions allow for a standing helix structure. Thus, homeotropic anchoring conditions will not ensure a long-term stable lying helix structure for the flexoelectro-optic effect.[15]

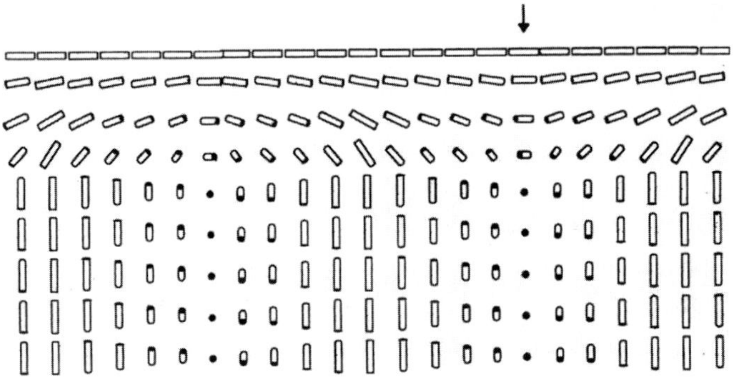

Fig. 7.9. Planar anchoring can allow for a lying helix structure but the elastic energy, especially in regions of opposite twist (marked by an arrow), leads to relaxation into the deformation-free standing helix texture.[15]

fully unwound.[10] Thus, the field-induced change in birefringence for non-zero dielectric anisotropy gives a strongly non-linear electro-optic response even though the tilt of the (distorted) indicatrix is still linear in the field. The problem with the stability of the lying helix alignment is also aggravated by the fact that the pitch is not really constant during switching. Furthermore, 'constant pitch materials' are not absolutely perfect and even a slight change in pitch value will lead to strains in the structure relative to the surface 'stripes', which might also lead to buckling instabilities. Thus more than twenty years after the pioneering work no final solution to this alignment problem of the uniformly lying helix structure for the flexoelectro-optic effect has been achieved. Neither has there been any real breakthrough in the dynamical performance beyond the work of Rudquist.[15]

### 7.2.6. *Standing helix geometry*

The only reasonable way to exploit the flexoelectro-optic effect is to work in the single stable configuration of a cholesteric in a sandwich cell, i.e. the Grandjean structure, in which the helix axis is perpendicular to the glass plates. In such a standing helix geometry the electric field should be applied in the plane of the cell. The Cambridge group[30] has recently demonstrated a theoretical and experimental analysis of this geometry. In a very short-pitch material ($p \ll \lambda$) the optical activity along the helix (optic axis) is vanishingly small[31] and a dark state was obtained between crossed polarizers. An applied field, at 45° to the polarizer axis, tilts the optic axis according to Eq. (7.4) giving a bright state with transmission controlled by the amplitude of $E$. This 'vertically aligned' flexoelectro-optic device is optically similar to a vertically aligned nematic (VAN) device. By using interdigitated electrodes the optic axis tilts in opposite directions in adjacent electrode gaps, reducing the asymmetry in the viewing-angle characteristics in the switched state. In this geometry the induced tilt is inherently small as the pitch must be very short to ensure a good dark state. This means that the brightness is limited unless the cell thickness or the birefringence are significantly increased to give a half-wave retardation and a fully bright state for small tilts. One drawback of using larger thicknesses, however, is that the viewing-angle range, for which a good dark state is obtained, is decreased. Another drawback is that the lateral field from the finger electrodes will not be constant in the bulk. It will be smaller close to the non-electroded surface, i.e. the tilt will vary along the cell normal. The small pitch also requires substantially higher driving fields than in the lying helix geometry and the field has to be applied over large distances, which increases the driving voltage further.

In Castles *et al.*[30] the driving voltage was limited to approximately 15 V/$\mu$m above which the structure became degraded by the field. At these field strengths the brightness was still relatively small. This device would need compensation foils to ensure an acceptable viewing angle in the dark state. We have found that on top of the indium-tin-oxide (ITO) electrodes, where the electric field lines are normal or tilted to the cell plane instead of parallel to it, the cholesteric helix can be reoriented into the plane of the device, probably due to $\Delta\epsilon \neq 0$. Considerable progress in materials development in terms of increasing the flexoelectric coefficient and the birefringence as well as in the electrode configurations is required before a useful device can be realized. Nevertheless, this geometry for the flexoelectro-optic effect is the natural way of circumventing the fundamental problem when it comes to alignment for the flexoelectro-optic effect.

The standing helix flexoelectro-optic geometry has also been suggested for use in some photonic applications.[32,33]

### 7.2.7. *Conclusions and outlook*

The original flexoelectro-optic effect is an *in-plane switching* working in conventional electrode geometry, i.e. with the field applied across the cell gap. It is capable of 100% modulation, has an inherent greyscale, a submillisecond response and is attractive for high-speed – even field-sequential-colour – active matrix displays. Unlike conventional nematic devices, switching between grey shades happens with the same time constants and is not dependent on the point of departure. The electro-optic effect is furthermore colour neutral (as in FLC) and finally, practically independent of temperature. This is a remarkable collection of properties, which should motivate an industrial effort.

In order to secure high transmission ($\lambda/2$ plate condition) the cell gap cannot be very thin, because of the low birefringence, which leads to relatively high driving voltages. For a powerful modulator this would not be a problem, but for TFT driving in displays lower voltages are required. This demands materials with higher values of the flexoelectric coefficient $e$ than presently available. Doping cholesterics with bent-core molecules with very high dipole moments would seem to be a promising solution. The bent-core nematic materials reported by Jákli and co-workers had extremely large flexoelectric coefficients[34] and could be an interesting alternative for such 'flexoelectric doping'. A drawback with such materials could be their high viscosity, which would reduce the switching speed relative to present materials. The condition $\Delta\varepsilon \approx 0$ also has to be kept. The weakness with the device structure is the stable alignment. However, in the last decade considerable progress has been made in polymer stabilization under electric fields such that this is now more or less a standard industrial process. This means that if a uniformly lying helix structure of high quality can be formed once, it should be possible to stabilize it.

The recent development in the vertically aligned flexoelectric cholesteric device is promising as this geometry circumvents the alignment problem. Instead of an unstable lying helix it uses the stable standing helix structure. There are still challenges in materials development and in device improvement, but it will be interesting to follow further developments in the years to come.

## 7.3. Early Attempts at Bistable Nematic Devices

### 7.3.1. *Nematic displays*

Prior to the development of thin film transistors (TFT)[35] and active matrix technology for liquid crystal displays, the maximum number of lines or rows in any display with an acceptable contrast was severely limited by the shallow voltage-transmission characteristics of, for instance, the twisted nematic mode. This inspired work to develop a liquid crystal device that could remain in either of two states (ideally black and white) after the removal of the electric field used to switch the liquid crystal into the selected state. With a memory in the liquid crystal an unlimited number of lines can be displayed using a simple passive matrix and the constraints are instead in the refreshment requirements.

### 7.3.2. *Bistability from bulk defects*

The challenge of creating bistable nematic devices is not only to stabilize two different optical states, but of course also to be able to switch between the states. Boyd *et al.* at Bell Labs used surfaces with a uniformly tilted but oppositely directed surface alignment that could stabilize a predominantly vertical (bend) or horizontal (splay) director structure.[36] These states are topologically distinct in the sense that they cannot continuously evolve into each other, and regions in different states are separated by disclinations of strength $s = \pm 1/2$, see Fig. 7.10. The two states are energetically degenerate when the pretilt angle $\theta_0 = \pi/4$, but the role of the disclinations in the switching makes both states sufficiently stable for deviations from this pretilt angle. The device used sets of interdigitated electrodes for producing predominantly horizontal or vertical fields, or alternatively conventional continuous electrodes and a two-frequency nematic material. An optical contrast was achieved by doping the liquid crystal with a pleochroic dye and limited to about 5:1. The Bell Lab team also realized that stability requires that the disclination lines must be pinned by surface discontinuities, either in alignment or in topography.[37] However, the device could not be switched practically and interest in bistable nematic displays decreased after the invention of the supertwisted nematic device,[38] the bistable ferroelectric liquid crystal device[39] and later of course after the success of TFT-technology.

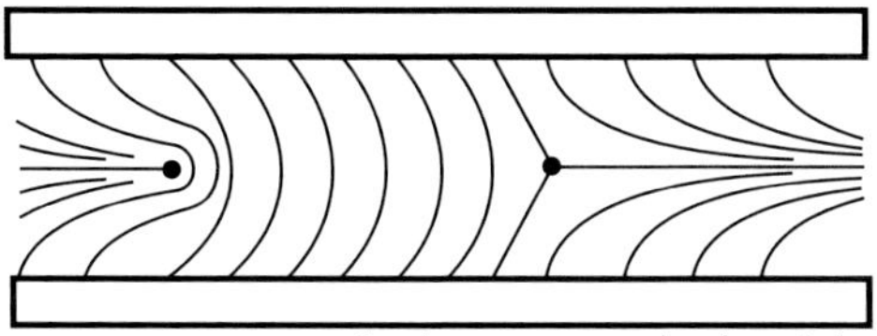


Fig. 7.10. Schematic illustration of the two stable states (essentially vertical and horizontal) in the geometry proposed by Boyd *at al.*[36]

### 7.3.3. *Bistability from multistate surfaces*

A different approach for bistable nematic devices is to create a surface that provides at least two stable surface director states. This was first achieved by the Orsay group through the oblique evaporation of SiO at the correct angle and film thickness.[40] A device using two such surfaces, each having one stable director state N in the plane of the surface, and two oblique metastable orientations O and O′ rotated by ±45° from N, was assembled to allow only two stable director configurations in the cell, azimuthally separated at about 45° (Fig. 7.11). The two director states are slightly splayed having flexoelectric polarization densities whose components orthogonal to the electrodes are oppositely directed. The electric field, normal to the boundary surfaces, breaks the surface anchoring through the dielectric anisotropy (and flexoelectric) coupling and the final state after removal of the field is governed by the sign of $E$ interacting with the normal components of the flexoelectric polarization.[41] The geometry of the electro-optic effect is similar to the surface-stabilized ferroelectric liquid crystal (SSFLC) device. The polarity-related selection of the final state was believed to be due to an ionic residual electric field, just after the pulse is switched off. The complicated method to obtain the multistate surface alignment layers for this flexoelectricity-based nematic mode was, however, hardly suitable for commercial production.

Another bistable device is the Binem display, which does not involve flexoelectric effects but is sufficiently similar for a brief mention. This device uses a chirally doped nematic in a cell with parallel alignment directions

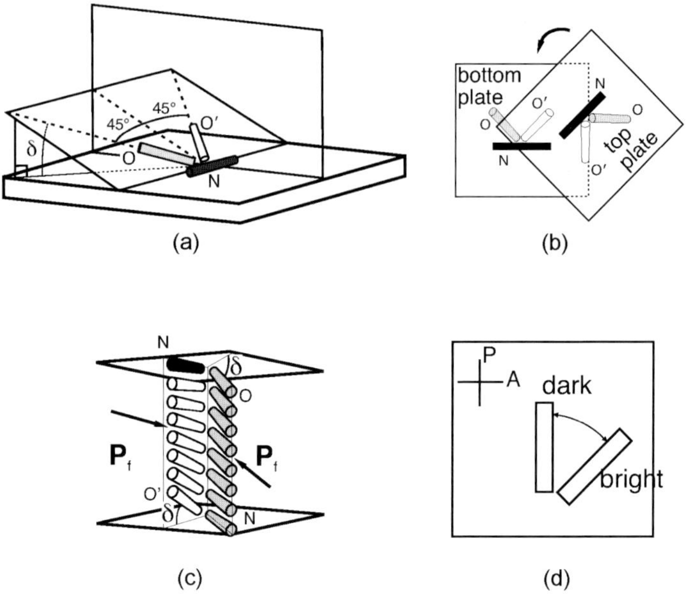


Fig. 7.11. The device described by Barberi *et al.*[41] (a) The tuned oblique evaporation of SiO gives three stable states for the surface director: N parallel to the surface and normal to the evaporation plane, and the symmetrically positioned O and O′, with polar tilt $\delta$ and azimuthal tilt $\pm 45°$. (b) Two such plates are assembled making the projections of N of the top plate and O′ of the bottom plate as well as O of the top plate and N of the bottom plate parallel. (c) This makes two splayed bulk director configurations possible, azimuthally separated by an angle of 45°. These states have opposite signs for the splay-induced flexoelectric polarization perpendicular to the surface. (d) The device can be switched between the stable dark and bright states depending on the polarity of the applied pulse.

at both surfaces. The cell thickness $d$ and the pitch $p$ are tuned such that both the non-twisted and the 180° twisted states have approximately equal energies. This is achieved when $p = 4d$. The state with equilibrium bulk pitch, i.e. having a 90° twist between the surfaces, is not allowed by the surface alignment directions. By using weak anchoring on one of the plates it is possible to break the surface anchoring with the applied electric field. The selection of the non-twisted or the 180° twisted state on relaxation after the field is switched off is governed by the shape of the addressing pulse.[42]

## 7.4. The Zenithal Bistable Device

### 7.4.1. *The ZBD surface grating*

The second application of flexoelectricity to be discussed in detail in this chapter is the so-called zenithal bistable device (ZBD).[43–46] The energy barrier between the topologically different states is, as in the Boyd device,[36] related to the creation and annihilation of line defects, while the selection of states is achieved using the flexoelectric coupling as in Barberi *et al.*[41] The basic principle of the ZBD mode, Fig. 7.12, is to use a non-reconfigurable surface that can support and stabilize both a homeotropic and a planar director structure in the bulk of a nematic liquid crystal cell.[47] First consider a flat surface promoting homeotropic anchoring of a nematic liquid crystal (Fig. 7.12a). Then introduce an undulation of the surface as in Fig. 7.12b. The director field of the nematic in contact with the surface will now be elastically deformed in the vicinity of the surface. The distortion elastic energy increases with the amplitude of the surface-grating relief, or rather with the ratio between the amplitude and periodicity of the relief (Figs 7.12b and c). Eventually the energy penalty of the heavily distorted but continuous director field becomes higher than for a new structure (Fig. 7.12d), in which the lines of discontinuity run along the ridges and valleys of the surface relief. These are disclination lines with a topological charge of $s = \pm 1/2$. The weaker the homeotropic anchoring the closer to the surface the disclinations form. If we compare the situation before and after the breakdown of the original director field we see that the bulk alignment of the director has changed from effectively homeotropic to effectively planar. By tuning the surface-grating amplitude $a$ and period $L$ we can make the two states energetically degenerate by choosing $a/L \approx 0.5$. For a shallow relief, i.e. for $a/L \ll 0.5$, the bulk director is normal to the cell plane, while for a deep relief, i.e. $a/L \gg 0.5$, the bulk director is essentially parallel to the cell plane. The role of the shape of the grating will be discussed later but the two states have been found to be simultaneously stable for $0.3 \leq a/L \leq 1$.[48] By using an asymmetric grating, Fig. 7.13, the director is given a pretilt $\delta$ relative to the cell plane.

### 7.4.2. *Switching between polarized states*

By replacing one or both aligning surfaces of a conventional liquid crystal device cell with a surface relief grating, as the one schematically depicted in Fig. 7.13, we can construct a device with two distinct optical states in

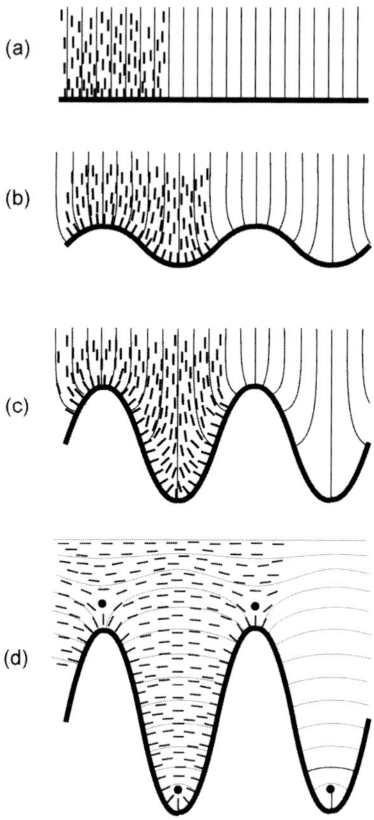


Fig. 7.12. (a) A flat homeotropic surface promotes a distortion-free director field perpendicular to the surface. When introducing an undulation of the surface (b) and (c), the director field is distorted in the vicinity of the surface. For large enough undulations the total elastic energy of the liquid crystal is minimized by adopting a new structure (d) with $s = \pm 1/2$ disclination lines. The discontinuous state (d) gives a bulk director parallel to the cell plane.

the absence of applied fields. But in order to get a useful electro-optic device it must be possible to electrically switch the liquid crystal between the states. A straightforward solution could be to use a two-frequency material, with $\Delta\varepsilon > 0$ at low frequency for selecting the homeotropic (here normally called the high-tilt) state, and $\Delta\varepsilon < 0$ for high-frequency driving fields for selecting the planar (or low-tilt) state. However, in addition to the high complexity of matrix multiplexing high information content displays using two-frequency materials are often slow, conductive and have strong

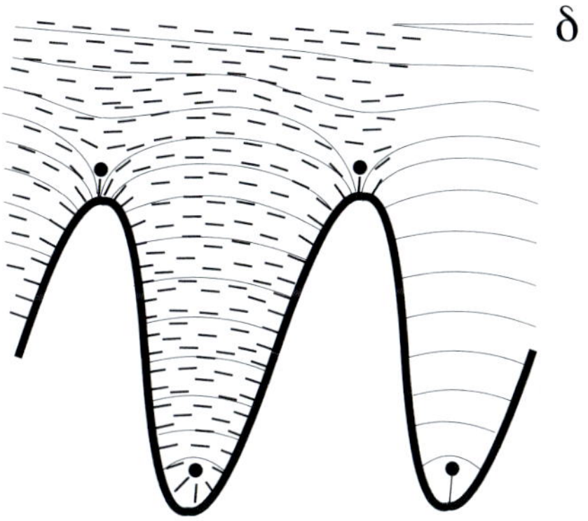


Fig. 7.13. An asymmetric, e.g. blazed, surface grating yields a pretilt $\delta$ of the bulk director in the discontinuous, planar state.

temperature dependence.[48] Instead the ZBD mode utilizes the flexoelectric polarization from the splay and bend deformations in the director field close to the grating surface for selection of either the homeotropic or the planar state at the grating surface. Figures 7.14a and b show the orientation of the flexoelectric polarization direction for splay and bend deformations for $e_1 > 0$ and $e_3 < 0$. If we look at the local flexoelectric polarization in the director fields of the homeotropic and planar states, Figs 7.14c and d, we can see that their flexoelectric polarization fields are very different, and that the effective flexoelectric polarization in the two states have opposite directions, i.e. the nematic is positively or negatively polarized, respectively. This facilitates the selection of either state using the polarity of an applied electric field perpendicular to the surface.

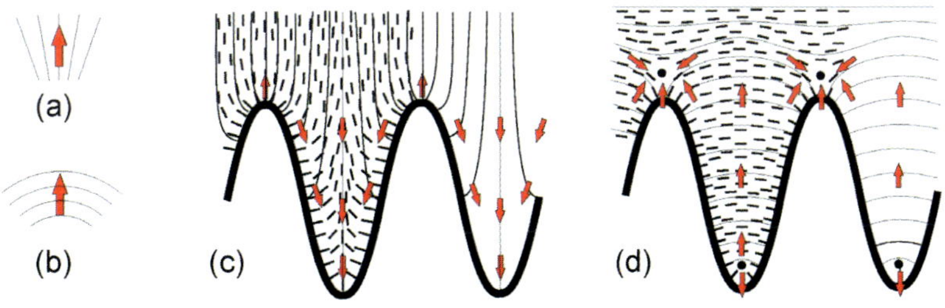


Fig. 7.14. The direction of flexoelectric polarization in pure (a) splay and (b) bend for $e_1 > 0$ and $e_3 < 0$. The (c) homeotropic continuous and (d) planar discontinuous states have different polarization fields and, importantly, opposite sign of the average flexoelectric polarization.

The first demonstrated zenithal bistable mode device used a cell with one conventional flat homeotropic surface and one grating surface (Fig. 7.15). In the high-tilt (homeotropic) director state, Fig. 7.15 left, the bulk director is perpendicular to the cell plane and the cell looks black between crossed polarizers. When switched to the low-tilt (planar) state, Fig. 7.15 right, the cell adopts a hybrid-aligned nematic configuration with the hybrid plane at 45° with respect to the polarizers, giving a bright state. A pulse of opposite polarity drives the liquid crystal back to the black state. The field threshold for switching between the homeotropic and planar states is often different from that for planar to homeotropic. This is no surprise as the two stable states are structurally different. In surface-stabilized ferroelectric devices, the two states are (ideally) identical except for the sign of **P** – which gives a 'true' or symmetric bistability – and the threshold is therefore the same for switching in either direction.

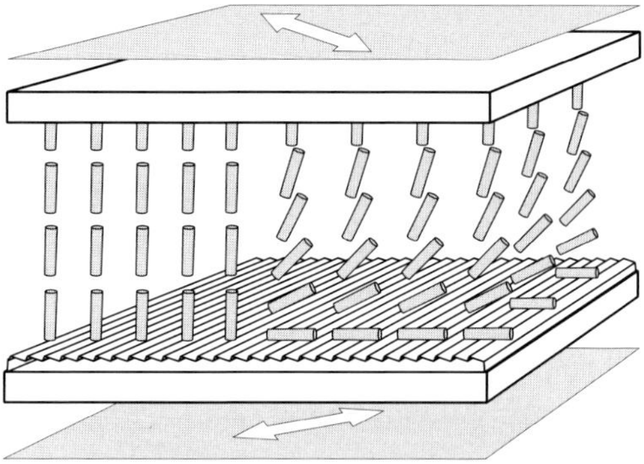

Fig. 7.15. The vertically aligned nematic–hybrid aligned nematic (VAN-HAN) zenithal bistable device geometry. The homeotropic state (left) is black while the hybrid state (right) is bright.

### 7.4.3. *Latching of the ZBD*

The latching between the two stable states is driven by electric fields of opposite polarity. Although it is the flexoelectric polarization that breaks the symmetry and allows for the polar switching, most of the switching process in fact involves the torque due to the dielectric anisotropy. For strongly positive $\Delta\varepsilon$ nematic materials the sample always latches the planar state with a positive field applied to the grating electrode. In order to reduce the unwanted effects of ions within the liquid crystal during switching, bipolar pulses are used[46] (similar to a FLC) and it is the trailing edge of the bipolar driving pulse that selects the final state through flexoelectric coupling. Although the basic polar mechanism can be explained qualitatively by the standard flexoelectricity described so far, the secondary flexoelectric effect (often called ordoelectricity), i.e. electric polarization emanating from gradients in the scalar part of the orientational order parameter $S$ close to the grating surface, is also assumed to play a part in the switching.

While the basic idea of creating two stable states with different polar configurations using a tuned grating surface profile is intuitively easy to understand, the latching mechanism of the ZBD is very complex. Not surprisingly the latching threshold, as in the FLC case, has some voltage-time area character. The complexity of the differential equations and the imposed boundary and switching conditions make analytical solutions for the ZBD behaviour very difficult to find. One-dimensional models[49,50] are not able to fully describe the behaviour. Numerical models assuming simplified geometries and zero dielectric anisotropy[51–53] gave latching voltages about one order of magnitude higher than those experimentally observed. Spencer and Care[54] built a more complete model using studies of the defect trajectories and annihilation times as functions of the material parameters for the switching from the homeotropic to the planar state. Recently, Spencer *et al.*[55] published a comprehensive comparative modelling and experimental study of the static and dynamic behaviour of ZBD cells, using a surface-grating profile computed from scanning electron microscope (SEM) images and the main results are summarized in the following. Further comparison between theory and experiment for the latching of ZBD devices is given in Jones and Amos.[56] The ZBD surface is assumed to have a finite homeotropic anchoring strength and the opposing flat surface infinite planar anchoring strength in a cell with a twisted nematic arrangement, i.e. where the planar states at the ZBD surface and at the opposing surface are rotated 90° with respect to each other. This is the twisted nematic–hybrid aligned nematic (TN-HAN) geometry shown in Fig. 7.16. The simulations show that the locations and trajectories of defects annihilating or nucleating strongly depend on the liquid crystal material parameters, the grating material and grating profile, but also on the magnitude, frequency and pulse length of the applied field.

When switching from the homeotropic to the planar state the first part of the bipolar pulse is negative. The nematic has a positive $\Delta\varepsilon$ and aligns with the applied field during this stage squeezing the splay and bend deformations towards the surface. This means that the homeotropic state is

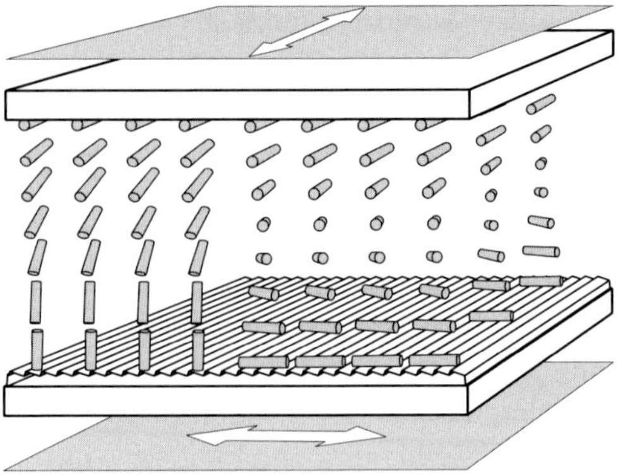

Fig. 7.16. The TN-HAN zenithal bistable device geometry. The HAN state (left) is black while the TN state (right) is bright.

enforced, i.e. a larger volume of the bulk director now aligns perpendicular to the cell plane. In the process the splay and bend deformations grow, increasing the local flexoelectric polarization density in the surface regions. The second part of the bipolar pulse (positive) couples to the flexoelectric polarization and a defect pair is nucleated on the steepest sidewall of the grating. These defects migrate to the energetically favoured regions of strong surface curvature. When applying a writing pulse with a $\tau V$ value considerably higher than the latching threshold a second pair of defects can be nucleated at the less steep sidewall. This extra pair is, however, annihilated after the field has been switched off and the final defect configuration state is independent of the detailed switching process. After the surface state has been switched from homeotropic to planar the liquid crystal relaxes into the final twisted nematic state, with a relaxation time governed by the viscoelastic properties of the material. When switching from the planar to the homeotropic state the first part of the bipolar pulse is positive. The field stabilizes the defect positions while the bulk director reorients into the field direction. Again the bulk alignment keeps the distortions close to the ZBD surface. The second (negative) pulse instead pushes the defects towards each other along the steep sidewall before they annihilate. Without the defects, the system relaxes to the homeotropic state when the field is turned off. Thus in both switching processes the ZBD cell 'relaxes' into the final state after the surface state has latched. While the latching pulse can be in the range of tens of microseconds but typically 500 $\mu$s, the total switching time of a pixel, including latching, is several milliseconds or tens of milliseconds. The relaxation part of the switching is independent of the pulse amplitude. The results of the theoretical analysis[55] also suggest that the weaker the homeotropic anchoring strength the smaller the latching threshold. In order to further lower the operating voltage the liquid crystal material should have high values for $(e_1 - e_3)$, $\Delta\varepsilon$ and $K_3$. The obtained results regarding latching voltages for the homeotropic to planar switching as a function of the cell gap agree well with experimental values. Figure 7.17 shows the pulse duration as a function of the pulse amplitude $V$ for a real ZBD display.[46]

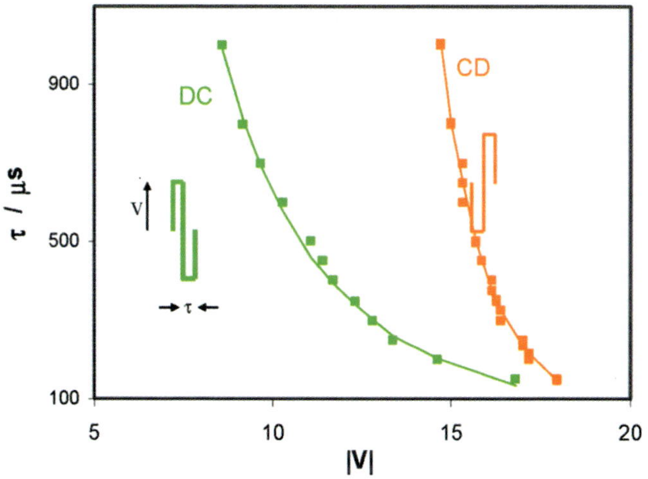


Fig. 7.17. Examples of $\tau V$ combinations for latching from the discontinuous to the continuous state (DC) and from the continuous to the discontinuous (CD) states of a ZBD display. From Jones,[46] reproduced with kind permission of the Society for Information Display.

### 7.4.4. *The grating structure*

In contrast to other bistable nematic modes the surface state of the director does not change in zenithal bistable mode devices, which gives them an important robustness. In other words, while the device is bistable, the director state at the ZBD surface is monostable. This feature gives a smaller temperature dependence than other modes utilizing weak anchoring and multiple surface states. Among the many similarities with FLC devices there is also the fact that the stable states are polarized, i.e. they have a non-zero polarization in bulk. Therefore we anticipate problems with image sticking (burn-in). Avoiding this requires perfectly controlled surface conditions and well-engineered materials (with ultra-low electrical conductivity). These problems seem to have been solved satisfactorily for the production of ZBD. Evidently the fact that the director does not switch at the surface simplifies the problem. In bistable SSFLC displays the device can also be tuned so as to have the switching occur only at the chevron interface, keeping the surface director fixed at both plates. This will, in a similar way as in the ZBD device, reduce the problem of image sticking. The image sticking problem has, however, been satisfactorily solved in, for instance, the bistable FLC technology developed by Citizen for ultra-high resolution displays or ultra-low power shelf-edge prize tags.[57]

The ZBD grating shape can be varied from sinusoidal to strongly blazed sinusoidal, Fig. 7.18. It is generally saw-tooth-shaped to give a proper bulk pretilt $\theta_D$ in the low-tilt state and the approximate relation with the distance $l_D$ between adjacent $+1/2$ and $-1/2$ disclinations and the grating period $L$ is $\theta_D = \pi/2 - l_D\pi/L$.

Gratings with pretilts ranging from 0° (symmetric) to 55° (strongly asymmetric) have been demonstrated and produced in practice.[58] For a twisted nematic ZBD (see Section 7.4.5) a pretilt of about 5° is ideal.

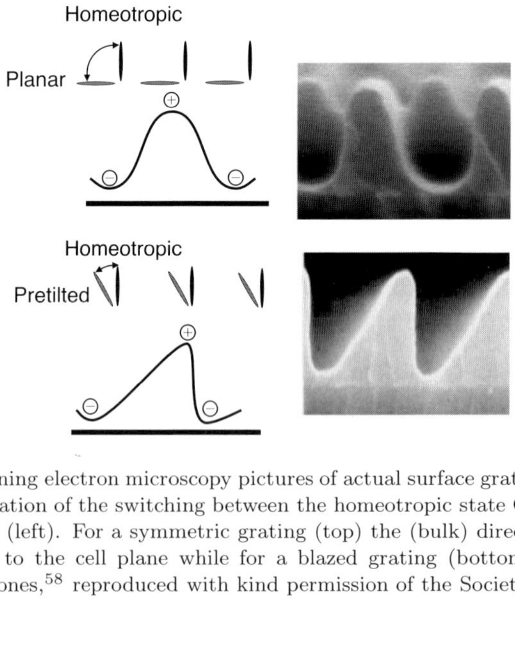


Fig. 7.18. Scanning electron microscopy pictures of actual surface gratings (right) and a schematic illustration of the switching between the homeotropic state C and the low-tilt (planar) state D (left). For a symmetric grating (top) the (bulk) director of the planar state is parallel to the cell plane while for a blazed grating (bottom) it is pretilted. Modified from Jones,[58] reproduced with kind permission of the Society for Information Display.

Common dimensions for device gratings are 0.8 $\mu$m pitch and 0.9 $\mu$m amplitude. More complex gratings with more ridges, like trapezoid gratings, can pin more disclination lines allowing multistable states for certain pretilts. The grating shape also has a large effect on the latching threshold. By variation of the grating shape within each pixel, separate latching thresholds can be achieved over the pixel area and hence shades of grey can be introduced.[45]

In a practical display device it is fundamental that adjacent domains with different states are stable in time and over wide temperature ranges at zero field even though the two states have different energies. As mentioned above, the ZBD is not symmetrically bistable as the two states are generally different with one state energetically more stable than the other. To prevent the domain with the lowest energy state growing at the expense of the other domain the grating is broken up into stripes, perpendicular to the grating ridges, with a 180° phase shift in the groove structure. In the slip regions there are vertical concave and convex surfaces and regions with negative Gaussian curvature. The slip regions make the $\pm 1/2$ disclination lines form a three-dimensional defect loop structure and act like pinning sites for the disclinations at zero field as well as a nucleation and annihilation site during electro-optic switching.[46,55,59] Such phase-shifted grating structures have been used to make devices with a temperature working range from $-10$ °C to above 70 °C with good shock stability and operating windows.

The grating surface reliefs are manufactured[60] in a process similar to the one used in CD and DVD production. First a photoresist layer, spin-coated on a substrate, is exposed to UV light through a chrome-on-glass mask, defining the grating pitch, mark-to-space ratio, slip periodicity and the structure of the desired grating. The grating shapes are further controlled by variations in the chrome pattern and process conditions, e.g. oblique incidence of the UV light through the mask results in a blazed grating after development. This resist grating is then transferred into an inverse nickel copy, which in turn is used in forming the shape into a suitable resin layer on a flexible film. The structured resin film is then used to emboss the final grating onto a photocurable polymer film on the final display substrate. The method allows for complex variations of the grating parameters and, hence, latching thresholds over the area of each pixel, Fig. 7.19, which allows for greyscale. The grating material can itself promote homeotropic anchoring of the nematic liquid crystal. Alternatively the grating can be subsequently covered with a homeotropic aligning agent. For more details of the ZBD grating manufacturing process see Jones and Bryan-Brown[60] and Jones and Beldon[61] and references therein.

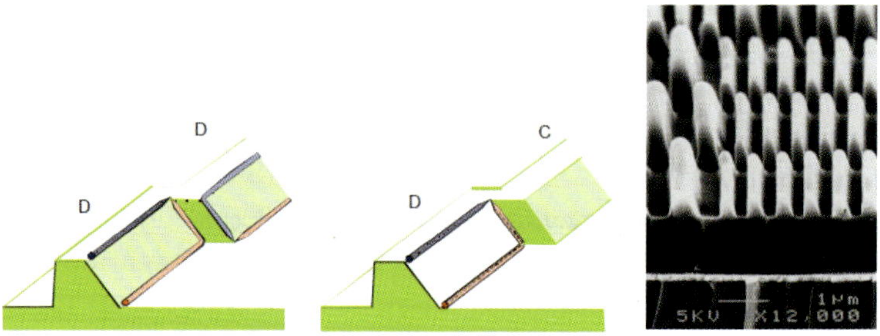


Fig. 7.19. Left: The 180° slips in the grating structure ensure stability of the selected states as they pin the disclination lines. Right: Scanning electron microscopy image of a surface grating showing the slips. The image also shows regions with different pitch gratings allowing for greyscale. From Jones,[46] reproduced with kind permission of the Society for Information Display.

### 7.4.5. *Material parameters and display modes*

In the first ZBD prototypes the pulse voltages required were quite high but after about 15 years of dedicated work in display analysis and material parameter optimization they are now more reasonable, about 20 V, still somewhat higher than for mainstream displays. The first ZBD displays used standard positive dielectric anisotropy nematics like **E7** from Merck; the most recent ones use materials in which the material parameters including the dielectric anisotropy and flexoelectric coefficients have been tuned. Today's ZBD displays can be driven by standard STN-drivers and have recently found the way to the market. The power consumption is extremely low in static ZBD displays showing the same information for years as the device only consumes power during updates. The power consumption is, however, about 10 times higher for ZBD displays that are refreshed every minute compared to reflective-mode TN displays.[46]

#### 7.4.5.1. *The VAN-HAN mode*

Zenithal bistable devices have been used in several types of electro-optic geometry. In the VAN-HAN mode, Fig. 7.15, one ZBD surface is combined with a flat homeotropic surface. In the planar ZBD state (Fig. 7.15 right), the nematic is hybrid aligned with the director plane rotated 45° with respect to the polarizer, giving a bright state. The homeotropic state is dark but elastic deformations close to the grating surface probably lead to some light leakage. Moreover, especially in transmissive mode the HAN cell has poor viewing-angle characteristics.

Fig. 7.20. Electronic shelf label using the TN-HAN ZBD display. From Jones,[46] reproduced with kind permission of the Society for Information Display.

### 7.4.5.2. *The TN-HAN mode*

A better contrast and viewing-angle performance is obtained in a TN-HAN configuration; see Fig. 7.16. Here the opposing flat surface has planar anchoring with the rubbing direction parallel to the grating grooves. The planar ZBD state results in a twisted nematic and a bright state between crossed polarizers. As in conventional TN cells the bright state has excellent viewing-angle properties and is essentially colour neutral through an appropriate choice of birefringence and cell thickness so that the cell is approximately between the first and second Gooch–Tarry minima.[60] The homeotropic ZBD state again gives a hybrid-aligned nematic but now with the hybrid plane parallel to the polarizer, which ensures a good dark state. The TN-HAN cell has been commercialized with a reflective geometry using ambient light in the form of electronic shelf labels. For the device shown in

Fig. 7.20 the resolution is greater than 100 dpi, there are 224 × 96 pixels, passively addressed with a line addressing time of 1.5 ms and a driving voltage of 19 V. The optical response is about 50 ms. The refractive index of the grating material is matched to the ordinary refractive index of the nematic liquid crystal to avoid diffractive losses from the grating.[46]

## References


1. I. Dozov, Ph. Martinot Lagarde and G. Durand, Flexoelectrically controlled twist of texture in a nematic liquid crystal, *J. Physique Lett.* **43**(10), L-365–L-369, (1982). doi: 10.1051/jphyslet:019820043010036500
2. R. Kiefer, B. Weber, F. Windscheid and G. Baur, In-plane switching of nematic liquid crystals. In *Proceedings of 12th IDRC Japan Display,* Society for Information Display, Hiroshima, 1992, pp. 547–550.
3. G. Durand, Flexoelectricity in nematics: Continuous distortions and instabilities, *Mol. Cryst. Liq. Cryst.* **113**, 237–246, (1984). doi: 10.1080/00268948408071686
4. R.B. Meyer and J.S. Patel, Flexoelectric liquid crystal device, US Patent Number 4,917,475 (April 17, 1990).
5. J.S. Patel and R.B. Meyer, Flexoelectrooptic electro-optics of a cholesteric liquid crystal, *Phys. Rev. Lett.* **58**(15), 1538–1540, (1987). doi: 10.1103/PhysRevLett.58.1538
6. J.S. Patel and S.-D. Lee, Fast linear electro-optic effect based on cholesteric liquid crystals, *J. Appl. Phys.* **66**(4), 1879–1881, (1989). doi: 10.1063/1.344369
7. S.-D. Lee, J.S. Patel and R.B. Meyer, Effect of flexoelectric coupling on helix distortions in cholesteric liquid crystals, *J. Appl. Phys.* **67**(3), 1293–1297, (1990). doi: 10.1063/1.345679
8. S.-D. Lee, J.S. Patel and R.B. Meyer, Switching characteristics of a field-induced structure near the isotropic-cholesteric transition, *Mol. Cryst. Liq. Cryst.* **209**(1), 79–84, (1991). doi: 10.1080/00268949108036180
9. P. Rudquist, *The Linear Electro-optic Effect in a Chiral Nematic Liquid Crystal*, Gothenburg Institute of Physics Report, No. 314, (1993).
10. P. Rudquist, L. Komitov and S.T. Lagerwall, Linear electrooptic effect in a cholesteric liquid crystal, *Phys. Rev. E.* **50**(6), 4735–4743, (1994). doi: 10.1103/PhysRevE.50.4735
11. L. Komitov, S.T. Lagerwall, B. Stebler and R. Strigazzi, Sign reversal of the linear electrooptic effect in the chiral nematic phase, *J. Appl. Phys.* **76**(6), 3762–3768, (1994). doi: 10.1063/1.357379
12. P. Rudquist, M. Buivydas, L. Komitov and S.T. Lagerwall, Linear electrooptic effect based on flexoelectricity in a cholesteric with sign change of dielectric anisotropy, *J. Appl. Phys.* **76**(12), 7778–7783, (1994). doi: 10.1063/1.358523

13. P. Rudquist, T. Carlsson, L. Komitov and S.T. Lagerwall, The flexoelectro-optic effect in cholesterics, *Liq. Cryst.* **22**(4), 445–449, (1997). doi: 10.1080/026782997209153
14. P. Rudquist and S.T. Lagerwall, On the flexoelectric effect in nematics, *Liq. Cryst.* **23**(4), 503–510, (1997). doi: 10.1080/026782997208082
15. P. Rudquist, *The Flexoelectrooptic Effect in Cholesteric Liquid Crystals*, PhD Thesis, Chalmers University of Technology, Göteborg, 1997.
16. P. Rudquist, L. Komitov and S.T. Lagerwall, The flexoelectrooptic effect, *Ferroelectrics* **213**, 447–456, (1998).
17. P. Rudquist, S.T. Lagerwall, N.A. Clark, J.E. Maclennan and D.M. Walba, Symmetric thresholdless electrooptic effects in liquid crystals. In *Proc. 20th IDRC, SID'00*, Palm Beach, 24–28 September 2000, pp. 59–63.
18. B. Musgrave, P. Lehmann and H.J. Coles, A new series of chiral nematic bimesogens for the flexoelectro-optic effect, *Liq. Cryst.* **26**(8), 1235–1249, (1999). doi: 10.1080/026782999204255
19. H.J. Coles, B. Musgrave, M.J. Coles and J. Willmott, The effect of the molecular structure on flexoelectric coupling in the chiral nematic phase, *J. Mater. Chem.* **11**(11), 2709–2716, (2001). doi: 10.1039/B105647F
20. H.J. Coles, M.J. Clarke, S.M. Morris, B.J. Broughton and A.E. Blatch, Strong flexoelectric behavior in bimesogenic liquid crystals, *J. Appl. Phys.* **99**(3), 034104/1–5, (2006). doi: 10.1063/1.2166643
21. S.M. Morris, M.J. Clarke, A.E. Blatch and H.J. Coles, Structure-flexoelastic properties of bimesogenic liquid crystals, *Phys. Rev. E* **75**(4), 041701/1–9, (2007). doi: 10.1103/PhysRevE.75.041701
22. R.B. Meyer, Effects of electric and magnetic fields on the structure of cholesteric liquid crystals, *Appl. Phys. Lett.* **12**(9), 281–282, (1968). doi: 10.1063/1.1651992
23. R.B. Meyer, Distortion of a cholesteric structure by a magnetic field, *Appl. Phys. Lett.* **14**(7), 208–209, (1969). doi: 10.1063/1.1652780
24. Y. Bouligand, Sur l'existence des pseudomorphoses cholesteriques chez divers organismes vivants, *J. Phys. Colloques* **30**(C4), C4-90–C4-103, (1969).
25. S.T. Lagerwall, Linear and non-linear polar effects in liquid crystals, *Mol. Cryst. Liq. Cryst.* **543**(1), 3–47, (2011). doi: 10.1080/15421406.2011.568326
26. S.T. Lagerwall, *Ferroelectric and Antiferroelectric Liquid Crystals*, Wiley VCH, Weinheim, 1999.
27. J. Pavel and M. Glogarova, A new type of layer structure defects in chiral smectics, *Liq. Cryst.* **9**(1), 87–93, (1991). doi: 10.1080/02678299108036768
28. R.F. Shao, P.C. Willis and N.A. Clark, The field induced stripe texture in surface stabilized ferroelectric liquid crystal cells, *Ferroelectrics* **121**(1), 127–136, (1991). doi: 10.1080/00150199108217616
29. P. Rudquist, L. Komitov and S.T. Lagerwall, Volume stabilized ULH structure for the flexoelectrooptic effect and the phase-shift effect in cholesterics, *Liq. Cryst.* **24**(3), 329–334, (1998). doi: 10.1080/026782998207127
30. F. Castles, S.M. Morris, D.J. Gardiner, Q.M. Malik and H.J. Coles, Ultrafast-switching flexoelectric liquid-crystal display with high contrast, *Journal of the SID* **18**(2), 128–133, (2010). doi: 10.1889/JSID18.2.128

31. H.L. de Vries, Rotatory power and other optical properties of certain liquid crystals, *Acta Cryst.* **4**(3), 219–226, (1951). doi: 10.1107/S0365110X51000751
32. B.J. Broughton, M.J. Clarke, A.E. Blatch and H.J. Coles, Optimized flexoelectric response in a chiral liquid-crystal phase device, J. Appl. Phys. 98(3), 34109/1–6, (2005). doi: 10.1063/1.2006227
33. H.J. Coles, M.J. Coles, B.J. Broughton and S.M. Morris, Flexoelectro-optic liquid crystal device, Patent No. WO/2006/003441 (2006).
34. J. Harden, B. Mbanga, N. Éber, K. Fodor-Csorba, S. Sprunt, J.T. Gleeson and A. Jákli, Giant flexoelectricity of bent-core nematic liquid crystals, *Phys. Rev. Lett.* **97**(15), 157802/1–4, (2006). doi: 10.1103/PhysRevLett.97.157802
35. T.P. Brody, The birth and early childhood of active matrix: A personal memoir, *Journal of the SID* **4**(3), 113–127, (1996). doi: 10.1889/1.1985000
36. G.D. Boyd, J. Cheng and P.D.T. Ngo, Liquid-crystal orientational bistability and nematic storage effects, *Appl. Phys. Lett.* **36**(7), 556–558, (1980). doi:10.1063/1.91578
37. J. Cheng and G.D. Boyd, Threshold and switching characteristics of a bistable nematic liquid-crystal storage display, *Appl. Phys. Lett.* **37**(12), 1072–1074, (1980). doi:10.1063/1.91881
38. C.M. Waters and E.P. Raynes, Liquid crystal devices, UK Patent GB 2123163 (1982).
39. N.A. Clark and S.T. Lagerwall, Submicrosecond bistable electro-optic switching in liquid crystals, *Appl. Phys. Lett.* **36**(11), 899–901, (1980). doi: 10.1063/1.91359
40. M. Monkade, M. Boix and G. Durand, Order electricity and oblique nematic orientation on rough solid surface, *Europhys. Lett.* **5**(8), 697–702, (1988). doi: 10.1209/0295-5075/5/8/006
41. R. Barberi, M. Giocondo and G. Durand, Flexoelectrically controlled surface bistable switching in nematic liquid crystals, *Appl. Phys. Lett.* **60**(9), 1085–1086, (1992). doi: 10.1063/1.106452
42. I. Dozov, A. Boissier and T. Laboureau, Nemoptic's bistable nematic liquid-crystal technology – Cholesterics or ferroelectrics are not necessary to make a bistable LCD, and using conventional nematics offers some substantial advantages, *Information Display* **18**(1), 10–13, (2002).
43. G.P. Bryan-Brown, C.V. Brown and J.C. Jones, Bistable nematic liquid crystal device, UK Patent GB 2318422 (9 October 1996).
44. G.P. Bryan-Brown, C.V. Brown, J.C. Jones, E.L. Wood, I. Sage, P. Brett and J. Rudin, Grating aligned bistable nematic device, *SID Int. Symp. Digest Tech. Papers* **28**, 37–40, (1997).
45. J.C. Jones, S.M. Beldon and E.L. Wood, Gray scale in zenithal bistable LCDs: The route to ultra-low-power color displays, *Journal of SID* **11**(2), 269–275, (2003). doi: 10.1889/1.1825653

46. J.C. Jones, The zenithal bistable display: From concept to consumer, *Journal of SID* **16**(1), 143–154, (2008). doi: 10.1889/1.2835021
47. Here we consider only homeotropic grating surface alignment. A zenithal bistable device surface can also be realized with planar surface anchoring on the grating surface.[46]
48. J.C. Jones, G. Bryan-Brown, E. Wood, A. Graham, P. Brett and J. Hughes, Novel bistable liquid crystal displays based on grating alignment, In eds. R. Shashidar and B. Gnade, *Liquid Crystal Materials, Devices, and Flat Panel Displays*, Proc. of SPIE. Vol. 3955, pp. 84–93, (2000). doi: 10.1117/12.379969
49. F.P. Da Costa, M. Grinfeld, N.J. Mottram and J.T. Pinto, A mathematical study of a bistable nematic liquid crystal device, *Math. Models Meth. Appl. Sci.* **17**(12), 2009–2034, (2007). doi: 10.1142/S0218202507002546
50. A.J. Davidsson and N.J. Mottram, Flexoelectric switching in a bistable nematic device, *Phys. Rev. E* **65**(5), 051710/1–10, (2002). doi: 10.1103/PhysRevE.65.051710
51. C. Denniston and J.M. Yeomans, Flexoelectric surface switching of bistable nematic devices, *Phys. Rev. Lett.* **87**(27), 275505/1–4, (2001). doi: 10.1103/PhysRevLett.87.275505
52. L.A. Parry-Jones and S.J. Elston, Flexoelectric switching in a zenithally bistable nematic device, *J. Appl. Phys.* **97**(9), 093515/1–7, (2005). doi: 10.1063/1.1885184
53. L.A. Parry-Jones, R.B. Meyer and S.J. Elston, Mechanisms of flexoelectric switching in a zenithally bistable nematic device, *J. Appl. Phys.* **106**(1), 014510/1–9, (2009). doi: 10.1063/1.3153971
54. T.J. Spencer and C.M. Care, Lattice Boltzmann scheme for modeling liquid-crystal dynamics: Zenithal bistable device in the presence of defect motion, *Phys. Rev. E* **74**(6), 061708/1–14, (2006). doi: 10.1103/PhysRevE.74.061708
55. T.J. Spencer, C.M. Clark, R.M. Amos and J.C. Jones, Zenithal bistable device: Comparison of modeling and experiment, *Phys. Rev. E* **82**(2), 021702/1–13, (2010). doi: 10.1103/PhysRevE.82.021702
56. J.C. Jones and R.M. Amos, Relating display performance and grating structure of a zenithal bistable display, *Mol. Cryst. Liq. Cryst.* **543**(1), 57–68, (2011). doi: 10.1080/15421406.2011.568330
57. S. Kondoh, Ultra high-resolution FLC display for PDA, *Displays* **25**(1), 31–36, (2004). doi: 10.1016/j.displa.2004.04.004
58. J.C. Jones, 51.2: Novel geometries of the zenithal bistable device, *SID Symp. Digest Tech. Papers* **37**(1), 1626–1629, (2006). doi: 10.1889/1.2433313
59. J.C. Jones, Patent application WO 04,070,465 (2003).
60. C. Jones, and G. Bryan-Brown, 15.2: Low cost zenithal bistable display with improved white state, *SID Symp. Digest Tech. Papers* **41**(1), 207–210, (2010). doi: 10.1889/1.3500407
61. J.C. Jones and S. Beldon, 10.4: High image content zenithal bistable devices, *SID Symp. Digest Tech. Papers* **35**(1), 140–143, (2004). doi: 10.1889/1.1825769

# Appendix A

## Measured Flexoelectric Coefficients of Nematic Liquid Crystals

Nándor Éber

*Institute for Solid State Physics and Optics*
*Wigner Research Centre for Physics*
*Hungarian Academy of Sciences*
*P.O. Box 49, H-1525 Budapest, Hungary*
*e-mail: eber@szfki.hu*

Since the discovery of flexoelectricity[1] in 1969 several direct and indirect methods for measuring the flexoelectric coefficients of nematic liquid crystals have been reported. These techniques, their advantages and drawbacks are discussed in detail in Chapters 2, 3 and 4 of this volume.[2–4] Despite the diversity of the available methods, measurement of the flexocoefficients can still not be regarded as a routine task; $e_1$ and $e_3$ have been measured for a few compounds only. Unfortunately, sometimes even the data for a specific compound exhibit substantial deviations (differences in the magnitude or in the sign), depending on the experimental technique selected or even on the method of evaluation of the measured data sets.

Comparison of data from various sources is encumbered by the fact that only a few techniques are capable of providing $e_1$ or $e_3$ directly; most methods provide combinations of them, sometimes without information about their sign. Moreover, earlier papers give the flexocoefficients in CGS units ($\text{dyn}^{1/2}$) instead of the SI ones ($\text{Cm}^{-1}$) used nowadays ($10^{-4}$ CGS units correspond to $3.33 \times 10^{-12}$ $\text{Cm}^{-1}$).

To make the situation even more complicated, there exist two naming conventions for the flexocoefficients, which can be distinguished by looking at the definition of flexoelectric polarization. According to the original definition of 1969 by Meyer[1]

$$\mathbf{P}_{\mathrm{fl}} = e_1\mathbf{n}(\nabla \cdot \mathbf{n}) - e_3\mathbf{n} \times (\nabla \times \mathbf{n}), \tag{A.1}$$

while the other convention, first used by Patel and Meyer[5] in 1987, also proposed by Rudquist and Lagerwall[6] in 1997 and now suggested again by Meyer in the Introduction to the present book, uses

$$\mathbf{P_{fl}} = e_s\mathbf{n}(\nabla \cdot \mathbf{n}) + e_b\mathbf{n} \times (\nabla \times \mathbf{n}). \tag{A.2}$$

It can be seen that $e_s = e_1$ but $e_b = -e_3$; i.e. the two conventions differ only in the sign of the bend flexocoefficient. Unfortunately, in some papers using the convention of Eq. (A.2) the flexocoefficients are still named $e_1$ and $e_3$ (instead of $e_s$ and $e_b$), which makes the sign of $e_3$ ambiguous.

In order to resolve these problems and help the comparison of data for the flexocoefficients on a wider scale than previous reviews,[7,8] data available in the literature have been collected and are given in Tables A.1–A.9 in the same format. All values are in SI units and use the more widespread (original) naming convention corresponding to Eq. (A.1).

The tables contain only an abbreviated name of each compound. The precise chemical names of the compounds are given in the relevant tables of Appendix B.

In addition to the compounds listed in the tables, several other molecules have also been tested, however, only as a small concentration of dopants in some host nematic material. A linear extrapolation from these data might provide values for the flexocoefficients of the neat dopant; however, the applicability of this procedure to many compounds is questionable (especially for bent-core molecules); therefore these data are not included in the tables.

Table A.1. Flexoelectric coefficients of nematic liquid crystals using naming convention in Eq. (A.1): **MBBA**.

| Compound | Flexoelectric coefficient | (Temperature) | Alignment | Method | Reference |
|---|---|---|---|---|---|
| MBBA | $\lvert e_3\rvert \approx 23.3$ pC/m[a] | ([b]) | homeotropic | in-plane field, induced bend, interpretation of data from Ref. 10 | 9 |
| MBBA | $e_3 \approx 1.2$ pC/m[a] | ($\sim 22$ °C) | homeotropic | in-plane field, induced bend | 11 |
| MBBA | $e_3 \approx 3.3$ pC/m[a] | ([b]) | homeotropic | in-plane field, induced bend, reinterpretation of data in Refs. 10 and 11 | 12 |
| MBBA | $e_3 \approx 17.3 \pm 1.7$ pC/m[a] | ([b]) | homeotropic | in-plane field, conoscopy | 13 |
| MBBA | $\lvert e_3\rvert \approx 1.3$ pC/m | (25 °C) | homeotropic | in-plane field, induced bend | 14 |
| MBBA | $e_1 - e_3 \approx 3.3 \pm 0.7$ pC/m[a] | ([b]) | hybrid | induced twist | 15, 16 |
| MBBA | $e_1 - e_3 \approx 14 \pm 1$ pC/m | (30 °C) | hybrid | in-plane field, induced twist | 7 |
| MBBA | $\lvert e_1 - e_3\rvert \approx 17$ pC/m | (25 °C) | hybrid | flexodomains | 17 |
| MBBA | $e_1 - e_3 = 6$ pC/m | (30 °C) | planar | fit of EC threshold | 18 |
| MBBA | $\lvert e_1 - e_3\rvert \approx 7 \pm 0.15$ pC/m[a] | ([b]) | planar | flexodomains | 19 |
| MBBA | $e_1 + e_3 \approx 10$ pC/m[a] | ([b]) | planar | flexoelectric oscillations | 20 |
| MBBA | $e_1 + e_3 \approx 23.6 \pm 4.6$ pC/m[a] | (35 °C) | homeotropic | interdigitated electrodes, diffraction | 21 |
| MBBA | $e_1 + e_3 > 6.7$ pC/m[a] | (room) | planar | flexodomains | 22 |
| MBBA | $e_1 + e_3 \approx -3.3$ pC/m[a] | (room) | homeotropic | in-plane field, conoscopy | 23, 16 |
| MBBA | $e_1 + e_3 \approx -24.8$ pC/m[a] | ($\sim 20$°C) | hybrid | birefringence | 24 |
| MBBA | $e_1 + e_3 \approx -3.3$ pC/m[a] | ($T_{NI} - 10$°C) | hybrid | birefringence | 24 |
| MBBA | $e_1 + e_3 \approx 0.17$ pC/m[a] | ([b]) | homeotropic | oscillating pressure | 25 |
| MBBA | $e_1 + e_3 \approx 0.5$ pC/m[a] | ([b]) | homeotropic | oscillating pressure | 25 |
| MBBA | $e_1 + e_3 \approx -15$ pC/m[a] | ([b]) | hybrid | birefringence | 26 |
| MBBA | $e_1 + e_3 \approx -15 \pm 2.3$ pC/m[a] | ([b]) | hybrid | birefringence | 27 |
| MBBA | $e_1 + e_3 \approx -2.5 \pm 0.2$ pC/m[a] | (room) | hybrid | pyroelectric | 28 |
| MBBA | $e_1 + e_3 \approx -2.5 \pm 0.2$ pC/m[a] | (room) | hybrid | pyroelectric | 29 |
| MBBA | $e_1 + e_3 \approx 40$ pC/m | ([b]) | homeotropic | acoustic tilt | 30 |
| MBBA | $e_1 + e_3 \approx 10$ pC/m[a] | (20 °C) | 72° tilt | ellipsometry | 31 |
| MBBA | $e_1 + e_3 \approx -54 \pm 1$ pC/m | (30 °C) | hybrid | birefringence | 7 |
| MBBA | $e_1 + e_3 \approx -24$ pC/m | (room) | hybrid | pyroelectric, reinterpretation of data in Ref. 28 | 32 |
| MBBA | $e_1 + e_3 \approx 88$ pC/m | ([b]) | hybrid | birefringence, reinterpretation of data in Ref. 24 | 33 |
| MBBA | $e_1 + e_3 \approx 1.6 \pm 0.6$ pC/m | ([b]) | hybrid | pyroelectric | 34 |
| MBBA | $e_1 + e_3 \approx -20 \pm 5$ pC/m | (30 °C) | hybrid | birefringence, reinterpretation of data in Ref. 7 | 35 |
| MBBA | $e_1 + e_3 = -35$ pC/m | (30 °C) | planar | fit of EC threshold | 18 |

[a]Converted from original data given in CGS units.
[b]Temperature of measurement was not specified.

Table A.2. Flexoelectric coefficients of nematic liquid crystals using naming convention in Eq. (A.1): cyanobiphenyls.

| Compound | Flexoelectric coefficient | (Temperature) | Alignment | Method | Reference |
|---|---|---|---|---|---|
| **5CB** | $e_1 \approx 5$ pC/m² [a] | (room) | hybrid | pyroelectric | 28 |
| **5CB** | $e_1 \approx -5$ pC/m² [a] | (room) | hybrid | pyroelectric | 29 |
| **5CB** | $\|e_3\| \approx 40$ pC/m | ($T_{NI} - 4$ °C) | planar | direct flexing | 36 |
| **5CB** | $(e_1 - e_3)/K_2 \approx 2.4$ C/(Nm) | (303 K) | hybrid | in-plane field, induced twist | 37 |
| **5CB** | $(e_1 + e_3)/K_3 \approx 1.08$ C/(Nm) | (303 K) | homeotropic | quadrupolar field, conoscopy | 37 |
| **5CB** | $e_1 + e_3 \approx -53$ pC/m | (25 °C) | hybrid | pyroelectric | 32 |
| **5CB** | $e_1 + e_3 \approx -45.6$ pC/m | (25 °C) | hybrid | pyroelectric | 32 |
| **5CB** | $e_1 + e_3 \approx -42$ pC/m | (25 °C) | hybrid | pyroelectric | 38 |
| **5CB** | $e_1 + e_3 \approx 30$ pC/m | (30 °C) | hybrid | half-leaky guided mode | 39 |
| **5CB** | $e_1 + e_3 \approx -13$ pC/m | (room) | hybrid | pyroelectric | 40 |
| **6CB** | $(e_1 - e_3)/K_2 \approx 2.51$ C/(Nm) | (301 K) | hybrid | in-plane field, induced twist | 37 |
| **7CB** | $(e_1 - e_3)/K_2 \approx 3.53$ C/(Nm) | (303 K) | hybrid | in-plane field, induced twist | 37 |
| **8CB** | $e_1 - e_3 K \approx 3.3$ pC/m² [a] | (36.3 °C) | hybrid | in-plane field, induced twist | 41, 16 |
| **8CB** | $(e_1 - e_3)/K \approx 0.9$ C/(Nm) [a] | (36.3 °C) | hybrid | in-plane field, induced twist | 41 |
| **8CB** | $(e_1 - e_3)/K \approx 3.3$ C/(Nm) [a] | (36.3 °C) | hybrid | in-plane field, induced twist | 41 |
| **8CB** | $(e_1 - e_3)/K_2 \approx 2.28$ C/(Nm) | (307 K) | hybrid | in-plane field, induced twist | 37 |
| **8CB** | $e_1 + e_3 \approx 480$ pC/m | ([b]) | homeotropic | acoustic tilt | 30 |
| **7OCB** [d] | $(e_1 - e_3)/K \approx 1$ C/(Nm) [c] | ($T_{NI} - 8$ °C) | homeotropic | cholesteric flexoelectro-optic effect | 42 |
| **8OCB** | $e_1 - e_3 \approx -16.7$ pC/m² [a] | (77 °) | hybrid | in-plane field, induced twist | 41, 16 |
| **8OCB** | $(e_1 - e_3)/K \approx -4.7$ C/(Nm) [a] | (77 °) | hybrid | in-plane field, induced twist | 41 |
| **8OCB** | $(e_1 - e_3)/K \approx 2.3$ C/(Nm) | ($T_{NI} - 8$ °C) | hybrid | in-plane field, induced twist | 43 |
| **8OCB** | $e_1 + e_3 \approx 21.3 \pm 5$ pC/m² [a] | (73 °C) | homeotropic | interdigitated electrodes, diffraction | 21 |
| **8OCB** | $e_1 + e_3 \approx 4.7$ pC/m² [a] | ($T_{NI} - 7$ ° C) | homeotropic | in-plane field, conoscopy | 23 |
| **8OCB** | $e_1 + e_3 \approx 7 \pm 1.5$ pC/m | (75 °C) | hybrid | pyroelectric | 44 |
| **8OCB** | $(e_1 + e_3)/K_3 \approx 0.29$ C/(Nm) | (343 K) | homeotropic | quadrupolar field, conoscopy | 37 |
| **5CT** | $(e_1 - e_3)/K_2 \approx 0.87$ C/(Nm) | (423 K) | hybrid | in-plane field, induced twist | 37 |
| **HBACPE** | $e_1 + e_3 \approx 270$ pC/m | ([b]) | homeotropic | acoustic tilt | 30 |
| **CBmix** | $e_1 + e_3 \approx 45 \pm 22$ pC/m | (22 °C) | hybrid | flexoelectric offset voltage | 45 |

[a] Converted from original data given in CGS units.
[b] Temperature of measurement was not specified.
[c] Original data were given using naming convention in Eq. (A.2).
[d] Chiral dopant added for the measurement.

Table A.3. Flexoelectric coefficients of nematic liquid crystals using naming convention in Eq. (A.1): cyclohexanes.

| Compound | Flexoelectric coefficient | (Temperature) | Alignment | Method | Reference |
|---|---|---|---|---|---|
| **PCH3** | $(e_1 - e_3)/K_2 \approx 0.36$ C/(Nm) | (310 K) | hybrid | in-plane field, induced twist | 37 |
| **PCH3** | $(e_1 + e_3)/K_3 \approx 0.11$ C/(Nm) | (310 K) | homeotropic | quadrupolar field, conoscopy | 37 |
| **PCH4** | $(e_1 - e_3)/K_2 \approx 1.55$ C/(Nm) | (311 K) | hybrid | in-plane field, induced twist | 37 |
| **PCH5** | $(e_1 - e_3)/K_2 \approx 1.66$ C/(Nm) | (303 K) | hybrid | in-plane field, induced twist | 37 |
| **PCH5** | $(e_1 + e_3)/K_3 \approx 0.12$ C/(Nm) | (303 K) | homeotropic | quadrupolar field, conoscopy | 37 |
| **PCH7** | $(e_1 - e_3)/K_2 \approx 1.2$ C/(Nm) | (303 K) | hybrid | in-plane field, induced twist | 37 |
| **PCH7** | $(e_1 + e_3)/K_3 \approx 0.23$ C/(Nm) | (303 K) | homeotropic | quadrupolar field, conoscopy | 37 |
| **CCH7** | $(e_1 - e_3)/K_2 \approx -1.29$ C/(Nm) | (343 K) | hybrid | in-plane field, induced twist | 37 |
| **CCH7** | $(e_1 + e_3)/K_3 \approx 0.51$ C/(Nm) | (343 K) | homeotropic | quadrupolar field, conoscopy | 37 |
| **CCH7** | $(e_1 + e_3)/S \approx 36$ pC/m[a] | $(0.93T_{NI})$ | hybrid | AC field, birefringence | 46 |
| **MPPC** | $(e_1 - e_3)/K_2 \approx 0.52$ C/(Nm) | (323 K) | hybrid | in-plane field, induced twist | 37 |
| **EPPC** | $(e_1 - e_3)/K_2 \approx 0.28$ C/(Nm) | (323 K) | hybrid | in-plane field, induced twist | 37 |
| **PPPC** | $(e_1 - e_3)/K_2 \approx 2.97$ C/(Nm) | (310 K) | hybrid | in-plane field, induced twist | 37, 47 |
| **PCHmix** | $e_1 + e_3 \approx 48 \pm 24$ pC/m | (22 °C) | hybrid | flexoelectric offset voltage | 45 |
| **Mix1** | $(e_1 - e_3)/K_2 \approx 0.33$ C/(Nm) | (303 K) | hybrid | in-plane field, induced twist | 37, 47, 48 |
| **Mix1** | $(e_1 + e_3)/K_3 \approx -0.60$ C/(Nm) | (303 K) | homeotropic | quadrupolar field, conoscopy | 37, 47, 48 |

[a]Converted from original data given in CGS units.

Table A.4. Flexoelectric coefficients of nematic liquid crystals using naming convention in Eq. (A.1): other calamitics.

| Compound | Flexoelectric coefficient | (Temperature) | Alignment | Method | Reference |
|---|---|---|---|---|---|
| **BHAOB** | $e_1 + e_3 \approx 16$ pC/m | ([b]) | homeotropic | acoustic tilt | 30 |
| **BMAOB** | $\|e_1 - e_3\| \approx 5.7$ pC/m[a] | (25 °C) | planar | flexodomains | 49 |
| **BMAOB** | $\|e_1 - e_3\| \approx 5.7$ pC/m[a] | ([b]) | planar | flexodomains | 50 |
| **BMAOB** | $\|e_1 - e_3\| \approx 5.3$ pC/m | (29 °C) | planar | flexodomains | 51 |
| **BMAOB** | $\|e_1 - e_3\| \approx 12.3 \pm 2$ pC/m[a] | (room) | planar | flexodomains | 52 |
| **BMAOB** | $\|e_1 - e_3\| \approx 5.8$ pC/m[a] | ([b]) | hybrid | flexodomains | 53 |
| **BBMBA** | $e_1 + e_3 \approx 8.3$ pC/m[a] | ($T_{NA} + 1$ °C) | homeotropic | interdigitated electrodes, diffraction | 54 |
| **CBOOA** | $\|e_1 + e_3\| \sim 3$ pC/m[a] | ([b]) | homeotropic | interdigitated electrodes, diffraction | 54 |
| **ROCP-7037** | $(e_1 - e_3)/K_2 \approx 4.36$ C/(Nm) | (319 K) | hybrid | in-plane field, induced twist | 37 |
| **ROCP-7037** | $(e_1 + e_3)/K_3 \approx 0.05$ C/(Nm) | (319 K) | homeotropic | quadrupolar field, conoscopy | 37 |
| **ROCP-7334** | $(e_1 - e_3)/K_2 \approx 2.52$ C/(Nm) | (373 K) | hybrid | in-plane field, induced twist | 37 |
| **ROCP-7334** | $(e_1 + e_3)/K_3 \approx 0.16$ C/(Nm) | (373 K) | homeotropic | quadrupolar field, conoscopy | 37 |
| **HOT** | $(e_1 + e_3)/K \approx 2.7$ C/(Nm)[a] | (99.3 °C) | homeotropic | interdigitated electrodes, diffraction | 21 |
| **3PBClB** | $(e_1 + e_3)/K \approx 0.9$ C/(Nm)[a] | (48.9 °C) | homeotropic | interdigitated electrodes, diffraction | 21 |
| **NPOOB** | $e_1 + e_3 \approx 98$ pC/m | ([b]) | homeotropic | acoustic tilt | 30 |
| **8/7** | $e_1 - e_3 = 2.9$ pC/m | (86 °C) | planar | fit of EC thresholds | 55 |
| **8/7** | $e_1 - e_3 = 1$ pC/m | (86 °C) | planar | fit of EC thresholds | 56 |
| **8/7** | $e_1 + e_3 = -59.5$ pC/m | (86 °C) | planar | fit of EC thresholds | 55 |
| **8/7** | $e_1 + e_3 = -20$ pC/m | (86 °C) | planar | fit of EC thresholds | 56 |
| unknown compound | $\|e_1 - e_3\| = 4.4$ pC/m | ([b]) | planar | domain walls | 57 |
| unknown compound | $e_1 + e_3 = 8$ pC/m | ([b]) | planar | domain walls | 57 |
| **DEmix** | $e_1 + e_3 \approx 46 \pm 23$ pC/m | (22 °C) | hybrid | flexoelectric offset voltage | 45 |
| **PB** | $\|e_1 - e_3\| \approx 6$ pC/m[a] | (25 °C) | planar | flexodomains | 49 |
| **MixA** | $e_3 \approx 0.3$ pC/m[a] | ([b]) | homeotropic | oscillating pressure | 58 |

[a]Converted from original data given in CGS units.
[b]Temperature of measurement was not specified.

Table A.5. Flexoelectric coefficients of nematic liquid crystals using naming convention in Eq. (A.1): commercial mixtures.

| Compound | Flexoelectric coefficient | (Temperature) | Alignment | Method | Reference |
|---|---|---|---|---|---|
| **E7** | $(e_1 - e_3)/K \approx 0.9$ C/(Nm)[c] | ($0.98T_{NI}$) | hybrid | in-plane field, induced twist | 59 |
| **E7** | $e_1 - e_3 \approx$ -10.6 pC/m | ([b]) | twisted | in-plane field, induced tilt, conoscopy | 60 |
| **E7**[d] | $e_1 - e_3 \approx 9.3 \pm 0.3$ pC/m | ([b]) | twisted | cholesteric flexoelectro-optic effect | 61 |
| **E7** | $e_1 - e_3 \approx 12.2 \pm 1$ pC/m | ([b]) | twisted | in-plane field, induced tilt, conoscopy | 60 |
| **E7** | $\lvert e_1 - e_3\rvert \approx 5$ pC/m | (57.6 °C) | 45° twist, 37° pretilt | bistable polar switching | 62 |
| **E7**[d] | $e_1 - e_3 \approx 12.2 \pm 1$ pC/m | ([b]) | twisted | cholesteric flexoelectro-optic effect | 63 |
| **E7**[d] | $e_1 - e_3 \approx 12.2 \pm 0.8$ pC/m | ([b]) | hybrid | cholesteric flexoelectro-optic effect | 64 |
| **E7**[d] | $e_1 - e_3 \approx 12.2 \pm 2$ pC/m | ([b]) | planar | chiral Grandjean texture | 64 |
| **E7** | $e_1 + e_3 \approx 15 \pm 2$ pC/m | ([b]) | hybrid | fully leaky guided mode | 65 |
| **E7** | $e_1 + e_3 \approx 15 \pm 3$ pC/m | ([b]) | hybrid | fully leaky guided mode | 66 |
| **E7** | $e_1 + e_3 \approx 20 \pm 5$ pC/m | ([b]) | homeotropic, grating | diffraction | 67 |
| **E7** | $e_1 + e_3 \approx 600$ pC/m | ([b]) | homeotropic ZBD | AC field, birefringence | 68 |
| **E7** | $e_1 + e_3 \approx 40 \pm 5$ pC/m | (~ 20 °C) | homeotropic ZBD | AC field, birefringence | 69 |
| **E7** | $e_1 + e_3 \approx 14 \pm 1$ pC/m | ([b]) | planar | interferometry, interdigitated electrodes | 70 |
| **E7** | $e_1 + e_3 \approx 45 \pm 22$ pC/m | (22 °C) | hybrid | flexoelectric offset voltage | 45 |
| **E7** | $e_1 + e_3 \approx 35 \pm 5$ pC/m | ([b]) | 180° twist, 35° pretilt | birefringence | 63 |
| **E7** | $e_1 + e_3 \approx 14 \pm 1$ pC/m | ([b]) | planar | interferometry, interdigitated electrodes | 71 |
| **E70A**[d] | $e_1 - e_3 \approx 3.5 \pm 0.2$ pC/m | ([b]) | twisted | cholesteric flexoelectro-optic effect | 61 |
| **E70A** | $e_1 + e_3 \approx 34 \pm 17$ pC/m | (22 °C) | hybrid | flexoelectric offset voltage | 45 |

[b]Temperature of measurement was not specified.
[c]Original data were given using naming convention in Eq. (A.2).
[d]Chiral dopant added for the measurement.

Table A.6. Flexoelectric coefficients of nematic liquid crystals using naming convention in Eq. (A.1): commercial mixtures.

| Compound | Flexoelectric coefficient | (Temperature) | Alignment | Method | Reference |
|---|---|---|---|---|---|
| **ZLI-4788-000** | $e_1 + e_3 \approx 22$ pC/m[c] | ([b]) | hybrid | fully leaky guided mode | 72 |
| **ZLI-4788-000** | $e_1 + e_3 \approx -7$ pC/m[c] | ([b]) | hybrid | fully leaky guided mode | 72 |
| **ZLI-4792** | $(e_1 - e_3)/K \approx 0.6$ C/(Nm)[c] | ($0.98T_{NI}$) | hybrid | in-plane field, induced twist | 59 |
| **ZLI-4792** | $\lvert e_1 + e_3 \rvert \leq 10 \pm 1$ pC/m | (30 °C) | hybrid | birefringence | 7 |
| **ZLI-4792** | $e_1 - e_3 \approx -15 \pm 1$ pC/m | (30 °C) | hybrid | in-plane field, induced twist | 7 |
| **TL-216**[d] | $e_1 - e_3 \approx 1.45 \pm 0.3$ pC/m | ([b]) | hybrid | cholesteric flexoelectro-optic effect | 64 |
| **TL-216**[d] | $e_1 - e_3 \approx 0.9 \pm 0.3$ pC/m | ([b]) | planar | chiral Grandjean texture | 64 |
| **TL-216** | $e_1 - e_3 \approx 0.4 \pm 0.25$ pC/m | ([b]) | twisted | in-plane field, induced tilt, conoscopy | 60 |
| **TL-216** | $e_1 + e_3 \approx 1$ pC/m | (22 °C) | hybrid | flexoelectric offset voltage | 45 |
| **MDA-02-2419**[d] | $e_1 - e_3 \approx 22 \pm 1.5$ pC/m | ([b]) | hybrid | cholesteric flexoelectro-optic effect | 64 |
| **MDA-02-2419**[d] | $e_1 - e_3 \approx 28.2 \pm 4$ pC/m | ([b]) | planar | chiral Grandjean texture | 64 |
| **MDA-02-2419** | $e_1 + e_3 \approx 102 \pm 51$ pC/m | (22 °C) | hybrid | flexoelectric offset voltage | 45 |
| **MDA-02-2419** | $e_1 - e_3 \approx 22 \pm 1$ pC/m | ([b]) | twisted | in-plane field, induced tilt | 64 |
| **MLC-7029** | $e_1 + e_3 \approx -7 \pm 3.5$ pC/m | (22 °C) | hybrid | flexoelectric offset voltage | 45 |
| **MLC-6437-000**[d] | $e_1 - e_3 \approx 3.2$ pC/m[c] | ($0.9T_{NI}$) | homeotropic | cholesteric flexoelectro-optic effect | 73 |
| **Phase4** | $e_1 - e_3 = -4.7$ pC/m | (20 °C) | planar | fit of EC thresholds | 74 |
| **Phase4** | $e_1 + e_3 = -31.5$ pC/m | (20 °C) | planar | fit of EC thresholds | 74 |
| **Phase5** | $e_1 - e_3 = -2.9$ pC/m | (30 °C) | planar | fit of EC thresholds | 18, 55 |
| **Phase5** | $e_1 + e_3 = -50.1$ pC/m | (30 °C) | planar | fit of EC thresholds | 18, 55 |
| **Phase5A** | $e_1 - e_3 = -2.9$ pC/m | (30 °C) | planar | fit of EC thresholds | 18, 55 |
| **Phase5A** | $e_1 + e_3 = -50.1$ pC/m | (30 °C) | planar | fit of EC thresholds | 18, 55 |

[b]Temperature of measurement has not been specified.
[c]Original data were given using naming convention in Eq. (A.2).
[d]Chiral dopant added for the measurement.

Table A.7. Flexoelectric coefficients of nematic liquid crystals using naming convention in Eq. (A.1): bent-core compounds.

| Compound | Flexoelectric coefficient | (Temperature) | Alignment | Method | Reference |
|---|---|---|---|---|---|
| **ClPbis10BB** | $\lvert e_3\rvert \approx 35000$ pC/m | $(T_{NI} - 2\ °\text{C})$ | planar | direct flexing | 36 |
| **ClPbis10BB** | $e_3 \approx 40000$ pC/m | $(T_{NI} - 4\ °\text{C})$ | planar | flexible electrode, interferometry | 75 |
| **ClPbis10BB** | $\lvert e_3\rvert \approx 15.8$ pC/m | (71.5 °C) | homeotropic | in-plane field, induced bend | 76 |
| **ClPbis10BB** | $\lvert e_1 - e_3\rvert \approx 6$ pC/m | (74 °C) | planar | flexodomains | 76 |
| **ClPbis10BB** | $\lvert e_1 - e_3\rvert \approx 11.2$ pC/m[a] | (75 °C) | hybrid | flexodomains | 53 |
| **CNRbis12OBB** | $e_3 \approx 3.9 \pm 1$ pC/m | (120 °C) | homeotropic | in-plane DC field, induced bend | 77 |
| **CNRbis12OBB** | $\lvert e_3\rvert \approx 5.4 \pm 0.4$ pC/m | (120 °C) | homeotropic | in-plane AC+DC field, induced bend | 77 |
| **CNRbis12OBB** | $\lvert e_1\rvert \sim 7.7$ pC/m | (120 °C) | homeotropic | in-plane AC+DC field, induced bend | 77 |
| **BEPC** | $\lvert e_1 - e_3\rvert \approx 5.95$ pC/m | (55 °C) | planar | flexodomains | 78 |
| **BCCB** | $e_1 \approx 17.6$ pC/m | (140 °C) | homeotropic | surface polar instability | 77 |
| **BCCB** | $e_3 \approx 17.2$ pC/m | (140 °C) | homeotropic | surface polar instability | 77 |
| **C6ban**[d] | $e_1 - e_3 \approx -17.4$ pC/m | $(T_{NI} - 11\ °\text{C})$ | homeotropic | cholesteric flexoelectro-optic effect | 79 |
| **Azpac** | $\lvert e_1 - e_3\rvert \approx 3.3$ pC/m | (100 °C) | planar | flexodomains | 80 |

[a]Converted from original data given in CGS units.
[d]Chiral dopant added for the measurement.

Table A.8. Flexoelectric coefficients of nematic liquid crystals using naming convention in Eq. (A.1): bimesogens.

| Compound | Flexoelectric coefficient | (Temperature) | Alignment | Method | Reference |
|---|---|---|---|---|---|
| **11Es5CB**[d] | $e_1 - e_3 \approx -10.4$ pC/m[c] | ([b]) | homeotropic | cholesteric flexoelectro-optic effect | 81 |
| **11Es5CB**[d] | $e_1 - e_3 \approx 10.4$ pC/m[c] | ([b]) | homeotropic | cholesteric flexoelectro-optic effect | 82 |
| **5Es11CB**[d] | $e_1 - e_3 \approx 9 \pm 1$ pC/m[c] | ([b]) | homeotropic | cholesteric flexoelectro-optic effect | 81 |
| **5Es11CB**[d] | $e_1 - e_3 \approx 10.6$ pC/m[c] | ([b]) | homeotropic | cholesteric flexoelectro-optic effect | 82 |
| **CBO8OCB**[d] | $(e_1 - e_3)/K \approx 1.2$ C/(Nm)[c] | ($T_{NI} - 22$ °C) | homeotropic | cholesteric flexoelectro-optic effect | 42 |
| **FFO5OCB**[d] | $e_1 - e_3 \approx 20.4$ pC/m[c] | ($0.84T_{NI}$) | homeotropic | cholesteric flexoelectro-optic effect | 83 |
| **FFO6OCB**[d] | $e_1 - e_3 \approx 12.6$ pC/m[c] | ($0.84T_{NI}$) | homeotropic | cholesteric flexoelectro-optic effect | 83 |
| **FFO7OCB**[d] | $e_1 - e_3 \approx 20.6$ pC/m[c] | ($0.84T_{NI}$) | homeotropic | cholesteric flexoelectro-optic effect | 83 |
| **FFO8OCB**[d] | $e_1 - e_3 \approx 14.6$ pC/m[c] | ($0.84T_{NI}$) | homeotropic | cholesteric flexoelectro-optic effect | 83 |
| **FFO9OCB**[d] | $e_1 - e_3 \approx 19.8$ pC/m[c] | ($0.84T_{NI}$) | homeotropic | cholesteric flexoelectro-optic effect | 83 |
| **FFO10OCB**[d] | $e_1 - e_3 \approx 15$ pC/m[c] | ($0.84T_{NI}$) | homeotropic | cholesteric flexoelectro-optic effect | 83 |
| **FFO11OCB**[d] | $e_1 - e_3 \approx 21.2$ pC/m[c] | ($0.84T_{NI}$) | homeotropic | cholesteric flexoelectro-optic effect | 83 |
| **FFO12OCB**[d] | $e_1 - e_3 \approx 13.2$ pC/m[c] | ($0.84T_{NI}$) | homeotropic | cholesteric flexoelectro-optic effect | 83 |
| **NSO9OFF**[d] | $e_1 - e_3 \approx 20.4$ pC/m[c] | ($0.84T_{NI}$) | homeotropic | cholesteric flexoelectro-optic effect | 83 |
| **NSO9OphO9OCB**[d] | $e_1 - e_3 \approx 11.6$ pC/m[c] | ($0.91T_{NI}$) | homeotropic | cholesteric flexoelectro-optic effect | 83 |
| **Mixture1**[d] | $(e_1 - e_3)/K \approx 0.94$ C/(Nm)[c] | ([b]) | homeotropic | cholesteric flexoelectro-optic effect | 84 |
| **Mixture2**[d] | $(e_1 - e_3)/K \approx 1.74$ C/(Nm)[c] | ([b]) | homeotropic | cholesteric flexoelectro-optic effect | 84 |
| **Mixture2**[d] | $(e_1 - e_3)/K \approx 1.88$ C/(Nm)[c] | ($0.84T_{NI}$) | homeotropic | cholesteric flexoelectro-optic effect | 83 |
| **Mixture3**[d] | $(e_1 - e_3)/K \approx 3.48$ C/(Nm)[c] | ($0.84T_{NI}$) | homeotropic | cholesteric flexoelectro-optic effect | 83 |
| **MixBimes**[d] | $(e_1 - e_3)/K \approx 2$ C/(Nm)[c] | ($T_{NI} - 22$ °C) | homeotropic | cholesteric flexoelectro-optic effect | 42 |

[b]Temperature of measurement was not specified.
[c]Original data were given using naming convention in Eq. (A.2).
[d]Chiral dopant added for the measurement.

Table A.9. Flexoelectric coefficients of chiral nematic liquid crystals using naming convention in Eq. (A.1).

| Compound | Flexoelectric coefficient | (Temperature) | Alignment | Method | Reference |
|---|---|---|---|---|---|
| **(S)-9PPP8dM** | $e_1 - e_3 \approx 2$ pC/m[a,c] | ([b]) | planar | cholesteric flexoelectro-optic effect | 5 |
| **8OCB/CC** | $\|e_1 - e_3\| \approx -3$ pC/m | (57.6 °C) | planar | Lehmann rotation | 85 |
| **TM-216** | $(e_1 - e_3)/K \approx 0.12$ C/(Nm) | ([b]) | homeotropic | cholesteric flexoelectro-optic effect | 86 |

[a]Converted from original data given in CGS units.
[b]Temperature of measurement was not specified.
[c]Original data were given using naming convention in Eq. (A.2).

## References


1. R.B. Meyer, Piezoelectric effects in liquid crystals, *Phys. Rev. Lett.* **22**(18), 918–921, (1969). doi: 10.1103/PhysRevLett.22.918
2. N.V. Madhusudana, Chapter 2. Flexoelectro-optics and measurements of flexo-coefficients. In eds. Á. Buka and N. Éber, *Flexoelectricity in Liquid Crystals. Theory, Experiments and Applications*, Imperial College Press, London, 2012. pp. 33–60.
3. A. Jákli, J. Harden and N. Éber, Chapter 3. Flexoelectricity of bent-core molecules, In eds. Á. Buka and N. Éber, *Flexoelectricity in Liquid Crystals. Theory, Experiments and Applications*, Imperial College Press, London, 2012. pp. 61-99.
4. Á. Buka, T. Tóth-Katona, N. Éber, A. Krekhov and W. Pesch, Chapter 4. The role of flexoelectricity in pattern formation. In eds. Á. Buka and N. Éber, *Flexoelectricity in Liquid Crystals. Theory, Experiments and Applications*, Imperial College Press, London, 2012. pp. 101–135.
5. J.S. Patel and R.B. Meyer, Flexoelectric electro-optics of a cholesteric liquid crystal, *Phys. Rev. Lett.* **58**(15), 1538–1540, (1987). doi: 10.1103/PhysRevLett.58.1538
6. P. Rudquist and S.T. Lagerwall, On the flexoelectric effect in nematics, *Liq. Cryst.* **23**(4), 503–510, (1997). doi: 10.1080/026782997208082
7. T. Takahashi, S. Hashidate and T. Akahane, Novel measurement method for flexoelectric coefficients of nematic liquid crystals, *Jpn. J. Appl. Phys.* **37**(4A), 1865–1869, (1998). doi: 10.1143/JJAP.37.1865
8. A.G. Petrov, Measurements and interpretation of flexoelectricity. In eds. D.A. Dunmur, A. Fukuda and G.R. Luckhurst,*Physical Properties of Liquid Crystals, Vol 1: Nematics*, Inspec, London, 2001. pp. 251–264.
9. W. Helfrich, A simple method to observe the piezoelectricity of liquid crystals, *Phys. Lett.* **35A**(6), 393–394, (1971). doi: 10.1016/0375-9601(71)90160-5
10. W. Haas, J. Adams and J.B. Flannery, New electro-optic effect in a room-temperature nematic liquid crystal, *Phys. Rev. Lett.* **25**(19), 1326–1327, (1970). doi: 10.1103/PhysRevLett.25.1326
11. D. Schmidt, M. Schadt and W. Helfrich, Liquid-crystalline curvature electricity: The bending mode of MBBA, *Z. Naturforsch.* **27A**(2), 277–280, (1972).
12. A.I. Derzhanski, A.G. Petrov and M.D. Mitov, One-dimensional dielectric-flexoelectric deformations in nematic layers, *J. Phys. France* **39**(3), 273–285, (1979). doi: 10.1051/jphys:01978003903027300
13. A.S. Vasilevskaya and A.S. Sonin, Transverse electro-optical effect in MBBA, *Fiz. Tverd. Tela* **21**(1), 196–200, (1979).
14. A.G. Petrov, A.Th. Ionescu, C. Versace and N. Scaramuzza, Investigation of flexoelectric properties of a palladium-containing nematic liquid crystal, Azpac, and its mixtures with MBBA, *Liq. Cryst.* **19**(2), 169–178, (1995). doi: 10.1080/02678299508031966
15. I. Dozov, Ph. Martinot Lagarde and G. Durand, Flexoelectrically controlled twist of texture in a nematic liquid crystal, *J. Physique Lett.* **43**(10), L-365–L-369, (1982). doi: 10.1051/jphyslet:019820043010036500

16. G. Durand, Flexoelectricity in nematics: Continuous distortions and instabilities, *Mol. Cryst. Liq. Cryst.* **113**(1), 237–246, (1984). doi: 10.1080/00268948408071686
17. V.A. Delev and O.A. Scaldin, Electrooptics of hybrid aligned nematics in the regime of flexoelectric instability, *Tech. Phys. Lett.* **30**(8), 679–681, (2004). [*Pis'ma Zh. Tekh. Fiz.* **30**(16), 36–40, (2004).]
18. T. Tóth-Katona, N. Éber, Á. Buka and A. Krekhov, Flexoelectricity and competition of time scales in electroconvection, *Phys. Rev. E* **78**(3), 036306/1–12, (2008). doi: 10.1103/PhysRevE.78.036306
19. H.P. Hinov, On the coexistence of the flexo-dielectric walls-flexoelectric domains for the nematic MBBA – A new estimation of the modulus of the difference between the flexoelectric coefficients of splay and bend $|e_{1z} - e_{3x}|$, *Mol. Cryst. Liq. Cryst.* **524**(1), 26–35, (2010). doi: 10.1080/15421400903568161
20. A.I. Derzhanski and A.G. Petrov, Piezoelectricity in nematic liquid crystals, *Acta Phys. Polon. A* **55**(5), 747–754, (1979).
21. J.P. Marcerou and J. Prost, The different aspects of flexoelectricity in nematics, *Mol. Cryst. Liq. Cryst.* **58**(3–4), 259–284, (1980). doi: 10.1080/00268948008082127
22. H.P. Hinov, Further experimental evidence for the wall structure of the flexoelectric domains in symmetrically weakly-anchored MBBA layers, *Mol. Cryst. Liq. Cryst.* **89**(1), 227–248, (1982). doi: 10.1080/00268948208074480
23. I. Dozov, I. Penchev, Ph. Martinot-Lagarde and G. Durand, On the sign of flexoelectric coefficients in nematic liquid crystals, *Ferroelectrics Letters* **2**(4), 135–141, (1984). doi: 10.1080/07315178408202434
24. N.V. Madhusudana and G. Durand, Linear flexo-electro-optic effect in a hybrid aligned nematic liquid crystal cell, *J. Physique Lett.* **46**(5), L-195–L-200, (1985). doi: 10.1051/jphyslet:01985004605019500
25. O.A. Skaldin, A.N. Lachinov and A.N. Chuvyrov, Two contributions to the piezoelectric effect in nematic liquid crystals, *Fiz. Tverd. Tela* **27**(4), 1220–1222, (1985).
26. G. Barbero, P. Taverna Valabrega, R. Bartolino and B. Valenti, Evidence for the flexo-electric effect in a nematic liquid crystal cell, *Liq. Cryst.* **1**(5), 483–490, (1986). doi: 10.1080/02678298608086271
27. B. Valenti, C. Bertoni, G. Barbero, P. Taverna-Valabrega and R. Bartolino, Flexoelectricity in the hybrid aligned nematic cell, *Mol. Cryst. Liq. Cryst.* **146**(1), 307–320, (1987). doi: 10.1080/00268948708071820
28. L.A. Beresnev, L.M. Blinov, S.A. Davidyan, S.G. Kononov and S.B. Yablonskii, Direct measurement of the flexoelectric polarization of nematic liquid crystals, *JETP Lett.* **45**(12), 755–758, (1987). [*Pisma Zh. Eksp. Teor. Fiz.* **45**(12), 592–594, (1987).]
29. L.M. Blinov, L.A. Beresnev, S.A. Davidyan, S.G. Kononov and S.B. Yablonskii, Flexoelectric effects in liquid crystals, *Ferroelectrics* **84**(1), 365–377, (1988). doi: 10.1080/00150198808016233
30. G. Barbero, A.N. Chuvyrov, A.P. Krekhov and O.A. Scaldin, Surface polarization and flexoelectric effect, *J. Appl. Phys.* **69**(9), 6343–6348, (1991). doi: 10.1063/1.348834

31. L.M. Blinov, G. Durand and S.V. Yablonsky, Curvature oscillations and linear electro-optical effect in a surface layer of a nematic liquid crystal, *J. Phys. II France* **2**(5), 1287–1297, (1992). doi: 10.1051/jp2:1992200
32. L.M. Blinov, M. Ozaki and K. Yoshino, Flexoelectric polarization in nematic liquid crystals measured by a field onoff pyroelectric technique, *JETP Lett.* **69**(3), 236–242, (1999). [*Pisma Zh. Eksp. Teor. Fiz.* **69**(3), 220–225, (1999).]
33. S. Ponti, P. Ziherl, C. Ferrero and S. Zumer, Flexoelectro-optic effect in a hybrid nematic liquid crystal cell, *Liq. Cryst.* **26**(8), 1171–1177, (1999). doi: 10.1080/026782999204192
34. L.M. Blinov, M.I. Barnik, H. Ohoka, M. Ozaki and K. Yoshino, Separate measurements of the flexoelectric and surface polarization in a model nematic liquid crystal p-methoxybenzylidene-p′-butylaniline: Validity of the quadrupolar approach, *Phys. Rev. E* **64**(3), 031707/1–7, (2001). doi: 10.1103/PhysRevE.64.031707
35. N.T. Kirkman, T. Stirner and W.E. Hagston, Continuum modelling of hybrid-aligned nematic liquid crystal cells: Optical response and flexoelectricity-induced voltage shift, *Liq. Cryst.* **30**(9), 1115–1122, (2003). doi: 10.1080/02678290310001594562
36. J. Harden, B. Mbanga, N. Éber, K. Fodor-Csorba, S. Sprunt, J.T. Gleeson and A. Jákli, Giant flexoelectricity of bent-core nematic liquid crystals, *Phys. Rev. Lett.* **97**(15), 157802/1–4, (2006). doi: 10.1103/PhysRevLett.97.157802
37. P.R. Maheswara Murthy, V.A. Raghunathan and N.V. Madhusudana, Experimental determination of the flexoelectric coefficients of some nematic liquid crystals, *Liq. Cryst.* **14**(2), 483–496, (1993). doi: 10.1080/02678299308027664
38. L.M. Blinov, M. Kozlovsky, T. Nagata, M. Ozaki and K. Yoshino, Influence of guest conformation (rod- or banana-like photo-isomers) on flexoelectric coefficients in nematic liquid crystals, *Jpn. J. Appl. Phys.* **38**(9A/B), L1042–L1045, (1999).
39. A. Mazzulla, F. Ciuchi and J.R. Sambles, Optical determination of flexoelectric coefficients and surface polarization in a hybrid aligned nematic cell, *Phys. Rev. E* **64**(2), 021708/1–6, (2001). doi: 10.1103/PhysRevE.64.021708
40. H. Ohoka, M. Ozaki, L.M. Blinov, Barnik, N.M. Shtykov and K. Yoshino, Polarization measurement in nematic liquid crystal based on the pyroelectric response to laser irradiation, *Mol. Cryst. Liq. Cryst.* **366**(1), 283–293, (2001). doi: 10.1080/10587250108023971
41. I. Dozov, Ph. Martinot Lagarde and G. Durand, Conformational flexoelectricity in nematic liquid crystals, *J. Physique Lett.* **44**(19), L-817–L-822, (1983). doi: 10.1051/jphyslet:019830044019081700
42. B. Musgrave, M.J. Coles, S.P. Perkins and H.J. Coles, Enhanced flexoelectric switching behaviour in the chiral nematic phase, *Mol. Cryst. Liq. Cryst.* **366**(1), 735–742, (2001). doi: 10.1080/10587250108024014
43. B. Kundu, A. Roy, R. Pratibha and N.V. Madhusudana, Flexoelectric studies on mixtures of compounds made of rodlike and bent-core molecules, *Appl. Phys. Lett.* **95**(8), 081902/1–3, (2009). doi: 10.1063/1.3213544

44. L.M. Blinov, M.I. Barnik, H. Ohoka, M. Ozaki and K. Yoshino, Surface and flexoelectric polarisation in nematic and smectic A phases of liquid crystal 4-octyloxy-4′cyanobiphenyl, *Jpn. J. Appl. Phys.* **40**(8), 5011–5018, (2001).
45. C. Kischka, S.J. Elston and E.P. Raynes, Measurement of the sum $(e_1 + e_3)$ of the flexoelectric coefficients $e_1$ and $e_3$ of nematic liquid crystals using a hybrid aligned nematic (HAN) cell, *Mol. Cryst. Liq. Cryst.* **494**, 93–100, (2008). doi: 10.1080/15421400802430158
46. S.R. Warrier and N.V. Madhusudana, An AC electrooptic technique for measuring the flexoelectric coefficient $(e_1+e_3)$ and anchoring energies of nematics, *J. Phys. II France* **7**(12), 1789–1803, (1997). doi: 10.1051/jp2:1997215
47. V.A. Raghunathan, Investigations on some field induced instabilities in nematic liquid crystals, PhD Thesis, Mysore University, Mysore, 1989. url: http://hdl.handle.net/2289/3518
48. N.V. Madhusudana and V.A. Raghunathan, Influence of flexoelectricity on electrohydrodynamic instabilities in nematics, *Liq. Cryst.* **5**(6), 1789–1812, (1989). doi: 10.1080/02678298908045689
49. M.I. Barnik, L.M. Blinov, A.N. Trufanov and B.A. Umanski, Flexo-electric domains in liquid crystals, *J. Phys. France* **39**(4), 417–422, (1978). doi: 10.1051/jphys:01978003904041700
50. Y.P. Bobylev, V.G. Chigrinov and S.A. Pikin, Threshold flexoelectric effect in nematic liquid crystal, *J. Phys. Coll. (France)* **40**(C3), C3-331–C3-333, (1979). doi: 10.1051/jphyscol:1979364
51. Y. Marinov, J. Kosmopoulos, W. Weissflog, A.G. Petrov and D.J. Photinos, Flexoelectricity of wedge-like molecules in nematic mixtures, *Mol. Cryst. Liq. Cryst.* **357**(1), 221–228, (2001). doi: 10.1080/10587250108028255
52. Y. Marinov, A.G. Petrov and H.P. Hinov, On a simple way for obtaining important material constants of a nematic liquid crystal: Longitudinal flexoelectric domains under the joint action of DC and AC voltages, *Mol. Cryst. Liq. Cryst.* **449**(1), 33–45, (2006). doi: 10.1080/15421400600580188
53. Y.G. Marinov and H.P. Hinov, On the threshold characteristics of the flexoelectric domains arising in a homogeneous electric field: The case of anisotropic elasticity, *Eur. Phys. J. E* **31**(2), 179–189, (2010). doi: 10.1140/epje/i2010-10560-0
54. J. Prost and P.S. Pershan, Flexoelectricity in nematic and smectic-A liquid crystals, *J. Appl. Phys.* **47**(6), 2298–2312, (1976). doi: 10.1063/1.323021
55. T. Tóth-Katona, N. Éber and Á. Buka, Flexoelectricity in electroconvection, *Mol. Cryst. Liq. Cryst.* **511**(1), 11–24, (2009). doi: 10.1080/15421400903048461
56. A. Krekhov, W. Pesch, N. Éber, T. Tóth-Katona and Á. Buka, Nonstandard electroconvection and flexoelectricity in nematic liquid crystals, *Phys. Rev. E* **77**(2), 021705/1–11, (2008). doi: 10.1103/PhysRevE.77.021705
57. S.J. Elston, Flexoelectricity in nematic domain walls, *Phys. Rev. E* **78**(1), 011701/1–7, (2008). doi: 10.1103/PhysRevE.78.011701

58. S.V. Yablonskii, L.M. Blinov and S.A. Pikin, Observation of the direct flexoelectric effect in nematic liquid crystals, *JETP Lett.* **40**(6), 995–998, (1984). [*Pis'ma ZhETF.* **40**(6), 226–228, (1984).]
59. J.H. Wild, K. Bartle, N.T. Kirkman, S.M. Kelly, M. O'Neill, T. Stirner and R.P. Tuffin, Synthesis and investigation of nematic liquid crystals with flexoelectric properties, *Chem. Mater.* **17**(25), 6354–6360, (2005). doi: 10.1021/cm051682y
60. E.K. Tidey, L.A. Parry-Jones and S.J. Elston, Determination of the difference of flexoelectric coefficients in a nematic liquid crystal using a conoscopic technique, *Liq. Cryst.* **34**(2), 251–255, (2007). doi: 10.1080/02678290601097342
61. R.A. Ewings, C. Kischka, L.A. Parry-Jones and S.J. Elston, Measurement of the difference in flexoelectric coefficients of nematic liquid crystals using a twisted nematic geometry, *Phys. Rev. E* **73**(1), 011713/1–6, (2006). doi: 10.1103/PhysRevE.73.011713
62. C.-J. Yu, H.W. Pae and S.-D. Lee, Physical mechanism for bistable polar switching of a nematic liquid crystal in a boundary layer configuration, *Jpn. J. Appl. Phys.* **47**(12), 8887–8891, (2008).
63. C. Kischka, L.A. Parry-Jones, S.J. Elston and E.P. Raynes, Measurement of the flexoelectric coefficients $e_1$ and $e_3$ in nematic liquid crystals, *Mol. Cryst. Liq. Cryst.* **480**(1), 103–110, (2008). doi: 10.1080/15421400701825706
64. P.S. Salter, C. Kischka, S.J. Elston and E.P. Raynes, The influence of chirality on the difference in flexoelectric coefficients investigated in uniform lying helix, Grandjean and twisted nematic structures, *Liq. Cryst.* **36**(12), 1355–1364, (2009). doi: 10.1080/02678290903233991
65. S.A. Jewell and J.R. Sambles, Fully leaky guided mode study of the flexoelectric effect and surface polarization in hybrid aligned nematic cells, *J. Appl. Phys.* **92**(1), 19–24, (2002). doi: 10.1063/1.1483392
66. S.A. Jewell and J.R. Sambles, Fully-leaky guided mode measurement of the flexoelectric constant $(e_{11} + e_{33})$ of a nematic liquid crystal, *Mol. Cryst. Liq. Cryst.* **401**(1), 67–73, (2003). doi: 10.1080/744815187
67. E.G. Edwards, C.V. Brown, E.E. Kriezis and S.J. Elston, Behaviour of a nematic liquid crystal cell containing a diffraction grating, *Mol. Cryst. Liq. Cryst.* **400**(1), 13–19, (2003). doi: 10.1080/15421400390242887
68. C.V. Brown, L. Parry-Jones, S.J. Elston and S.J. Wilkins, Comparison of theoretical and experimental switching curves for a zenithally bistable nematic liquid crystal device, *Mol. Cryst. Liq. Cryst.* **410**(1), 417–425, (2004). doi: 10.1080/15421400490433451
69. H.M. Sykulska, L.A. Parry-Jones and S.J. Elston, Measurement of flexoelectric coefficients in nematic liquid crystals using shallow grating devices, *Mol. Cryst. Liq. Cryst.* **436**(1), 267–279, (2005). doi: 10.1080/15421400590955839
70. C.L. Trabi, A.A.T. Smith and C.V. Brown, Interferometric method for determining the sum of the flexoelectric coefficients $e_1 + e_3$ in an ionic nematic material, *Appl. Phys. Lett.* **92**(22), 223509/1–3, (2008). doi: 10.1063/1.2938722

71. C.L. Trabi, A.A.T. Smith and C.V. Brown, Interferometric method for the measurement of the flexoelectric polarisation using nematic liquid crystal phase gratings, *Mol. Cryst. Liq. Cryst.* **509**(1), 378–392, (2009). doi: 10.1080/15421400903066125
72. S.L. Cornford, T.S. Taphouse and J.R. Sambles, Analysis of the sign-dependent switching observed in a hybrid aligned nematic cell, *New J. Phys.* **11**(1), 013045/1–14, (2009). doi: 10.1088/1367-2630/11/1/013045
73. N. Aziz, S.M. Kelly, W. Duffy and M. Goulding, Rod-shaped dopants for flexoelectric nematic mixtures, *Liq. Cryst.* **36**(5), 503–520, (2009). doi: 10.1080/02678290903031643
74. M. May, W. Schöpf, I. Rehberg, A. Krekhov and Á. Buka, Transition from longitudinal to transversal patterns in an anisotropic system, *Phys. Rev. E* **78**(4), 046215/1–9, (2008). doi: 10.1103/PhysRevE.78.046215
75. J. Harden, R. Teeling, J.T. Gleeson, S. Sprunt and A. Jákli, Converse flexoelectric effect in a bent-core nematic liquid crystal, *Phys. Rev. E* **78**(3), 031702/1–5, (2008). doi: 10.1103/PhysRevE.78.031702
76. K.V. Le, F. Araoka, K. Fodor-Csorba, K. Ishikawa and H. Takezoe, Flexoelectric effect in a bent-core mesogen, *Liq. Cryst.* **36**(10–11), 1119–1124, (2009). doi: 10.1080/02678290902854086
77. P. Kumar, Y.G. Marinov, H.P. Hinov, U.S. Hiremath, C.V. Yelamaggad, K.S. Krishnamurthy and A.G. Petrov, Converse flexoelectric effect in bent-core nematic liquid crystals, *J. Phys. Chem. B* **113**(27), 9168–9174, (2009). doi: 10.1021/jp903241z
78. P. Kumar and K.S. Krishnamurthy, Gradient flexoelectric switching response in a nematic phenyl benzoate, *Liq. Cryst.* **34**(2), 257–266, (2007). doi: 10.1080/02678290601111192
79. P.S. Salter, C. Tschierske, S.J. Elston and E.P. Raynes, Flexoelectric measurements of a bent-core nematic liquid crystal, *Phys. Rev. E* **84**(3), 031708/1–5, (2011). doi: 10.1103/PhysRevE.84.031708
80. N. Scaramuzza and M.C. Pagnotta, Evidence for flexoelectric effect in pallated metallorganic liquid crystals, *Mol. Cryst. Liq. Cryst.* **239**(1), 263–267, (1994). doi: 10.1080/10587259408047188
81. B. Musgrave, P. Lehmann and H.J. Coles, A new series of chiral nematic bimesogens for the flexoelectro-optic effect, *Liq. Cryst.* **26**(8), 1235–1249, (1999). doi: 10.1080/026782999204255
82. B. Musgrave, P. Lehmann and H.J. Coles, Flexoelectro-optic properties of a series of novel chiral nematic liquid crystals, *Mol. Cryst. Liq. Cryst.* **328**(1), 309–316, (1999). doi: 10.1080/10587259908026072
83. S.M. Morris, M.J. Clarke, A.E. Blatch and H.J. Coles, Structure-flexoelastic properties of bimesogenic liquid crystals, *Phys. Rev. E* **75**(4), 041701/1–9, (2007). doi: 10.1103/PhysRevE.75.041701
84. M.J. Clarke, A.E. Blatch and H.J. Coles, The effects of introducing ester linking groups on the flexoelectro-optic properties of symmetric bimesogens, *Mol. Cryst. Liq. Cryst.* **434**(1), 39–47, (2005). doi: 10.1080/15421400590955271

85. A. Dequidt and P. Oswald, Does the electric Lehmann effect exist in cholesteric liquid crystals? *Eur. Phys. J. E* **24**(2), 157–166, (2007). doi: 10.1140/epje/i2007-10226-0
86. P. Rudquist, M. Buivydas, L. Komitov and S.T. Lagerwall, Linear electro-optic effect based on flexoelectricity in a cholesteric with sign change of dielectric anisotropy, *J. Appl. Phys.* **76**(12), 7778–7783, (1994). doi: 10.1063/1.358523

# Appendix B

# Abbreviations

## B.1. Abbreviations of Chemical Names

Table B.1. Nematic compounds.

| Abbreviation | | Chemical name |
|---|---|---|
| **BHAOB** | : | n-butyl-n-heptanoil azoxy benzene |
| **BMAOB** | : | 4-n-butyl-4′-methoxyazoxybenzene |
| **BBMBA** | : | 4-butoxybenzal-4′-($\beta$-methylbutyl)aniline |
| **CBOOA** | : | 4-cyanobenzylidene-4′-octyloxyaniline |
| **MBBA** | : | N-(4-metoxybenzylidene)-4′-butylaniline |
| **nCB** | : | 4-alkyl-4′-cyanobiphenyl |
| **5CB** | : | 4-pentyl-4′-cyanobiphenyl |
| **6CB** | : | 4-hexyl-4′-cyanobiphenyl |
| **7CB** | : | 4-heptyl-4′-cyanobiphenyl |
| **8CB** | : | 4-octyl-4′-cyanobiphenyl |
| **7OCB** | : | 4-n-heptyloxy-4′-cyanobiphenyl |
| **8OCB** | : | 4-n-octyloxy-4′-cyanobiphenyl |
| **5CT** | : | 4′-pentyl-4″-cyano-4-terphenyl |
| **PCHn** | : | 4-cyanophenyl-trans-4′-n-alkylcyclohexane |
| **PCH3** | : | 4-cyanophenyl-trans-4′-n-propylcyclohexane |
| **PCH4** | : | 4-cyanophenyl-trans-4′-n-butylcyclohexane |
| **PCH5** | : | 4-cyanophenyl-trans-4′-n-pentylcyclohexane |
| **PCH7** | : | 4-cyanophenyl-trans-4′-n-heptylcyclohexane |
| **CCHn** | : | 4-alkyl-4′-cyanocyclohexylcyclohexanes |
| **CCH7** | : | 4-heptyl-4′-cyanocyclohexylcyclohexane |
| **MPPC** | : | 4-methoxyphenyl trans-4-pentylcyclohexylcarboxylate |
| **EPPC** | : | 4-ethoxyphenyl trans-4-pentylcyclohexylcarboxylate |
| **PPPC** | : | 4-pentylphenyl trans-4-pentylcyclohexylcarboxylate |
| **PEC** | : | 1-propyl-4(4-ethoxyphenyl)cyclohexane |
| **DE3** | : | 4′-cyanophenyl 4-propylcyclohexanecarboxylate |
| **DE5** | : | 4′-cyanophenyl 4-pentylcyclohexanecarboxylate |
| **DE7** | : | 4′-cyanophenyl 4-heptylcyclohexanecarboxylate |
| **ROCP-7037** | : | 5-heptyl-2-(4-cyanophenyl)pyrimidine |
| **ROCP-7334** | : | 5-(4-butylphenyl)-2-(4-cyanophenyl)pyrimidine |
| **6OO8** | : | 4-n-octyloxy-phenyl-4-n-hexyloxybenzoate |
| **8/7** | : | 4-n-octyloxy-phenyl-4-n′-heptyloxy-benzoate |
| **BEPC** | : | butyl p-(p-ethoxyphenoxycarbonyl)phenyl carbonate |

Table B.1. Nematic compounds (continued).

| Abbreviation | | Chemical name |
|---|---|---|
| **HBACPE** | : | n-heptyl benzonoic acid n-cyanobiphenyl ester |
| **NPOOB** | : | 4-nytrophenyl-4-n-octyloxy benzoate |
| **HOT** | : | 4,4′-dihexyloxytolan |
| **3PBClB** | : | 4,n-pentylphenyl-4,(pentylbenzoyloxy)-3-chloro benzoate |
| **MBB** | : | 1,4-di-(4-butylbenzoyloxy)-2-methylbenzene |
| **CC**[a] | : | cholesteryl chloride |
| **MHPOBC**[a] | : | 4-(1-methylheptyloxycarbonyl)phenyl 4′-octyloxybiphenyl-4-carboxylate |
| **(S)-9PPP8dM**[a] | : | S-4-n-nonyloxyphenyl-4′-(3″,7″-dimethyloctyloxybenzoyloxy) benzoate |

[a]Chiral compound

Table B.2. Bent-core nematic compounds.

| Abbreviation | | Chemical name |
|---|---|---|
| **ClPbis10BB** | : | 4-chloro-1,3-phenylene bis 4-[4′-(9-decenyloxy) benzoyloxy] benzoate |
| **10DClPBBC** | : | 4,6-dichloro-1,3-phenylene bis[4-(9-decen-1-yloxy)-1,1′-biphenyl-4-carboxylate] |
| **11DClPBBC** | : | 4,6-dichloro-1,3-phenylene bis[4-(10-undecen-1-yloxy)-1,1-biphenyl-4-carboxylate] |
| **10DClPBCP** | : | 2,4-dichloro-5-[(4-decyloxy-1,1′-biphenyl)carbonyloxy]phenyl -4′-[(9-decene-1-yloxy)-1,1′-biphenyl]-4-carboxylate |
| **BC12** | : | 1,3-phenylene bis[4-(3-methylbenzoyloxy)] 4′- n-dodecylbiphenyl 4′-carboxylate |
| **CNRbis12OBB** | : | 4-cyanoresorcinol bis[4-(4-n-dodecyloxybenzoyloxy) benzoate] |
| **C6ban** | : | 4-cyano-1,3-phenylene bis [4-[4′-(hexyl) benzoyloxy] benzoate] |
| **Azpac** | : | acetylacetonate derivative of the cyclopalladated 4,4-bis(hexyloxy) azoxybenzene |

Table B.3. Bimesogens.

| Abbreviation | | Chemical name |
|---|---|---|
| **BCCB** | : | 4-((3-(4-(4-(decyloxy)benzoyloxy)benzoyloxy) phenylimino) methyl) -3-hydroxyphenyl 4-(6-(4′-cyanobiphenyl-4-yloxy)hexyloxy) benzoate |
| **11Es5CB** | : | 3-(10-Undecenyloxy)estra-1,3,5(10)-triene-17β-yl 5-(4′-cyanobiphenyl-4-yloxy)pentanoate |
| **5Es11CB** | : | 3-(4-Pentenyloxy)estra-1,3,5(10)-triene-17β-yl 10-(4′-cyanobiphenyl-4-yloxy)undecanoate |
| **FFOnOFF** | : | α-(2′,4-difluorobiphenyl-4′-yloxy)-ω-(2′,4-difluorobiphenyl-4′-yloxy)alkanes |
| **FFO9OFF** | : | α-(2′,4-difluorobiphenyl-4′-yloxy)-ω-(2′,4-difluorobiphenyl-4′-yloxy)nonane |
| **FFO11OFF** | : | α-(2′,4-difluorobiphenyl-4′-yloxy)-ω-(2′,4-difluorobiphenyl-4′-yloxy)undecane |
| **FFEnEFF** | : | α-(2′,4-difluorobiphenyl-4′-ester)-ω-(2′,4-difluorobiphenyl-4′-ester)alkanes |
| **FFE9EFF** | : | α-(2′,4-difluorobiphenyl-4′-ester)-ω-(2′,4-difluorobiphenyl-4′-ester)nonane |
| **FFE11EFF** | : | α-(2′,4-difluorobiphenyl-4′-ester)-ω-(2′,4-difluorobiphenyl-4′-ester)undecane |

Table B.3. Bimesogens (continued).

| Abbreviation | | Chemical name |
|---|---|---|
| **CBO8OCB** | : | $\alpha$-(4-cyanobiphenyl-4′-yloxy)-$\omega$-(4-cyanobiphenyl-4′-yloxy)octane |
| **FFOnOCB** | : | $\alpha$-(2′,4-difluorobiphenyl-4′-yloxy)-$\omega$-(4-cyanobiphenyl-4′-yloxy)alkanes |
| **FFO5OCB** | : | $\alpha$-(2′,4-difluorobiphenyl-4′-yloxy)-$\omega$-(4-cyanobiphenyl-4′-yloxy)pentane |
| **FFO6OCB** | : | $\alpha$-(2′,4-difluorobiphenyl-4′-yloxy)-$\omega$-(4-cyanobiphenyl-4′-yloxy)hexane |
| **FFO7OCB** | : | $\alpha$-(2′,4-difluorobiphenyl-4′-yloxy)-$\omega$-(4-cyanobiphenyl-4′-yloxy)heptane |
| **FFO8OCB** | : | $\alpha$-(2′,4-difluorobiphenyl-4′-yloxy)-$\omega$-(4-cyanobiphenyl-4′-yloxy)octane |
| **FFO9OCB** | : | $\alpha$-(2′,4-difluorobiphenyl-4′-yloxy)-$\omega$-(4-cyanobiphenyl-4′-yloxy)nonane |
| **FFO10OCB** | : | $\alpha$-(2′,4-difluorobiphenyl-4′-yloxy)-$\omega$-(4-cyanobiphenyl-4′-yloxy)decane |
| **FFO11OCB** | : | $\alpha$-(2′,4-difluorobiphenyl-4′-yloxy)-$\omega$-(4-cyanobiphenyl-4′-yloxy)undecane |
| **FFO12OCB** | : | $\alpha$-(2′,4-difluorobiphenyl-4′-yloxy)-$\omega$-(4-cyanobiphenyl-4′-yloxy)dodecane |
| **NSO9OFF** | : | (E)-2,4′-difluoro-4-((9-(4-(4-nitrostyryl)phenoxy)nonyl)oxy)-1,1′-biphenyl |
| **NSOnOphOnOCB** | : | (E)-4′-((9-(3-((9-(4-(4-nitrostyryl)phenoxy)nonyl)oxy)phenoxy)nonyl)oxy)-[1,1′-biphenyl]-4-carbonitrile |

Table B.4. Mixtures.

| Abbreviation | | Composition |
|---|---|---|
| **8OCB/CC** | : | compensated cholesteric mixture of **8OCB** (50%) and **CC** (50%) |
| **CBmix** | : | mixture of **3CB** (30%), **5CB** (40%) and **7CB** (30%) |
| **DEmix** | : | mixture of **DE3** (30%), **DE5** (40%) and **DE7** (30%) |
| **Mix1** | : | mixture of **MBB** (46%), **PEC** (50%) and **5CB** (4%) |
| **MixA** | : | mixture of azoxy compounds and 8CB |
| **PB** | : | mixture of phenylbenzoates |
| **PCHmix** | : | mixture of **PCH3** (30%), **PCH5** (40%) and **PCH7** (30%) |
| **FFO9/11OFF** | : | mixture of **FFO9OFF** (50%) and **FFO11OFF** (50%) |
| **Mixture3** | : | mixture of **FFE9EFF** (50%) and **FFE11EFF** (50%) |
| **Mixture1** | : | mixture of **FFO9OFF** (40%) and **FFO11OFF** (60%) |
| **Mixture2** | : | mixture of **FFO9OFF** (50%) and **FFO11OFF** (50%) |
| **MixBimes** | : | mixture of substituted bimesogens, composition unspecified |
| **E7** | : | commercial nematic mixture from Merck |
| **E70A** | : | commercial nematic mixture from Merck |
| **MDA-02-2419** | : | commercial nematic mixture from Merck |
| **MLC-6437-000** | : | commercial nematic mixture from Merck |
| **MLC-7029** | : | commercial nematic mixture from Merck |
| **Phase4** | : | commercial nematic mixture from Merck |
| **Phase5** | : | commercial nematic mixture from Merck |
| **Phase5A** | : | commercial nematic mixture from Merck |
| **TL-216** | : | commercial nematic mixture from Merck |
| **TM-216** | : | commercial thermochromic mixture from Merck |
| **ZLI-4788-000** | : | commercial nematic mixture from Merck |
| **ZLI-4792** | : | commercial nematic mixture from Merck |

## B.2. Common Abbreviations

| | | |
|---|---|---|
| AC | : | Alternating current |
| AFM | : | Atomic force microscopy |
| BC | : | Bent-core |
| BCLCE | : | Bent-core liquid crystalline elastomer |
| BCN-LCE | : | Bent-core nematic swollen in a liquid crystal elastomer |
| BLM | : | Black lipid membranes |
| DC | : | Direct current |
| EC | : | Electroconvection |
| EHD | : | Electrohydrodynamic |
| FLC | : | Ferroelectric liquid crystal |
| HAN | : | Hybrid-aligned nematic |
| HEK | : | Human embryonic kidney |
| I | : | Isotropic |
| IPS | : | In-plane switching |
| ITO | : | Indium-tin-oxide |
| LC | : | Liquid crystal |
| LCD | : | Liquid crystal display |
| LCE | : | Liquid crystalline elastomer |
| N | : | Nematic |
| ns-EC | : | Non-standard electroconvection |
| OHC | : | Outer hair cell |
| PVDF | : | Polyvinylidene fluoride |
| PZT | : | Pb-zirconate-titanate |
| SAM | : | Self-assembled monolayer |
| SAXS | : | Small-angle X-ray scattering |
| s-EC | : | Standard electroconvection |
| SEM | : | Scanning electron microscopy |
| SmA | : | Smectic A |
| SmC | : | Smectic C |
| SSFLC | : | Surface-stabilized ferroelectric liquid crystal |
| ST | : | Splay-twist |
| STN | : | Supertwisted nematic |
| TEM | : | Transmission electron microscopy |
| TN | : | Twisted nematic |
| TFT | : | Thin film transistor |
| ULH | : | Uniform lying helix |
| UV | : | Ultraviolet |
| VAN | : | Vertically aligned nematic |
| WEM | : | Weak electrolyte model |
| ZBD | : | Zenithal bistable device |

# Author Index

# Subject Index